To my mother, Sarah.

PREFACE

The purpose of this book is to show how mathematics can be applied to improve cancer chemotherapy. Unfortunately, most drugs used in treating cancer kill both normal and abnormal cells. However, more cancer cells than normal cells can be destroyed by the drug because tumor cells usually exhibit different growth kinetics than normal cells. To capitalize on this last fact, cell kinetics must be studied by formulating mathematical models of normal and abnormal cell growth. These models allow the therapeutic and harmful effects of cancer drugs to be simulated quantitatively. The combined cell and drug models can be used to study the effects of different methods of administering drugs. The least harmful method of drug administration, according to a given criterion, can be found by applying optimal control theory.

The prerequisites for reading this book are an elementary knowledge of ordinary differential equations, probability, statistics, and linear algebra. In order to make this book self-contained, a chapter on cell biology and a chapter on control theory have been included. Those readers who have had some exposure to biology may prefer to omit Chapter 1 (Cell Biology) and only use it as a reference when required. However, few biologists have been exposed to control theory. Chapter 7 provides a short, coherent and comprehensible presentation of this subject. The concepts of control theory are necessary for a full understanding of Chapters 8 and 9. For readers not already familiar with control theory, the time required for the mastery of Chapter 7 will be well spent since this powerful tool is applicable to many other branches of biology.

The appendices provide a brief description of topics which are not treated in detail in this book. These short topic outlines are more helpful in choosing a new research direction than scattered references throughout the book.

Biology has become a quantitative science. It is hoped that this book will interest biologists in mathematics and mathematicians in biology. A biologist will find that mathematical models are absolutely essential for research in modern cell kinetics. A mathematician will discover that there are many exciting, unsolved mathematical problems in cell biology.

The author is grateful to Tom Slook and George Swan for their helpful comments on Chapter 7 and 8 respectively, to Werner Duchting for contributing Appendix G, to Simon Levin for his editorial work and Alan Perelson for his constructive suggestions for improving the manuscript. I am glad to record my thanks to Gerry Sizemore and Mittie Davis for typing the book and Rae Ballou for tracing many obscure references and obtaining copies of many papers. The author is indebted to his wife, Carole, for helping with all phases of the manuscript and his daughter, Debby, for proofreading the book and preparing the index.

Martin Eisen
University of Maryland School of Medicine
Temple University

TABLE OF CONTENTS

INTRODUCTION

This year, approximately 350,000 Americans will die from cancer. The number of new cases which develop yearly is twice this figure. Cancer strikes 7,000 children each year. The purpose of this monograph is to indicate how mathematicians can play an important role in irradicating this disease.

Two well known methods used in cancer treatments are removal of the tumor (surgery) and destruction of the tumor in situ by an external attacking agent (chemotherapy, radiation) or an internal attacking agent (immunotherapy, endocrinotherapy). A third, more subtle method, described in Chapter 1, tumor regression, has not yet been used. However, sometimes a tumor which remains after surgery, chemotherapy or immunotherapy disappears. There have even been instances of tumor regression without any form of treatment. These spontaneous tumor regressions may be the reason for the successes cited in controversial forms[1] of therapy such as laetrile, vitamin C, meditation and so on.

The most frequently used treatment for cancer is surgery. Surgery is most successful when combined with early diagnosis. Perhaps the most spectacular example of surgical success is the reduction in deaths from cancer of the cervix due to early detection by the Pap test.

Unfortunately, surgery has its limitations. The primary tumor may be inoperable or may be so large that it is only partially removable. Moreover, if metastases have occurred it is usually impossible to remove all of the secondary growths.[2] Several forms of cancer are disseminated and therefore cannot be surgically removed (e.g. the leukemias which cause about 10% of all cancer deaths). Radiation treatment is also limited for the same reasons.

Utilizing the body's own defense mechanism is a theoretically possible treatment for disseminated cancer. However, much basic research is still required in immunotherapy before this form of treatment becomes practical (See Appendix D.).

---

[1] These treatments are controversial since they have not been tested in properly designed trials on a <u>single</u> form of cancer.

[2] More than 50% of present cancer deaths are due to nonlocalized or diffuse neoplasms.

Endocrinotherapy is only helpful for tumors in hormone dependent organs. The use of hormones in cancer is described in Carbone [4], McKerns [10] and Segaloff [14].

This leaves only chemotherapy. The modern era of chemotherapy began in 1948. S. Farber et al. noted that aminopterin could produce clinical remissions in acute leukemia [7]. Spurred on by this success, early workers tried to find a biochemical difference between normal and malignant cells that could be exploited. The first malignant tumor cured by chemotherapy was announced by Li et al. [9]. In the 1960's it was announced that certain leukemic cells needed external asparagine to grow, while normal leukocytes could synthesize their own. Treatment of animal leukemias with L-aspariginase (an enzyme which breaks down asparagine) cured the animals. Unfortunately, this method was ineffective for the treatment of human leukemias. To date, no exploitable biochemical differences between normal and malignant cells in humans have been found[3]. A history of chemotherapy can be found in Burchenal [2]. Despite this setback, combinations of drugs, sometimes coupled with other forms of treatment have produced statistically substantiated cures[4] in acute childhood (lymphocytic) leukemia [4], Hodgkin's disease (Mopp protocol [17]), choriocarcinoma and Burkitts lymphoma. With the coupling of chemotherapy to surgery young patients with osteogenic sarcoma of bone, who almost always died of their disease, are living normal disease-free lives. Young children with Wilms' tumor of the kidney are now expected to be cured after being treated with surgery, radiotherapy and chemotherapy.

Aside from these successes, the past twenty years has brought little progress in treatments designed specifically for cancer. The increase in the number of cures is thought, by some authorities, to be due to improved surgical technique [6].

Readers interested in obtaining pharmacological information about chemotherapy can consult [1,3,8,11,12,13]. However, such knowledge is not required to read

---

[3] Differences, however, have been found in their membrane permeability to certain drugs.

[4] The patient is presumed cured if no trace of the disease can be found for five years after treatment.

this monograph. Mathematical analysis of cancer chemotherapy as well as other treatment modalities can be found in Swan [16].

The major difficulty in chemotherapy is that most cancer drugs kill normal as well as tumor cells. Therefore drugs are usually given on a cyclical basis. Periodic rest periods allow the patient's normal cells time to recover. It is well known that optimal[5] scheduling can make the difference between success and failure in chemotherapy [15].

In this book we will describe some mathematical techniques for devising "optimal" schedules in cancer chemotherapy. Research is still in its formative stages and a complete theory does not exist. The cornerstone of optimal scheduling is that tumor cells usually exhibit different growth kinetics than critical normal tissue. To take advantage of this fact requires appropriate models of cells. These are presented in the first part of the book. Various techniques for obtaining kinetic parameters necessary to describe cellular proliferation are also described. Finally, it is shown how optimal control theory can be applied to obtain drug schedules. Realistically, such an approach must have a sound biological basis. Hence we begin with a chapter on cell biology

References

1. Becker, F.F. (editor). Cancer, A Comprehensive Treatise, Vol. 5, Chemotherapy, Plenum Press, New York, 1977.

2. Burchenal, J.H. The historical development of cancer chemotherapy, Seminars in Oncol., 4, 135-146, 1977.

3. Busch, H. and Lane, M. Chemotherapy, Year Book Medical, Chicago, 1967.

4. Carbone, P.P. Systemic treatment of breast cancer: past, present, and future, Int. J. Radiat. Oncol. Biol. Phys., 1, 759-767, 1976.

5. Clarkson, B. and Fried, J. Changing concepts of the treatment of acute leukemia, Med. Cli. N. Amer., 55, 561-600, 1971.

6. Enstrom, J.E. and Austin, D.F. Interpreting cancer survival rates, Science, 195, 847-851, 1977.

---

[5] Optimal is used in the sense of the best procedure according to some criterion.

7. Farber, S., Diamond, L.K., Mercer, R.D., Sylvester, R.F., and Wolfe, J.A. Temporary remissions in acute leukemia in children produced by folic acid antagonist, 4-aminopteroylglutamic acid (aminopterin), New England J. Med., 238, 787-793, 1948.

8. Krakoff, I.H. Cancer chemotherapeutic agents. Ca-A Cancer J. for Clinicians, 27, 130-143, 1977.

9. Li, M.C., Hertz, R., Spencer, D.B. Effects of methotrexate therapy upon choriocarcinoma and chorioadenoma, Proc. Soc. Exp. Biol., New York, 93, 361-366, 1956.

10. McKerns, K.W. (editor). Hormones and Cancer, Academic Press, New York, 1974.

11. Pratt, W.B. Fundamentals of Chemotherapy, Oxford University Press, Oxford, 1973.

12. Sartorelli, A.C. (editor). Cancer Chemotherapy, Am. Chem. Soc., Washington (D.C.), 1976.

13. Sartorelli, A.C. and Johns, D.G. (editors). Antineoplastic and Immunosuppressive Agents, Part 1. Springer-Verlag, New York, 1974.

14. Segaloff, A. The treatment of human malignancies with steroids. In Methods of Hormone Research (Steroidal Activity in Experimental Animals and Man. Part C) Vol. 5, 203-213, edited by R.I. Dorfman, Academic Press, New York, 1966.

15. Skipper, H. Combination therapy: some concepts and results, Canc. Chemo. Reports, Pt. 2, Vol. 4, No. 1, 137-145, 1974.

16. Swan, G. Some Current Mathematical Topics in Cancer, Monograph Publishing, University Microfilms, Ann Arbor, 1978.

17. Weed, R. (editor). Hematology for Internists, Little, Brown and Co., Boston, 1971.

I. CELLS

1.1 Introduction

Just as atoms are the fundamental units of elements, the cell is the fundamental unit of structure of all living things except viruses (see Appendix B). Simplistically, the cell consists of a collection of substances surrounded by a membrane. One of the most important of these substances is the genetic (or nuclear) material which directs the activities of the cell.

Cells are classified into two major types according to the organization of their nuclear material. In the eukaryotic cell, the genetic material is enclosed by a membrane. In the prokaryotic cell, the genetic material is not surrounded by a membrane but instead is densely coiled in the nuclear zone. [See Fig. 1.1]

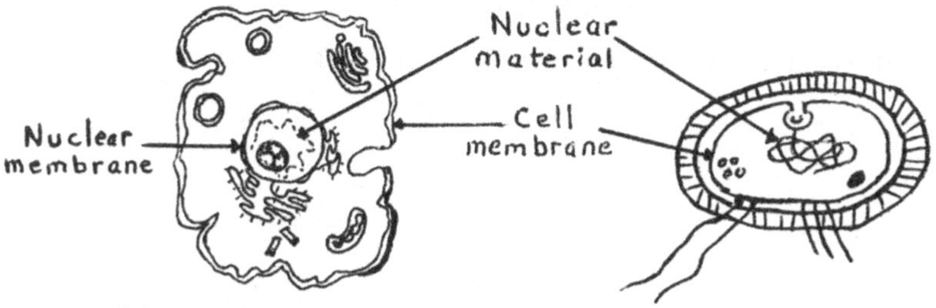

(a) A eukaryote (b) A prokaryote

Figure 1.1 The two major cell types.

Prokaryotic cells may be the first cells that arose in biological evolution. They are very small, simple cells having no membrananous organelles. In fact, their only membrane is the cell membrane. Prokaryotic cells contain only 1 chromosome. This chromosome contains a single molecule of double helical DNA. [See Appendix A.] The prokaryotes include the eubbacteria, the blue-green algae, the spirochetes, the rickettsiae, and the mycoplasma or pleuropneumonialike organisms.

Eukaryotic cells, on the other hand, are much larger and more complex then the prokaryotic cells. The cell volume of most eukaryotes is from 1000 to 10,000 times larger than that of prokaryotes. Unlike prokaryotic cells, eukaryotes have membranous organelles and their nuclear material is divided into chromosomes. Nearly all cells of higher organisms in the animal and plant kingdom are eukaryotic. In addition several simple organisms such as fungi, protozoa and many algae also have eukaryotic cells.

Further details about the structure and function of the cell is given in the next two sections. The events that occur in the living cell between the time it is first born after the division of the mother cell and the time it divides itself is outlined in Section 1.4. Sections 1.5 and 1.6 are concerned with cancer.

1.2 Cell Organization-Protoplasm

A typical eukaryote cell has two parts: the nucleus, which is filled with nucleoplasm, and the cytoplasm. The nucleus is separated from the cytoplasm by a nuclear membrane; the cytoplasm is separated from the extracellular fluids by a cell membrane.

The different substances that make up a cell are collectively called protoplasm. Protoplasm is chiefly composed of five basic substances. The function and location of these five substances are summarized in Table 1.1.

TABLE 1.1 THE FIVE BASIC SUBSTANCES IN PROTOPLASM.

SUBSTANCE	PERCENTAGE OF CELL	LOCATION	FUNCTION
Water	70-85%	Throughout cell	(1) Contains dissolved cellular chemicals and particles (2) Necessary for certain chemical reactions and aids other chemical reactions (3) Allows diffusion and the transportation of material from one part of the cell to another
Electrolytes	Trace amounts	Dissolved in water	(1) Provides inorganic chemicals for cellular reactions (2) Necessary for cellular operations
Proteins	10-20%	Throughout cell	(1) Structural proteins hold the cell together (2) Enzymes catalyze chemical reactions (3) Chromosomes are formed from nucleoproteins (See Appendix A)
Lipids	2-3%	Dispersed throughout cell, especially in membranes	Combines with structural proteins to form membranes which separate the different water compartments of the cell from each other
Carbohydrates	1%	Small amount stored in cell. Glucose always present in extracellurar fluid.	(1) Major role in nutrition (2) Part of a cell membrane (3) Stored in minute amounts as glycogen, a polymere of glucose

Notes on Table 1.1

1. Some examples of electrolytes are potassium, magnesium, phosphate, sulfate, bicarbonate and small quantities of sodium and chloride ions.
2. Important examples of electrolytic action are:
   (a) transmission of electrochemical impulses in nerve and muscle fibres due to changes in permeability of membrane to electrolytes.
   (b) determination of the activity of enzymatically catalyzed reactions

3. Most structural proteins are fibrillar proteins. The individual protein molecules are polymerized into long fibrous threads providing tensile strength. They are usually not soluble in cell fluid.

   Most enzymes are globular proteins. Such proteins are aggregates of protein molecules organized in a globular form. They are usually soluble in cell fluid. Frequently they are adsorbed on the surfaces of membranes.

4. Lipids are substances which are grouped together because of the common property of being soluble in fat solvents and so insoluble or partially soluble in water. Some examples of lipids are neutral fat, phospholipids and cholesterol.

1.3 Cellular Structures and their Function

The cell is not merely a bag of fluid, enzymes and other chemicals. It also contains highly organized physical structures called organelles which are described below.

Almost all of the physical structures of a eukaryotic cell are lined by a membrane. These membranes can be composite structures. The simplest basic unit of the structure is called the unit membrane. The unit membrane is about 80 Angstroms thick. Its composition is approximately: 55% proteins, 40% lipids and 5% polyaccharides.[1]

The precise molecular organization of the unit membrane is unknown. However, many experiments point to the structure schematically illustrated in Fig. 1.2. In the interior of the unit membrane, a central layer of lipids is covered by a layer of proteins. The outside surface of the unit membrane has a thin mucopolysaccharide[2] layer. This jellylike, slippery or sticky layer provides intercellular lubrication and acts as a flexible cement. The small knobbed structures lying at the base of the protein molecules are called phospholipid molecules. The fat portion of the phospholipid molecule is attracted to the

---

[1] A macromolecule composed of a number of similar or identical subunits which are sugars.

[2] A macromolecule composed of disaccharide units with attached substances.

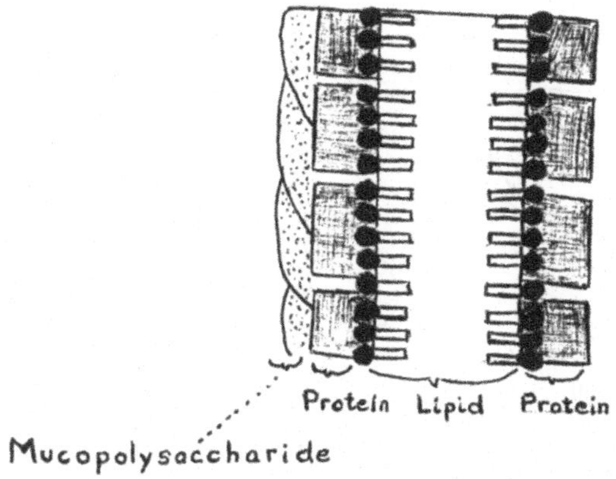

Figure 1.2 Molecular organization of the unit membrane.

central lipid phase of the cell membrane. The polar (ionized) portion of the phospholipid molecule protrudes toward the surface and is bound electrochemically with the inner and outer layers of proteins.

This model of the unit membrane explains many experimentally observed facts. The surface of the membrane is hydrophilic, which means that water adheres easily to the membrane. This is due to the presence of proteins and mucopolysaccharides on the surface. The membrane is impervious to lipid-insoluble substances. This fact is easily explainable by the lipid center. The chemical reactivity of the cell's outer surface is different from that of the inner surface. The mucopolysaccharide coating on the outer surface of the cell accounts for this phenomena.

The unit membrane is believed to have many minute pores passing through it. These pores have never been seen, even with an electron microscope. However, the study of movement of molecules of different sizes between the extra and intracellular fluid show that free diffusion of molecules up to a size of approximately 8 Angstroms occur. It is postulated that lipid-insoluble substances of very small sizes, such as water molecules, pass through these pores between the interior and exterior of the cell.

Table 1.2 and Fig. 1.3 summarize the different parts of the cell.

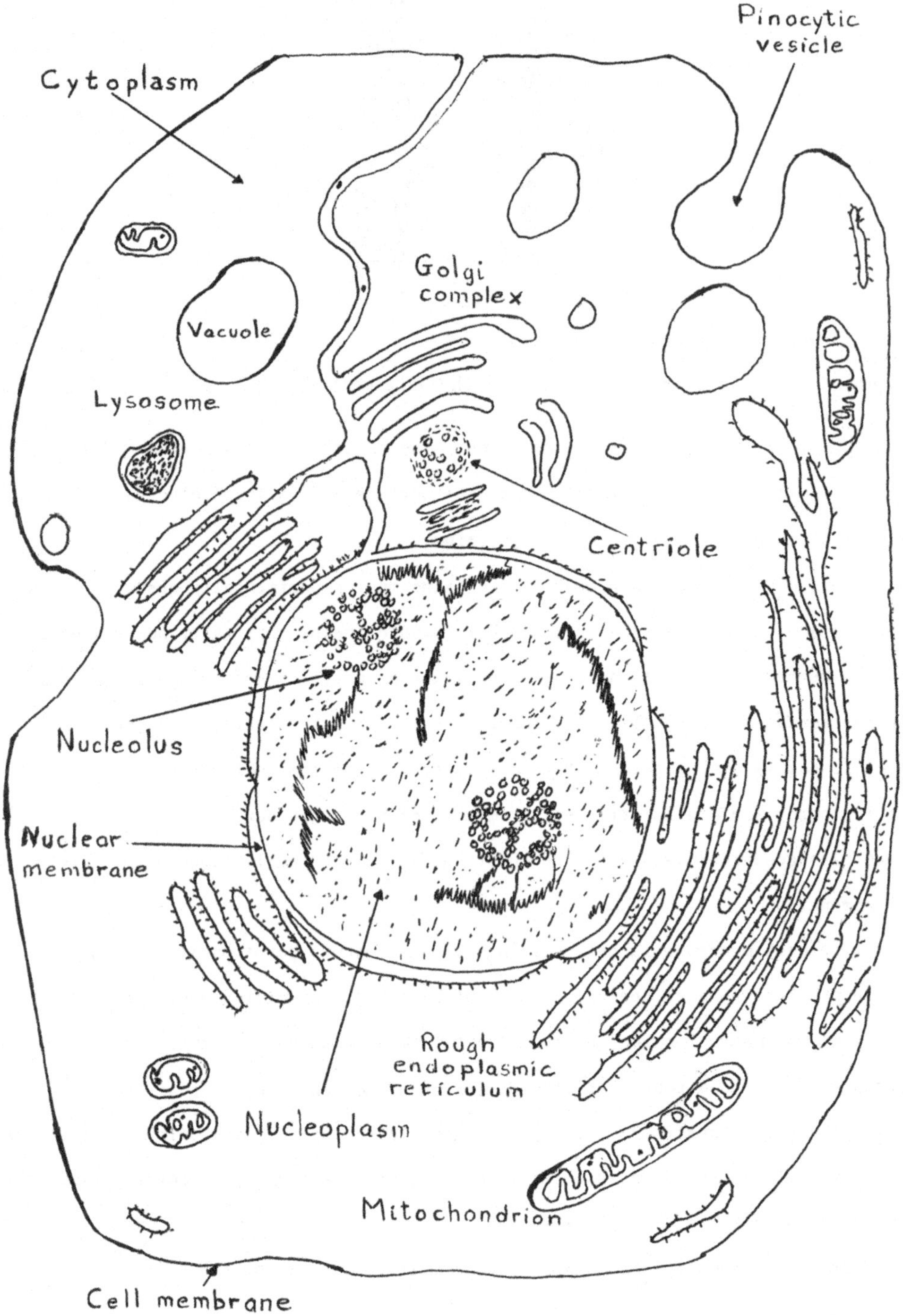

Figure 1.3 Diagram of a typical eukaryotic cell.

TABLE 1.2 FUNCTIONS OF CELLULAR ORGANELLES

ORGANELLE	FUNCTIONS
Plasma-membrane	Differentially permeable membrane. Admits certain extracellular substances and secretes certain cell products.
Cell wall (plants only)	Thick cellulose wall surrounding the cell membrane. Provides strength and rigidity.
Nucleus:	Regulates growth and reproduction.
Nuclear envelope	Differentiably permeable membrane. Allows transfer of materials between nucleus and cytoplasm.
Nucleoplasm (nuclear sap)	Contains materials for building DNA and messenger molecules.
Chromosomes	Provides hereditary instructions; regulates cellular processes
Nucleolus	Synthesizes rRNA
Cytoplasm:	Contains machinery for carrying out the nuclear instructions.
Hyaloplasm	Contains enzymes for glycolysis and structoral materials.
Endoplasmic reticulum	Isolates and conducts material to Golgi complex. Rough endoplasmic reticulum contains ribosomes.
Ribosomes	Sites of protein synthesis
Golgi complex	Production of cellular secretions
Mitochondria	Energy production
Lysosome (animals only)	Contains intracellular digestive enzymes.
Inclusions	Storage depots for various materials
Plastids (plants only)	Structures for storage of starch, pigments and other cellular products. Photosynthesis occurs in chloroplasts.

The different parts of the cell which are summarized in Fig. 1.3 and Table 1.2 will now be discussed in greater detail.

A. The Nucleus.

The nucleus is separated from the cytoplasm by the nuclear envelope (or membrane). Actually the nuclear membrane is composed of two unit membranes (without the polysaccharide coats) separated by a wide space (about 150 Angstroms). Large pores of several hundred Angstrom diameter are present in the membrane so that almost all dissolved substances can move with ease between the fluids of the nucleus (or nucleoplasm) and cytoplasm. The nuclear membrane is synthesized by the endoplasmic reticulum (see below). In antibody producing cells the nuclear envelope contains antibodies [3].

The nucleus is the control center of the cell. It controls both the chemical reactions that occur in the cell and the reproduction of the cell. Briefly, the nucleus contains large quantities of deoxyribonucleic acid (DNA, see Appendix A) which is complexed with nucleoproteins to form chromosomes. Chromosomes contain linear collections of specific genetic factors called genes. Each gene is responsible for the synthesis of a specific sequence of amino acids to form a polypeptide. Since proteins, in particular enzymes, are formed from polypeptides, genes control the production of enzymes. In turn, enzymes control the biochemical reactions occuring in the cell. To control reproduction, the chromosomes reproduce themselves. Then the cell splits by a special process called mitosis (see Section 4) to form two daughter cells, each of which receives one of the two sets of chromosomes. Further biochemical details appear in Appendix A.

Remark on Chromosome Number

In higher organisms, each somatic cells (i.e. any cell except a sex cell) contains a set of chromosomes inherited from the maternal parent and a comparable set of chromosomes (homologous chromosomes) from the paternal parent. The total number (2n) of chromosomes is called the diploid number. Sex cells or gametes, which contain half the number (n) of chromosomes found in somatic cells, are called haploid cells. The suffix "ploid" refers to chromosome sets while the prefix indicates the degree of ploidy. A genome is a set of chromosomes corresponding to the haploid set of a species. For instance, the somatic cells of a fruit fly contain 8 chromosomes, the garden pea has 14, humans 46, tobacco 48, cattle 60, etc.

---

[3] Globulin molecules formed primarily by plasma cells in the lymph nodes that are capable of attacking the invading agent or antigen.

A nucleus is present in nearly all metazoan cells. However, it is absent in mammalian erythrocytes (red blood cells) and a few other end line types (i.e., cells which have stopped evolving). Some cells may have more than one nucleus - for example, hepatocytes (2 or more); renal tubular cells (binucleate). There are giant cells (osteoclasts or foreign body giant cells) with 100 or more nuclei.

The nuclei of many cells contain one or more lightly staining structures called nucleoli. The nucleolus does not have a limiting membrane, but is simply an aggregate of loosely bound granules composed mainly of ribonucleic acid (RNA, see Appendix A) with some proteins. Its major activity is the synthesis of RNA for the ribosomes (an organelle of the cytoplasm which produces protein). Actively secreting cells, such as pancreatic acinar cells, have large and multiple nucleoli. On the other hand, in cells showing a low level of protein synthesis (e.g. muscle cells) certain small lympocytes) nucleoli may be small or absent.

B. Annulate Lamellae.

Annulate lamellae are stacks of membranes which resemble nuclear envelopes and are found in nucleoplasm or cytoplasm. These structures are common in germ cells. Their function is unknown but they may exercise nuclear control in parts of the cytoplasm distant from the nucleus by transferring information.

C. The Cytoplasm.

The cytoplasm surrounds the nucleus and is bounded by the plasma membrane (plasmalemma). It consists of a ground substance (hyaloplasm) and organelles.

The plasma membrane is a unit membrane. Besides the usual constituents of a unit membrane it contains

(a) enzymes: These permit discriminatory passage of materials across the membrane

(b) energy sources: The carbohydrate coating on the outer surface of the cell plays a part in cell homing, in cell recognition in cell-to-cell interaction, in the characterization of antigens and is also useful in the immunologic sorting of cells.

Organelles of Cytoplasm.

1. Endoplasmic Reticulum.

This organelle is a network of tubular and vesicular structures formed from unit membranes. It is filled with endoplasmic matrix, a fluid medium that is different from the fluid which is outside the endoplasmic reticulum. The space inside the endoplasmic reticulum is connected with the space between the two membranes of the double nuclear membrane. Hence, substances in the space between the two nuclear membranes can be conducted to other parts of the cell through the endoplasmic reticular tubules. The nuclear envelope is probably derived from the endoplasmic reticulum.

Ribosomes. Attached to the outer surface of many parts of the endoplasmic reticulum are large numbers of granular particles (about 250 Å in diameter) called ribosomes. When ribosomes are present, the reticulum is called the rough (granular) endoplasmic reticulum, or ergastoplasm. Ribosomes have molecular weights of about 3×10^6 and are approximately half protein and half ribosomal RNA (rRNA). About 60 different proteins are found in a single ribosome and two different rRNA chains. They are always constructed from a large (molecular weight $\sim 2 \times 10^6$) and a small (molecular weight $\sim 10^6$) subunit. Ribosomes synthesize protein (see Appendix A) and so occur in cells that secrete protein or isolate it within membrane bound sacs.

The endoplasmic reticulum isolates synthesized material composed of peptides and larger molecules made by ribosomes. The reticulum then permits the assembly of some of these peptides into larger molecules and facilitates the complexing of molecules with other compounds. It then channels the products into the Golgi Complex (see 2, below).

Part of the endoplasmic reticulum has no attached ribosomes and is called smooth (agranular) endoplasmic reticulum. It contains various enzymes involved in steroid biosynthesis, drug detoxification and glycogen metabolism. The nuclear membrane is reformed from the smooth endoplasmic reticulum in telophase[4]. The agranular reticulum probably acts as a channel for transporting secretory

---

[4] A stage in the division of a cell into two daughter cells.

substances into the exterior of the cell.

2. Golgi Complex.

The Golgi complex is probably a specialized portion of the endoplasmic reticulum. It is composed of four or more layers of thin flattened vesicles formed from unit membranes. There are direct connections between the endoplasmic reticulum and parts of the Golgi complex.

The Golgi complex is very prominent in secretory cells and is located on the side of the cell from which substances will be secreted. Some functions of the Golgi complex are as follows.

(a) Protein and other material made elsewhere in the cell are sometimes transported to the Golgi complex for condensation in membrane bound packets. This allows the transportation of material out of the cell or the storage of material within the cell.

(b) Carbohydrates in the Golgi complex are sometimes synthesized and complexed with proteins coming from the endoplasmic reticulum. For example, antibodies have carbohydrate moieties attached; mucoproteins such as mucous contain carbohydrates.

(c) The Golgi complex may also be active in lipoprotein synthesis.

3. Mitochondria.

Mitochondria are present in all cells. However the number of mitochondria per cell varies from a few hundred to many thousand, depending on the amount of energy required by the cell. For example, hepatic cells have many mitochondria while erythrocytes have relatively few. Mitochondria also vary in size and shape. Some are only a few hundred millimicrons in diameter and globular in shape while others are as large as 1 micron in diameter and are tubular in shape.

Mitochondria are composed of inner and outer unit membranes. Typically the inner membranes is corrugated to form shelves (cristae) which extend to the center of the mitochondrion. Knob-like repeating units, called elementary particles, are attached by slender stalks, to the inner membrane. The space enclosed by the inner membrane contains a fine granular material, called matrix, which has dense granules.

Mitochondria are the major sites of energy production in the cell since they contain many important enzymes. For example, most of the TCA[5] enzymes (See Appendix C) are present in the matrix while the cytochrome electron transport system and oxidative phosphorylation enzymes lie on the cristae. When nutrients and oxygen come in contact with the oxidative enzymes on the cristae, they eventually combine to form carbon dioxide and water. The liberated energy is used to synthesize a substance called adenosine triphosphate (ATP) (See Appendix C). The ATP then diffuses throughout the cell and releases its stored energy wherever it is needed.

Mitochondria are probably self-replicative whenever there is a need for increased amounts of energy. Indeed, the mitochondria contain a special type of DNA and RNA. It is known that DNA controls the replication of the entire cell. Therefore this substance might play a similar role in mitrochondria.

4. Lysosomes.

The lysosomes are 50-80 Å in diameter and are bounded by unit membranes. They contain about 35 different acid hydrolases. Hydrolases or hydrolytic enzymes are capable of splitting an organic compound into two or more parts by hydrolysis [6]. In hydrolysis, a hydrogen from a water molecule is combined with a part of a compound. The hydroxyl portion of the water molecule is then combined with the remaining part of the compound. For instance, protein is hydrolyzed to form amino acids and glycogen is hydrolyzed to form glucose.

Ordinarily, the membrane surrounding the lysosome prevents self digestion of the cell. However, many different conditions of the cell will cause the membranes of some lysosomes to break, allowing the release of the enzymes (frequently into pinocytic vesicles). These enzymes are then free to split the organic substances with which they come into contact. In this way, lysosomes act as digestive

[5] tricarboxylic acid cycle or citric acid cycle or Krebs cycle

[6] Reactions occur best in an acid environment.

organelles. They also remove damaged or senescent cell substances and are probably responsible for regression of tissues.

In several diseases such as Chediak-Higashi disease and chronic granulomatous disease of childhood, the lysosomal membrane in leukocytes (white blood cells) is highly resistant to breakdown. Because of this resistance, the control of bacteria is impaired and the victims die of infection.

Recent research suggests that a loss in potency of a certain lyosomal enzyme leads to muscular dystrophy.

5. Perioxisomes or Microbodies.

These membrane bound granules are somewhat larger than lysosomes. They contain the enzymes:

(a) oxidase, which catalyzes reactions that produce H_2O_2

(b) catalase, which catalyzes the reactions $2H_2O_2 \rightarrow 2H_2O+O_2$

6. Multivesicular Bodies.

These bodies consist of a number of small vesicles inside a larger vesicle and are larger than lysosomes. They are found in hepatic cells and a few other cell types. Their relationship to lysosomes and microbodies is unknown.

7. Microtubules.

Many cells have hollow nonbranching tubules that are approximately 250 Å in diameter and whose length varies from one to many microns. These microtubules are the major components of centrioles (see 9) and cilia. Their functions are:

(a) to support and shape the cell in the cytoskeleton

(b) to contract portions of the cell

(c) to transport material in the cell

(d) to move organelles (mitrochondria, for example, exhibit saltatory movement along tubules)

(e) to form axons

Note on Cilia

The surface of cells in the respiratory tract and in some portions of the digestive tract are covered by many minute, projecting hair-like processes which are 3 to 4 microns long. These processes, called cilia, are covered by the cell membrane. Each cilium is supported by 11 microtubular filaments - 9 double tubular filaments located around the periphery of the cilium and 2 single tubular filaments in the center of the cilium. Each cilium is an outgrowth of a structure that lies immediately beneath the cell membrane called the basal body.

The beating motion of the cilia serves to propel fluid from one part of a surface to another. For example, mucus can be moved out of the lungs and an ovum can be moved along the fallopian tube.

8. Microfilaments.

There are many types of microfilaments. Some are merely a dispersed form of microtubules containing specialized forms of proteins. Others have skeletal functions.

One of the most important classes of microfilaments is 50 Å in diameter and is made from the contractile protein actin. These microfilaments lie beneath the plasma membrane and account for its ruffling as well as pinocytosis,[7] phagocytosis[8] and ameboid movement. Microvilli also contain this type of microfilament.

Another class of thicker filaments (tonofilaments) which are about 100 Å in diameter act to strengthen and support cell.

9. Centrioles.

A centriole is a minute cylinder about .2 microns in diameter and about .4 microns long. Its walls are composed of nine parallel microtubules. Centrioles are found in all animal cells; none are found in higher plants. Multinucleate

[7] A mechanism by which the membrane actually engulfs some of the extracellular fluid and its contents. The invaginated portion of the membrane breaks away from the surface of the cell forming a vesicle which moves deeper into the cytoplasm.

[8] Ingestion of large particulate matter by the cell.

cells may contain many centrioles.

Diploid cells are cells with two sets of homologous chromosomes. One set is donated by the father; the other set is donated by the mother. Diploid cells have two pairs of centrioles lying close to each other near one pole of the nucleus. The two centrioles of each pair lie at right angles to each other. The two pairs of centrioles remain dormant until shortly before mitosis (cell division) takes place. Then microtubules connecting the pairs of centrioles begin to grow. This pushes the centrioles apart. At the same time other microtubules grow radially away from each pair. Some of these microtubules penetrate the nucleus. The set of microtubules connecting the centriole pair is called the spindle. The entire set of microtubules together with the two pairs of centrioles is called the mitotic apparatus.

10. Inclusions.

There are many different inclusions found in cells. Examples of inclusions are:

(a) secretory granules (located in secreting epithelial cells)

(b) lipid droplets (common in adipose tissue)

(c) glycogen particles (found in the liver)

(d) stored food such as protein and carbohydrates

(e) pigment. Pigment is classified into two types:

(i) exogenous - e.g., carbon, lead, silver

(u) endogenous - e.g., melanin, hemoglobin

Prokaryote cells have the same general structure as eukaryotic cells but have no membrane - surrounded nucleus and no membranous organelles such as mitochondria or endoplasmic reticulum. The cytosol (a uniform structureless, viscous fluid high in protein concentration) contains most of the cell enzymes as well as the metabolic intermediates and inorganic salts.

Photosynthetic leaf cells of higher plants contain most of the distinctive organelles and structures observed in the eukaryotic cells of animals. However, there are some distinctive differences. Cells of higher plants for example, characteristically contain plastids, membrane bounded organelles, some of which possess DNA different from that of the nucleus. Plastids containing pigment are called chromoplasts - e.g., carotenoid pigment or chlorophyll. Other plastids are unpigmented (leucoplasts). Common leucoplasts are the amyloplasts which store starch granules.

Chloroplasts (chlorophyll containing plastids) are about 8 times the size of large mitochondria. There are several to many chloroplasts per cell and they are variable in shape. The inner membranes form regular stacks of disk-like flattened vesicles called thylakoids, in which chlorophyll and the photosynthetic electron carriers are located. Chloroplasts are the site of photosynthetic formation of ATP, reducing power, and oxygen. The reduction of CO_2 to form glucose and starch also takes place in chloroplasts.

There are fewer mitochondria in plants than in the liver cells of animals. In the light, leaf cells obtain most of their energy from photosynthetic phosphorylation in chloroplasts. However, in the dark, mitochondrial respiration is the main energy supply.

Plant cells usually have larger vacuoles and have thicker and more rigid cell walls than animal cells have.

1.4 The Life Cycle of Cells

A cell is born by the division of a parent cell. This process is called mitosis. In mitosis each part of the parent cell is doubled. The parent cell then divides equally, on the average, into two daughter cells. These new cells are identical copies of their parent cell.

The metabolic activities of cells in the long interval between two successive mitosis is called the interphase. In recent years, these activities have been investigated using radioactive precursors. It was discovered that the interphase includes a series of metabolic events that are of great importance in the overall process of cell division. The orderly sequence of these metabolic activities, from the midpoint of one mitosis to the midpoint of the next mitosis, is called the cell cycle.

The four phases now used to express the time table of the cell cycle were first defined in 1953 by Howard and Pelc [6]. The cell cycle is shown diagramatically in Fig 1.4. The four phases and the corresponding time intervals they designate are described below. Some important associated metabolic events are also given.

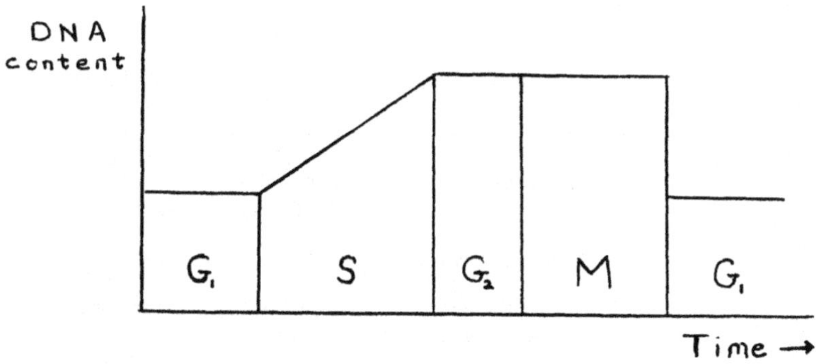

Figure 1.4 Life cycle of cells.

G_1: interval of time between the birth of the cell at division and the beginning of DNA replication.

(i) RNA synthesized

(ii) protein synthesized.

S: period of DNA replication

(i) DNA duplicated

(ii) RNA and protein synthesis continues with an increase in protein synthesis

G_2: interval between the end of S and the onset of cell division.

RNA and protein synthesis continues.

M: mitosis

(i) protein synthesis is reduced to a minimum

(ii) no DNA is synthesized

(iii) RNA synthesis occurs in early prophase[9] and late telophase[9]

The duration of all the phases may vary, but by far the greatest variation is found in the G_1 phase. When a cell cycle is long, most of the prolongation is in

[9] Divisions of mitosis which will be described below.

the G_1 phase; when a cell cycle is short, as it is in egg cells, there may be no measurable G_1 phase. The variability of the cell cycle for some tissues[10] is shown in the following table.

TABLE 1.3 RENEWAL TIMES

No renewal	Slow Renewal	Fast Renewal
nervous tissue	skin	g.i. tract (1-2 days)
muscle tissues	thyroid	hematopoietic system (e.g. erythrocytes - 120 days)
	liver (normally)	liver (post hepatectomy)
	T-lymphocytes (no antigen)	T-lymphocytes (antigen)

Our description of the cell cycle, must now be modified to account for the pattern of slow growth followed by fast growth shown by some cells and the fact that some cells do not grow at all. In the absence of growth - promoting factors (mitogens) most cells enter a resting state in the G_1 phase of the life cycle and so contain a diploid amount of DNA. Lajtha [23] and Quastler [33] both suggested that some cells merely do not remain longer in G_1 but that there is actually a resting state G_0. A cell in G_0 can return to the G_1 phase of the cell cycle. Once a cell has begun to replicate its DNA, it normally continues through the G_2 phase and then passes into mitosis. However, Gelfant [19] has noted the existence of extended periods of G_2 (as long as 16 days). These extended periods can also be interpreted as resting periods where the DNA content is tetraploid (twice the DNA content of diploid cells). Once into G_2 the presence of a new mitogenic stimulus will start another cell cycle. The recent version of the cell cycle is shown diagramatically in Fig. 1.5.

After mitosis, each of the two daughter cells can either (i) commence a cycle with the same characteristics as the mother cell (ii) differentiate to form a different kind of cell and commence a cycle with different characteristics

[10] Groups of cells that work together to perform a role in structure and functions of the body.

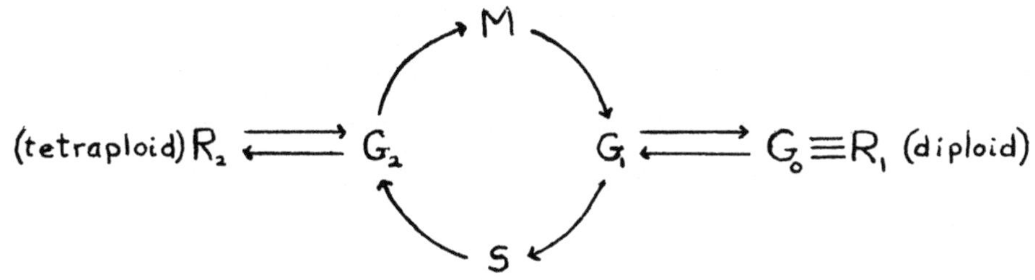

Figure 1.5 The cell cycle.

(iii) enter G_1 or prolonged G_0[11]; (iv) die or be physically removed from the population.

Remarks on Cell Differentiation

A quiescent state often leads to a differentiation process that may create a highly specialized cell (e.g. nerve cell) unable to divide. For instance, fibroblasts, epithelial cells and lymphoblasts are differentiated cells which divide. It is only when the physiological role of a given cell makes further division unwanted that the capacity to divide is lost.

Many forms of differentiation require specific differentiation factors. These not only promote differentiation, but also lead to the selective proliferation of partially differentiated "blast" cells. Some examples are given in Table 1.4.

TABLE 1.4 FACTORS IN DIFFERENTIATION AND PROLIFERATION

Example	Factor
red blood cell production	erythropoietin
nerve cell differentiation	nerve growth factor
epidermal cell differentiation	epidermal growth factor
uterine growth	estradiol

There are also forms of cell differentiation for which specific inducing molecules have not been found. The formation of a fat cell from a fibroblast seems to be

[11] The cell can be halted in G_0 if protein synthesis is stopped.

triggered by a rich food supply coupled with an inability to divide. The spontaneous fusion of myoblasts in culture yields striated muscles.

The life history of a cell can also be described at the chemical level by means of enzymes. Mitchison [30] has found that enzymes fall into groups with three kinds of history:

(a) continuous enzymes: These increase exponentially in a predictable way as the cell grows.

(b) step enzymes: These increase abruptly at certain points in the cell cycle; different step enzymes increase at different points.

(c) peak enzymes: These increase at certain points in the cell cycle and decrease thereafter. e.g. enzymes involved in the synthesis of DNA peak during the S phase.

The cell's production of an enzyme is the end result of a "readout" of a gene or genes for the enzyme. Hence, in light of Mitchison's observations, it is logical to ask whether genes determine the advance of the cell cycle from one phase to the next. This question has been answered affirmatively by several geneticists who experimented with one - celled organisms. They identified specific genes that control specific steps in the cell cycle. When the genes are made unworkable by mutation, the cell cannot advance beyond a certain point in the cell cycle. L.H. Hartwell of the University of Washington has described a number of such gene mutations in yeasts.

Remarks on Mitosis

Early light microscopists were fascinated by mitosis, since this was the only phase of the cell cycle in which notable changes occured. On the basis of these changes, mitosis was divided into four phases.

(a) Prophase

(i) In this phase, chromosomes first become thicker and more tightly coiled. Secondary coiling is visible with the light microscope.

(ii) Later, each chromosome appears to split longitudinally in half forming chromatids. This splitting actually occurs in the S and G_2 phases preceding mitosis but only becomes visible in late prophase.

(iii) After maximal contraction of the chromosomes the nuclear membrane and nucleoli disappear.

(iv) Each pair of centrioles moves to opposite poles of the cell. (The pair of centrioles was duplicated in earlier interphase). A spindle develops between the pairs of centrioles. Some fibres of the spindle attach to the chromosomes. The pair of centrioles and their radiating fibres is called an aster.

(v) Finally the nuclear membrane begins to degenerate and disappears by metaphase.

(b) Metaphase

(i) In this phase, the chromosomes first arrange themselves in an equatorial place and form an equatorial plate.

(ii) Each chromosome is then constricted at one plane along its length. An unstained zone, the centromere or kinetochore lies there. The two chromatids are free of one another except at the centromere. The spindle fibres are also attached at the centromere.

(iii) At the end of the metaphase, the centromeres divide and each chromatid is attached to a spindle by its own centromere.

(c) Anaphase

In the anaphase, sister chromatids move to the centrioles at opposite poles. At this stage, the chromatids can be designated as chromosomes.

(d) Telophase

(i) In this phase, membranes begin to form about groups of chromosomes. Later, when chromosomes unite into a single zone to become the nucleus, these membranes merge to form the nucleolemma. An identical set of chromosomes is assembled at each pole of the cell.

(ii) Nucleoli appear at satellite[12]- bearing chromosomes.

(iii) Segments of chromosomes uncoil to become euchromatin[13].

(iv) In animals, cytoplasmic division (or cytokinesis) is accomplished by the formation of a cleavage furrow which deepens and eventually pinches the cell

---

[12] A tiny terminal extension of chromatin material.

[13] Extended chromatin which is not visible by the light microscope. Dense coiled chromatin which is visible is called heterochromatin.

in two.

Cytokinesis in most plants involves the construction of a cell plate of pectin originating in the center of the cell and spreading laterally to the cell wall. Later, cellulose and other reinforcing materials are added to the cell plate, converting it into a new cell wall.

The four phases of mitosis were mainly described above in terms of events occuring in nuclear division (or karyokinesis). Definite changes can also be observed in cytoplasmic division. In the anaphase, bubbling of the cytoplasm can be observed as daughter nuclei separate. This implies a change in the solvation and the gelation of the cytoplasm. Nuclear mass separation is accompanied by cytoplasmic partition. Mitochondria, lysosomes, ribosomes and cytoplasmic membranes are distributed in approximately equal amounts about two newly formed nuclei. Although there is no assurance of equal distribution of cytoplasmic organelles to daughter cells, they do contain exactly the same type and number of chromosomes. The spindle may persist as a transient bridge between daughter cells. In the latter part of telophase a cytocentrum (cell center which may be gelated, contains centrioles, and is surrounded by Golgi elements) and Golgi elements are formed.

Remarks on Other Forms of Cell Division

(a) Amitosis

This process occurs in terminal or highly transient cell types (e.g. placenta, blood, multinucleated or giant cells of connective tissue). The nuclear membrane constricts and a single nucleus is pinched in two. However, the chromosomal material is usually not equally distributed to daughter cells.

(b) Polyteny

Polyteny occurs in the salivary gland cells of diptera[14]. DNA replication occurs without nuclear division. Replicates remain together and form giant chromosomes.

(c) Polyploidy

Here DNA replication and chromosome duplication occur but there is no

---

[14] The order of insects containing the true or winged flies.

karyokinesis. Hence, the nucleus contains more than the diploid number of chromosomes - e.g., hepatocytes, megakaryocytes.

(d) Meiosis

Sexual reproduction involves gametogenesis, the manufacture of gametes (sex cells) and their union (fertilization). Gametogenesis occurs only in specialized germ line cells of the reproductive organs. Gametes contain the haploid number (n) of chromosomes, but originate from diploid (2n) cells of the germ line. Hence, the number of chromosomes must be halved during gametogenesis. The reductional process is called meiosis. Two divisions occur in meiosis. The first division (meiosis I) produces two haploid cells from a single diploid cell. The second meiotic division (meiosis II) separates the sister chromatids of the haploid cells.

A more detailed discussion of the cell cycle is presented in Baserga [2,3] and Mitchison [29].

1.5 Control of Cell Proliferation

One of the goals of biologists is to discover how cells of multicellular organisms control their growth. Armed with this knowledge,new insight will be gained in embryology, developmental diseases and cancer.

At first sight, it would seem that if the ratio of the cytoplasmic mass to the nuclear mass exceeds a certain limit, some instability sets in and triggers cell division[15]. This nucleus cytoplasm relation theory (or critical mass theory) is untenable for all types of cells. Individual cells can exhibit large variations in mass at the time of division. Moreover, cells can be made to divide before they have doubled in size. D. Mazia and M. Mitchison starved yeast cells, which divide by division, by depriving them of nitrogen. Such cells could not grow because they could not make the proteins required for growth due to the absence of nitrogen. However, they converted some of their own protein into the kinds needed for cell division. The yeast cells divided into abnormally small daughter cells.

Although the critical mass theory may be invalid, the size of a cell appears to be limited by the capacity of the single nucleus to support growth. A cell

---

[15] This theory was proposed by a biologist R. von Hertwig in 1908.

containing two nuclei or one nuclei with two sets of chromosomes can grow to twice its normal adult size. For example, mammalian cultured cells when exposed to appropriate doses of ionizing radiation may reproduce their genetic material repeatedly. However, cytokinesis does not occur. As the amount of genetic material in the cell's nucleus increases, so does the cell's size.

It also has been hypothesized that total cellular volume is a major factor in controlling cell division. This seems to be true in several prokaryotes and lower eukaryotes. However, in mammalian cells the process is more complicated. These cells follow a series of well coordinated events which are required for cell division. Some current research into these events are described below. Several hypotheses for regulation of these events already exist. They range from a master clock [30], to a mitotic oscillator [42], to the interaction of parallel pathways controlled by means of checks and balances [30]. Mathematical volume models will be presented in Chapter 3.

One productive method of studying cell proliferation utilizes the growth of cells in culture. The cells excised from an animal to start a culture are called the primary cells. Their descendants which are present several division cycles later are called secondary cells. Most secondary cell cultures fail to multiply more than 20 to 50 times. Rarely, a cell multiplies indefinitely to form a cell line. The reason for this unlimited growth is unknown. It is easier to obtain cell lines from tumor cells than from their normal equivalents. Hence, some cells which produce cell lines may do so because they have acquired by mutation many of the essential properties of cancer cells. Cell death may be related to the aging phenomena in higher vertebrates.

Possession of well-defined cell lines led to the discovery of the detailed nutritional requirements of cultured cells. However, division is not entirely dependent on a well-defined collection of amino acids and vitamins. As pointed out in the first paragraph of this section, starved cells can proliferate but are usually smaller. There are other factors involved.

One of these factors seems to be a group of polypeptide hormones that are present in blood serum. These act by binding to the surface of their target cells.

As a result, the activity of the membrane bound enzymes, adenyl cyclase and guanyl cyclase are changed. This decreases the level of cAMP[16] and increases the level of cGMP[17]. For some unknown reason, these changes trigger cell division in some cells. The secret of cell cycle control may be revealed when a deep understanding of the cell membrane is attained.

Other agents which can effect the growth of cells are the chalones[18] investigated by W.S. Bullough of the University of London and promine and retine studied by A. Szent-Gyorgyi at the Marine Biological Laboratory in Woods Hole, Mass.

Bullough [8] suggested that a defective chalone mechanism may lead to instability of tissue size and so to the uncontrolled growth of some cancers. Glass [20] tried to quantify this idea and also study the effect of different geometries on the stability of growth [40]. Bell [4] also formulated a mathematical model to study mitotic inhibitors.

Remission of cancer has been achieved by administering chalones in the experiments of Bichel and Barfod [5], Mohr et al. [31] and Rytomoa and Kiviniemi [36].

Further details on chalones can be found in [8,17,21].

D. Mazia caused unfertilized eggs of sea urchins to replicate their chromosomes by adding a little ammonia to the sea water containing the eggs. The treated eggs remain unfertilized and can even be fertilized later. Ammonia is not a substitute for fertilization but only mimics something that normally happens after fertilization. Other experiments show that simple factors in the extracellular environment (e.g. ions of hydrogen, potassium and calcium) can affect the cell cycle.

---

[16] Cyclic adenosine monophosphate. Its structure is similar to adenosine 5-monophosphate except that the phosphate group is joined to the 3-carbon by an ester linkage (see Appendix A).

[17] Cyclic guanine monophosphate. Its structure is similar to cAMP except adenosine is replaced by guanine (see Appendix B).

[18] A chalone is a mitotic inhibitor produced by a tissue.

Recent evidence suggests that the transition from one phase of the cell cycle to the next phase requires a distinct event. The existence of "points of no return"[19] in the cell cycle have been known for a long time. For example, division in the protozoan tetrahymena can be prevented by raising the environmental temperature before a definite time in the cell cycle. However, once this point is passed, exposure to high temperature can not prevent division. This point in the protozoan cycle is a point of no return. It is just as if switches exist at various points in the cell cycle. Once one of these switches has been thrown, the cell must move from one phase to the next. The technique of cell fusion can be used to answer the question of whether such switches exist.

Sendai viruses alter the membranes of properly cultured cells in such a way that two or more adjacent cells can fuse into one cell. This newly-formed cell contains the cytoplasm and the nuclei of all the fused cells within a common membrane. Even cells from different species of animals can be fused by the action of the virus. These hybrids behave functionally as cells and reproduce as hybrids[20].

Consider the following cell fusion experiments:

	Investigator	Cell type 1	Fused type 2	Result
1.	P.N. Rao and R.L. Johnson University of Colorado Medical Center, Denver	S phase	G_1 phase	Nuclei of G_1 cells began DNA synthesis long before normal
2.	J. Graves University of California, Berkeley	hamster	mouse	(i) Both nuclei began DNA synthesis at the same time (ii) The shift to S phase occured after the shorter G_1-interval of hamster cells
3.	Same as 2	hamster	mouse	Each nuclei followed its own characteristic replication time table after entering the S phase.

[19] This descriptive phrase is due to D. Mazia.

[20] Sometimes one set of chromosomes dominates another in the sense that succeeding generations contain cells in which some dominated chromosomes are no longer present. Eventually, all the dominated chromosomes drop out.

[21] Normally have a shorter G_1 phase than a mouse cell.

Experiments 1 and 2 show that there is something which can start the replication of chromosomes. This substance can force any nucleus, ready or not, to enter the S phase. Experiment 3 implies that the signal for starting the S phase has no control over what happens once the phase has started. At present, the nature of the signal is not known. It may be a special molecule that can switch on replication or a simple change in the external environment which causes a change in the internal environment of the cell.

Virus-induced cell fusion has also been used to study entrance into the M phase. This period of cell division is characterized by condensation of the chromosomes making them visible under the light microscope. The following experiments were also performed by Johnson and Rao. They fused M phase cells with cells in other phases as shown below.

Phase of other Cell	Result
1. G_2	Normal looking condensed chromosomes in G_2 nucleus. Chromosomes are double.
2. G_1	Condensed single chromosomes.
3. S	Chromosomes condense in small fragments. This effect is called pulverization.

The conclusion from these experiments is that there is something in the M phase that forces chromosomes to condense even when they are obviously not ready to do so. Whatever this substance is,it appears to work on any cell. Hybrids of remotely related animal cells (e.g. men and toads) give the same experimental results.

In the body, some cells are cycling while others are noncycling. For instance, some cells may be in the G_o state. Although the aforementioned cell culture experiments indicates some possible biochemical explanation for cellular proliferation, it is not certain how cells become noncycling. There is some evidence that lack of nutrition may cause some cells to become noncycling. For some types of cells this may be the main factor, since many types of cells can be made to reenter the cycle by merely culturing them in the laboratory. However, other types of cells will not return to the cycle when cultured,for example, noncycling lymphocytes. The noncycling lymphocytes can be transformed into cycling ones by exposing them to lectins (a plant protein). It is therefore possible,that the internal environment of such cells has been changed by a hormonelike substance or by changes in the extracellular concentration of ordinary ions and molecules.

The influence that a cell society exercises over the cell cycle of its individual members is observable in mammalian cell cultures. Normally, cells grow and multiply on the bottom of a flask. Their growth ceases when the substrate has become covered by a single layer of cells. This phenomena is called contact inhibition.

Harry Rubin of the University of California at Berkeley has shown that contact inhibition can be overcome by infecting the cell culture with a cancer virus. The infection transforms the cells in a number of ways:

(i) The cells may change in appearance.

(ii) They will grow on top of each other to form more than a single layer.

(iii) If put into a host animal, they will grow in its body as cancer cells.

Contact inhibition is the effect of the cell society as a whole on its individual members. This can be demonstrated experimentally. If a single layer sheet of cells is punctured, the cell cycle resumes. This happens not only along the edges of the wound but in all parts of the culture. Reproduction persists until the empty space is filled. The cell cycle is then shut down.

The hypothesis that gentle caresses between adjacent cells will shut down the cell cycle is inconsistent with the following observations. Different kinds of cells will grow to different degrees of crowding under different conditions before the cycle is halted. Contact inhibition can be overcome by making changes in the environment; for instance, by adding large amounts of blood serum. Moreover, some kinds of cells are not subject to contact inhibition at all.

Some recent experiments indicate that the cell membrane may play a key role in contact inhibition. One group of workers in the Department of Embryology of the Carnegie Institution of Washington has found that the membrane potential changes during the cell cycle. The cell cycle will be halted by changes in the external environment that induce changes in the membrane potential.

Werner R. Loewenstein's group at the University of Miami School of Medicine has made electrode studies of the junctions between cells in contact with each other. The channels that the junctions provide, allow ions and small molecules to pass between the cells. They have found evidence suggesting that contact-inhibitable cells make junctions, while junctions are not formed between non contact-inhibitable cells.

Max M. Burger of Princeton University was the first to discover that the membrane of contact-inhibitable cells reacts with those plant proteins that induce cycling in noncycling lymphocytes. The reaction is easy to see. When the protein is present, the cell surface becomes sticky and the cells clump together. Cells that are not contact-inhibitable are unaffected by the presence of the protein.

All of the previously described processes or agents either block or stimulate cell division, but will not do both. Electric current can block the growth of cells. Moreover, by applying electric current at the desired site of growth, tissues have been regenerated in animals. Dr. Robert O. Becker of the Veterans Administration Hospital, Syracuse, New York has studied the electrical control of growth processes since the late 1950's. He has successfully treated fractures which would not normally heal due to inadequate bone growth by means of electrical stimulation.

The explanation of the above results on the cellular level may lie in the recent work of Dr. Clarence D. Cone at the Veterans Administration Hospital at Hampton, Virginia. He discovered a method of altering the concentrations of sodium, potassium and chloride ions in a cell. This changes the electrical potential difference across the cell membrane. Potential differences from -70 to -90 millivolts will block cell divisions, while those from -10 to -20 millivolts will stimulate cell growth.

Dr. Cone's early experiments were conducted on cultured hamster cells. Recently, nerve cells from the spinal cord have been induced to divide. These nerve cell experiments might lead to methods of combating senility and treating brain or spinal cord injuries. In the future, the regeneration of damaged tissue such as myocardium and kidney may well be standard clinical practice. Further, it might be possible to stop the uncontrolled proliferation of cancer cells by manipulating their ionic concentration.

1.6 Cancer

A popular misconception is that cancer cells grow faster than normal cells. Although the length of the cell cycle varies in different kinds of cells, it is shorter in certain animal cells than in some of the fastest growing tumors [3]. Later, we shall learn that tumor growth involves other kinetic parameters besides the speed of cell proliferation. Actually, the common characteristics of the many kinds of malignant growth that are called cancer is that such cells pursue the cell cycle without restraint. A cancer cell has lost the ability to control growth and division.

Cancer cells do not suddenly appear; an evolutionary process is involved. Most tumors develop as a clone[22] from a single cell [32]. This means that even though a large number of cells may be affected by a cancer-promoting agent, the macroscopic tumor that ultimately develops usually represents the progeny of a single cell, or at most, a few cells. Other precancerous cells in the exposed tissue never successfully proliferate or they are destroyed before progressing to a fully developed tumor.

The few exceptions to the unicellar concept of tumor origin are: tumors of viral etiology and growths with a strong hereditary component.

In tumors of viral etiology, there is a possibility of infection from adjacent cells; in tumors caused by a hereditary factor the gene defect involves every cell and greatly increases each cells susceptibility to cancerous change.

It had been thought that cancer strikes first in differentiated cells. This viewpoint assumes that the cause of cancer is connected with the genes that make

---

[22] A population of cells descended from a single cell.

a differentiated cell what it is. Recent evidence suggests that some cancers originate in normal stem cells. Stem cells are those which produce the differentiated cells. When the stem cells become malignant, they produce malignant undifferentiated cells. Such cells are observed in the early stages of cancer.

The three main stages in tumor development are:

(a) Initiation of Neoplasia

An induced change in a single previously normal cell makes it neoplastic. For the purpose of this model neoplasia is defined as a degree of escape from normal growth control. This escape provides the cell with a selective growth advantage over the normal cell from which it was born. Many studies indicate that changes in the external membrane causes defeciencies in normal growth control mechanisms mediated by cell to cell contact.

The specific gene products or genetic events that produce neoplasia usually cannot be seen by present methods. These facts, coupled with the reversibility of the transformation in some culture systems, suggests that initiation usually involves altered gene expression rather than a structural mutation[23]. In a few forms of cancer, the initial neoplastic event is visible at the chromosome level in nearly all cases of the disease - e.g. the Philadelphia chromosome is present in 85-90% of chronic granulocytic leukemia cases.

Specific agents that may initiate neoplasia are:

(i) ionizing radiation (causes genetic changes in cellular DNA)

(ii) carcinogenic (cancer producing) chemicals [The fact that many carcinogenic chemicals are general mutagens[24] makes it likely that mutation in a large number of genes can lead to the cancerous phenotype]

(iii) oncogenic viruses [see Appendix B]

(iv) inherited genetic errors - for instance, defective DNA repair in xeroderma pigmentosum leads to multiple skin cancer.

(v) unknown genetic deficits

---

23 Another supporting fact is that the fusion of a tumor and normal cell yields a nontumorous cell.

24 Substances which increase the mutability of a gene.

(b) Genetic Instability

Evidence indicates that neoplastic populations have a higher frequency of mitotic errors and other genetic changes. Experiments suggest that increased mitotic activity in tumors only partially accounts for this difference. Chromosome studies of solid tumors show that genetic instability may increase as the neoplasm evolves.

Some factors which act after tumor initiation to produce genetic instability are: radiation, viruses, inherited and acquired genetic defects, nutrition (e.g. deficiencies in a single amino acid can crease the frequency of nondisjunction[25] in cell culture) and therapy.

The specific type of genetic alteration may range from point mutations to major chromosomal aberrations (e.g. changes in chromosome number). Often it is difficult to demonstrate these changes or to find new products. Once again, alterations in genetic expression may be responsible.

(c) Emergence of Variant Subpopulations

Genetic instability often leads to the production of variants. Most variants die, because of metabolic or immunologic disadvantages. Occasionally a variant has a selective advantage. Its progeny becomes the predominant subpopulation until the appearance of a more favorable variant. This stepwise sequence differs in each tumor and results in a different, aneuploid karyotype (chromosomal constitution)[26] in each fully developed malignancy (see Fig. 1.6).

Some factors influencing the emergence of variant subpopulations are: inherited immunodeficiency, nutrition, infection, immune status of patient and therapy.

---

[25] Chromosomes do not separate.

[26] To determine the karyotype, cells are cultured in an artificial medium. Their growth is arrested in metaphase by exposing them to colchicine solution. After a brief treatment with a hypotonic salt solution, the cells are fixed, spread on a slide, stained and photographed. Individual chromosomes are cut out from enlarged photographs, matched in pairs of homologous chromosomes and arranged in order of descending length.

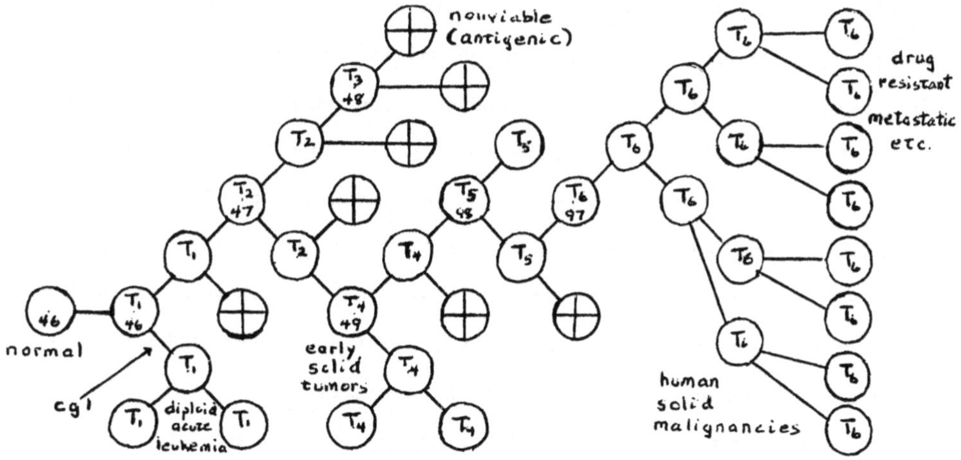

Figure 1.6 Model of clonal evolution in neoplasia.

Biological characteristics of tumor progression (e.g., morphological and metabolic loss of differentiation, invasion, metastases and resistance to therapy) parallel the stages of genetic evolution. Human tumors with minimal chromosome change (diploid acute leukemia, chronic granulocytic leukemia) are considered to be early in clonal evolution. Human solid cancers, typically highly aneuploid, are viewed as late in the developmental process.

The above model has several implications for therapy. The unicellular origin for most tumors suggests that a common membrane-related metabolic alteration or antigen may arise at the time of the initial neoplastic change. By taking advantage of this possible difference in membrane structure, one may be able to eradicate the clone and effect a cure.

Success in such a project is more likely in those human neoplasms that occur early in the evolutionary process and show the least evidence of genetic alteration. At present, consistent alterations either antigenic or metabolic which are useful for therapy have not been found. The possibility remains that each patient's cancer may require individual specific chemotherapy; even this may be thwarted by the emergence of genetically variant sublines. In such situations, it might be more feasible to produce specific cytotoxic antiserums or lymphocytes against a particular tumor than to design a specific chemotherapeutic agent for each neoplasm.

Finally, it might also be possible to provide an environment which forces the tumor cell population to cease unlimited proliferation and move into a state of controlled differentiation. This approach has been demonstrated to work for a few tumors. In general, this result seems more probable when the karyotype of the tumor is normal or near-normal - that is, in the early stages of the disease. Recently, scientists at the Sloan-Kettering Institute are investigating an intriguing substance they have named tumor necrosis factor (TNF). It is found in the body and inhibits the growth of human melanoma cells implanted in mice. TNF also curbs the proliferation of leukemic cells in cultured mice cells.

The paper of Armstrong and Doll [1] is one of the earliest scientific articles studying the influence of diet and environment on a number of human cancers. Several recent books [7,18,39] discuss modern hazards in the etiology of cancer.

The approach of molecular biology to cancer research is described in Watson [43].

Mathematical theories of the development cancer will not be presented in this book. Readers interested in pursuing this topic should consult Appendix E.

1.7 Metastasis and Invasion

Two fundamental concepts for understanding the spread of cancer are metastasis and invasion. A metastasis is a secondary growth center of cancer cells. The cells of the metastasis were detached from a primary tumor located at another site. Invasion is the infiltration of adjacent tissue by the malignant tumor.

Tumors exhibit different degrees of invasion and metastasis. Some tumors are localized (giant fibroadenoma) metastatic (thyroid); others are invasive (basal skin cancer); still others have a combination of these three properties. The reason why some tumors are invasive is not known. No enzyme or substance has been discovered to be present in all invasive tumors. Some possible causes of invasiveness are given below:

(i) The force due to the internal pressure of the tumor.

(ii) Tumor cells exhibit motility; they move about 1 cm. in 10 days. Normal cells move more slowly and exhibit contact inhibition.

(iii) Tumor cells can be separated much more easily than normal cells.

(iv) The tumor destroys normal tissue at its periphery by means of certain enzymes:

(a) The stromal ground substance contains hyaluronic acid; several tumors possess hyaluronase, an enzyme which disrupts hyaluronic acid.

(b) Many tumors possess fibrinolytic activity (i.e. lyse fibrin)

(c) Peptidase activity is greatest at the periphery of the tumor where the growth rate is greatest.

Moreover, some tumor cells produce lactic acid due to anaerobic[27] glycolysis. An acidic environment provides optimal conditions for the action of lyosomal enzymes.

(v) Trypsin (a protealytic enzyme may increase the invasiveness of certain tumors. Treatment of cells in culture with this substance increase their growth. (Tumors of the kidney treated with antiprotealytic enzymes are 50% less invasive.

(vi) The death of normal tissues. This may be due to lytic or toxic products secreted by the tumor; however, these products also destroy tumor cells. The normal cells may also die due to starvation since cancer cells utilize large amounts of glucose.

Three conditions must be fulfilled for metastases to occur.

(a) Invasion

Some tumors generally spread to regional lymph nodes (e.g. epithelial carcinomas) while others tend to spread to blood vessels (e.g. sarcomas). Alternately, blood vessels may send branches to the tumor. Folkman et al. [11] isolated a diffusible factor from tumors, called tumor-angiogenesis-factor (TAF). This factor induces endothelial cells to divide and causes vascular infiltration. Further details can be found in [12,13,14,15,16]. It has been shown by Rifkin et al. [35], that some human cancer cells produce plasminogen activator. This also leads to greater vascularization of the tumor. The increased vascularization not only increases the nutrient supply of the tumor but provides the cancer cells with an entrance to the circulatory

---

[27] able to survive without oxygen

system of the host. Experiments have demonstrated that tumor cells can reach afferent lymphatics from blood vessels and vica versa without direct invasion of the vessel's wall.

(b) Dissemenation to a different site.

The cancer cells enter the vessels and are carried by the lymphatic or circulatory system. Usually, a small number of cancer cells in the blood does not lead to metastases - thousands of cancer cells are required. Some walls of vessels may be impenetrable to cancer cells. For example, tumor emboli (from breast, liver or kidney) can occlude pulmonary arteries but have little capacity to penetrate the arterial walls and no metastases occur in the vicinity of these arteries.

(c) Attachment and growth in a new site

Circulating cancer cells may become arrested in the small vessels of some organ. These cells penetrate the organ tissue creating new sites of growth.

Metastases in certain organs can sometimes be explained by the anatomy of the circulation. However, not all sites of metastases can be explained in this manner. For example, carcinoma of the breast yields 15% metastases in the ovaries while carcinoma of the lungs does not yield any there. Muscles, lung and spleen are rarely involved in metastases. Such facts have led to the "fertile soil hypothesis". This hypothesis states that certain tissues are more suitable than others for supporting the growth of a particular cancer cell cell. Traumatized tissue is more fertile for the growth of cancer cells than normal tissue. (Hence, tissue injury may be one of the reasons that cancer is sometimes spread by massage or surgery). The following statistics also supports the fertile soil hypothesis. More metastases spread to the liver than to the lungs. However more metastases arise from a primary tumor in the lungs than a primary tumor in the liver.

The following brief discussion of mathematical models for some of the processes discussed in this section is presented for the interested reader. An early discussion of the diffusion of oxygen into a spherical cell can be found in Rashevsky [34]. Blumenson and Bross [6] applied this spherical diffusion equation to study the growth of avascular tumor cells in the anterior chamber of the eye. Deakin [9] formulated a model for vascular development in a melanoma transplant.

The first deterministic model describing the entire process of tumor transplantation up to the development of metastases was developed by Liotta et al [25]. The model was refined by Saidel et al [37].

Recall that while great quantities of tumor cells are released into the circulation very few (less than 0.1 percent) survive to form metastatic foci. Consequently a stochastic description of metastases is more appropriate. Such a model was developed by Liotta et al. [26] to provide a framework for predicting the development of metastatic foci from clumps in the pulmonary vessels. (Also see [27] and [38].)

The reader interested in obtaining a more detailed knowledge of cell biology can consult Leeson [24], Mazia [28], Mitchison [29] and Watson [49].

REFERENCES

1. Armstrong, B. and Doll, R. Environmental factors and cancer incidence and mortality in different countries with special references to dietary practices, Int. J. Cancer, 15, 617-631, 1975.

2. Baserga, R. Multiplication and Division in Mammalian Cells, M. Dekker, New York, 1976.

3. Baserga, R. The relationship of the cell cycle to tumor growth and control of cell division: a review, Cancer Res., 25, 581-590, 1965.

4. Bell, G.I. Models of carcinogenesis as an escape from mitotic inhibitors, Science, 192, 569-572, 1976.

5. Bichel, P. and Barfod, N.M. Specific chalone inhibition of the regeneration of the JB-1 ascites tumor studied by flow microfluorometry, Cell Tissue Kinet., 10, 183-193, 1977.

6. Blumenson, L.E. and Bross, I.D.J. A possible mechanism for enchancement of increased production of tumor angiogenic factor. Growth, 40, 205-209, 1976.

7. Braun, A.C. The Biology of Cancer, Addison-Wesley, Reading, 1974.

8. Bullough, W.S. Mitotic and functional homeostasis: a speculative review, Cancer Res., 25, 1683-1727, 1965.

9. Bullough, W.S. Mitotic control in adult mammalian tissues, Biol. Rev. Cambridge Phil. Soc., 50, 99-127, 1975.

10. Deakin, A.S. Model for initial vascular patterns in melanoma transplants, Growth, 40, 191-201, 1976.

11. Folkman, J., Merler, E., Abernathy, C., and Williams, G. Isolation of a tumor factor responsible for angiogenesis, J. Exp. Med., 133, 275-288. 1971.

12. Folkman, J. Tumor angiogenesis: therapeutic implications, New. Eng. J. Med., 285, 1182-1186, 1971.

13. Folkman, J. Anti-angiogenesis: new concept for therapy of solid tumors, Ann. Surg., 175, 409-416, 1972.

14. Folkman, J. Tumor angiogenesis factor, Cancer Res., 34, 2109-2113, 1974.

15. Folkman, J. and Klagsbrun, M. Tumor angiogenesis: effect on tumor growth and immunity. In Fundamental Aspects of Neoplasia, 401-412, edited by Gottlieb, A.A., Plescia, O.J. and Bishop, D.H.L., New York: Springer-Verlag, 1975.

16. Folkman, J. and Cotran, R. Relation of vascular proliferation to tumor growth, Intl. Rev. Exptl. Pathol., 16, 207-248, 1976.

17. Forscher, B.K. and Houck, J.C. (editors). Chalones: Concepts and Current Researches, Monograph 38, National Cancer Institute, Bethesda, 1973.

18. Fraumeni, J.F. (editor). Persons at high risk of cancer. An Approach to Cancer Etiology and Control, Academic Press, New York, 1975.

19. Gelfant, S. Patterns of cell division: the demonstration of discrete cell populations. In Methods of Cell Physiology, Vol. 2 (D.M. Prescott, Ed.),Academic Press, New York, 1966.

20. Glass, L. Instability and mitotic patterns in tissue growth, Trans. ASME: J. Dyn. Syst. Measure Contr. Ser. G, 95, 324-327, 1973.

21. Houck, J.C. (editor). Chalones, Elsevier, New York, 1976.

22. Howard, A. and Pelc. S.R. Synthesis of deoxyribonuncleic acid in normal and irradiated cells and its relation to chromosome breakage, Heredity (Suppl), 6, 261-273, 1953.

23. Lajtha, L.G. On the concept of the cell cycle, J. Cell Comp. Physiol. Suppl. 1, 62, 143-145, 1963.

24. Leeson, T.S. Histology, 3rd ed., Saunders, Philadelphia, 1976.

25. Liotta, L.A., Kleinerman, J. and Saidel, G.M. Quantitative relationships of intravascular tumor cells, tumor vessels and pulmonary metastases following tumor implantation, Cancer Res., 34, 997-1004, 1974.

26. Liotta, L.A., Saidel, G.M. and Kleinerman, J. Stochastic model of metastases formation, Biometrics 32, 535-550, 1976.

27. Liotta, L.A., Saidel, G.M. and Kleinerman, J. Diffusion model of tumor vascularization and growth, Bull. Math. Biol., 39, 117-128, 1977.

28. Mazia, D. The Cell Cycle, Scientific American, 230, 54-64, 1974.

29. Mitchison, J.M. The Biology of the cell cycle, Cambridge University Press, Cambridge, 1971.

30. Mitchison, J.M. Sequences, pathways and timers in the cell cycle, in Cell Cycle Controls (Podilla, G., Ed.) Academic Press, New York, 1974.

31. Mohr, R., Althoff, J., Kinzel, V., Suss, R. and Volm, M. Melanoma regression induced by "chalone": a new tumor inhibiting principle acting in vivo, Nature, 220, 138-139, 1968.

32. Nowell, P.C. The clonal evolution of tumor cells, Science, 195, 847-851, 1976.

33. Quastler, H. The analysis of cell population kinetics, in Cell Proliferation (Lamerton L.G. and Fry, R.M.J. (editors) , Blackwell Scientific, Oxford, 18-36, 1963.

34. Rashevsky, N. Mathematical Biophysics. Univ. of Chicago Press, Chicago, 1948.

35. Rifkin, D.B., Loeb, J.N., Moore, G. and Reich, E. Properties of plasminogen activators formed by neoplastic human cell cultures, J. Exp. Med., 139, 1317-1328, 1974.

36. Rytomaa, T. and Kiviniemi, K. Regression of generalized leukemia in rat induced by the granulocyte chalone, European J. of Cancer, 6, 401-410, 1970.

37. Saidel, G.M., Liotta, L.A. and Kleinerman, J. System dynamics of a metastatic process from an implanted tumor, J. Theor. Biol. 56, 417-434, 1975.

38. Saidel, G.M., Liotta, L.A. and Kleinerman, J. System dynamics of a metastatic process from an implanted tumor, J. Theor. Biol., 55, 417-434, 1976.

39. Searle, C.E. (editor). Chemical Carcinogens, Am. Chem. Soc., Washington (D.C.), 1976.

40. Shymko, R.M. and Glass, L. Cellular and geometric control of tissue growth and mitotic instability, J. Theor. Biol., 63, 355-374, 1976.

41. Suss, R., Kingel, F. and Scribner, J.D. Cancer: Experiments and Concepts, Springer-Verlag, New York, 1973.

42. Tyson, J. and Kauffman, S. Oscillator control of mitosis, J. Math, Biology, 1, 289-310, 1975.

43. Watson, J.D. Molecular biological approach to the cancer problem. In the Biological Revolution. Social Good or Social Evil, edited by Fuller, W., Doubleday, New York, 169-184, 1972.

44. Watson, J.D. Molecular Biology of the Gene, 3rd ed., W.A. Benjamin, Menlo Park, Cal., 1976.

II. MODELLING AND CELL GROWTH

2.1 Introduction

In this chapter, mathematical models of cell growth will be studied. It has been observed that cells can exhibit no growth (nerve cells), constant growth (bone marrow stem cells, intestinal crypt cells and epithelial cells) and variable growth (liver cells). Some cell populations just decay, like adult ovary cells which never divide but are released periodically. Section 3 contains some of these growth laws and their generalization.

In Section 4, a model for normal and abnormal tissue growth is defined by a system of differential equations using a generalized growth law. The model consists of a proliferating and a nonproliferating compartment. The compartments are also parametrized by their death and transfer rates. A stability analysis of the system shows in a precise way how abnormal growth may result from a perturbation of the parameters leading to a change in the equilibrium of tissue size. This analysis sheds some light on why normal cells having a faster growth rate than tumor cells do not grow wildly but hold to their steady state.

The results described in the preceding paragraph are based on a deterministic model. However, many biological phenomena exhibit randomness for instance, the length of the cell cycle is not constant. It seems natural to describe such random processes by using a stochastic model. In the next section we will try to reconcile the deterministic and stochastic approaches.

2.2 Modelling Philosophy

A plethora of cell models exist in the literature. These range from continuous to discrete; from stochastic to deterministic. Frequently the same phenomena is described by two different models. Which of these is correct?

The applied mathematician is not necessarily concerned with absolute precision nor a perfect description of the underlying phenomena. Some models exhibit nearly perfect correlation with all the physical parameters of the process being modelled. Others exhibit complete ignorance of the underlying mechanism (e.g. the black box approach via imput-output analysis in engineering). The criterion used by the applied mathematician is a practical one. If the results predicted by his model are sufficiently close to the observed data, the model is suitable. Hence, two models which give different results can both be satisfactory if they are both close to the observed data when biological variability and experimental errors are taken into account.

Some simple linear models will be used to illustrate the relation between the deterministic and corresponding stochastic models. They are only concerned with predicting cell number[1] at a particular time. More sophisticated models are required for studying the phases of the cell cycle and will be presented later.

(a) discrete models

(i) deterministic model.

Assume that every cell has a fixed life length or cycle time c and that exactly c seconds after it is born it splits into two identical cells. It is often said that the cell dies and produces 2 descendants. These two daughter cells also live c seconds and divide forming 2 cells.

The following notation will be used:

n_o = the number of cells in the 0th generation (i.e. the number of original or ancester cells just born at time $t = 0$)

n_1 = the number of cells in the 1st generation (i.e. all daughter cells or descendants of cells of the 0th generation)

n_k = the number of cells in the kth generation (i.e. cell descendants of the (k-1)th generation)

Since each cell divides into 2 cells

$$n_1 = 2\,n_o$$

$$n_2 = 2n_1 = 2^2\,n_o$$

$$n_3 = 2n_2 = 2^2\,n_1 = 2^3\,n_o$$

and generally

$$n_k = 2n_{k-1}$$

[1] The cell number is determined in a variety of ways - for example, by using the number of cells per unit volume in a sample, the mass of a cell, or the biochemical activity of the cells.

It follows that

(1) $$n_k = 2^k n_o \qquad k = 0,1,2,3,\ldots$$

$$= n_o 2^{t/c} \qquad t = kc$$

The last formula can be written as

(2) $$n(t) = n_o e^{(\frac{\ln 2}{c})t} \quad .$$

For comparison with the stochastic model two other modes of growth will be considered. First assume that all cells die; then $n_k = 0$. Next suppose that one of the daughter cells dies while the other lives and each generation exhibits this behavior. The population does not increase nor decrease and so $n_k = n_o$.

The results in this section can be summarized by the formula

(3) $$n_k = \mu^k n_o \qquad k = 0,1,2,3,\ldots$$

where μ is the number of live cells produced by each cell.

(ii) stochastic model (Galton-Watson Process)

In this model the cell cycle time c is assumed to be constant. However, each cell independently of every other cell gives birth to a random number 0,1 or 2 offspring with probabilities p_o, p_1, p_2, respectively. Note that $p_o + p_1 + p_2 = 1$. The mean[2] number of descendants is

(4) $$\mu = 0 \cdot p_o + p_1 + 2\, p_2 \ .$$

Unlike the deterministic process the number of descendants Z_k of a single cell (or ancestor) in the kth generation is not a constant but a random variable. This randomness is illustrated by the diagram in Figure 2.1. Cells are represented by dots; the number beside the point indicates the number of cells represented by the dot. Daughter cells are connected to mother cells by a line segment. The subscripted letters on the line segments represent the probability of the mother cells giving birth to the indicated number of daughter cells. For example, the line segment connecting 1 in generation 0 to 2 in generation 1 is labelled p_2

[2] Average and expected are also used.

to represent the probability of a cell having 2 offspring. Two cells can have a total of two offspring only if the following events occur

E_1: cell 1 has no descendants and cell 2 has 2 daughter cells.

E_2: cell 1 has two daughter cells and cell 2 has no descendants.

E_3: cell 1 has 1 daughter and cell 2 has 1 daughter.

Hence, probability that 2 cells have a total of two offsprings $p_0 p_2 + p_0 p_2 + p_1^2$. Therefore, the line segment joining 2 in generation 1 to 2 in generation 2 is labelled $2p_0 p_2 + p_1^2$.

From Fig. 2.1 it is evident that the number of offspring in the first generation is not determined with certainty, but only that $Z_1 = 0,1,2$ with probabilities p_0, p_1, p_2 respectively. The situation is more complicated for generation 2. However a careful examination of Fig. 2.1 reveals that

$$P(Z_2 = 0) = p_0 + p_1 p_0 + p_2 p_0^2$$

$$P(Z_2 = 1) = p_1^2 + 2\, p_2 p_1 p_0$$

$$P(Z_2 = 2) = p_1 p_2 + 2\, p_2^2 p_0 + p_2 p_1^2$$

$$P(Z_2 = 3) = 2\, p_1 p_2^2$$

$$P(Z_2 = 4) = p_2^3$$

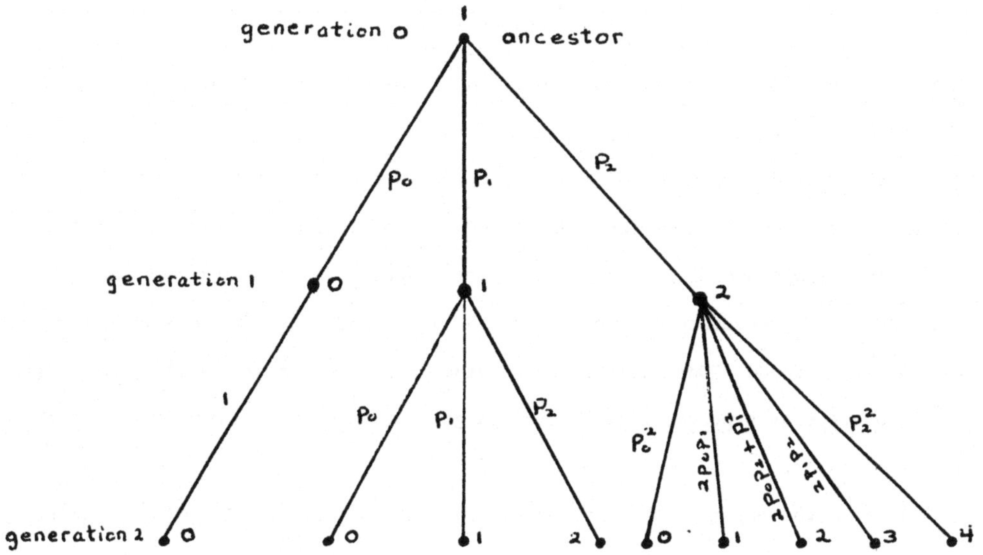

Figure 2.1 Number of daughter cells in each generation.

The distribution of the number of cells in later generation is difficult to find directly. One technique for finding these probabilities is the theory of branching processes (See Harris [7].) In particular, the generating function (See Eisen and Eisen [2].)

(6) $$f(s) = p_0 + p_1 s + p_2 s^2$$

for the number of offspring of a cell is required. Note that the coefficient of s^k is p_k $(k = 0,1,2)$. Theorem 4.1 in Chapter 1 of Harris [8] shows that the probability that i cells are alive in the kth generation, $P(Z_k = i)$, is the coefficient of s^i in $f_k(s)$ where the functions f_k are defined iteratively as follows:

$$f_0(s) = s$$

$$f_1(s) = f(s)$$

$$f_2(s) = f(f_1(s)) = p_0 + p_1(f(s)) + p_2(f(s))^2$$

$$= p_0 + p_1\ (p_0 + p_1 s + p_2 s^2) + p_2(p_0 + p_1 s + p_2 s^2)^2$$

and generally

(7) $$f_k(s) = f(f_{k-1}(s)). \qquad k = 1,2,3,\ldots$$

The reader should verify that the coefficients of powers of s in $f_2(s)$ give the probabilities in (5).

The number of cells in the kth generation is a random variable in the stochastic model. On the other hand, the number of cells in the kth generation is a well defined number in the deterministic model. To compare these two models a number representing the size of the kth generation for the stochastic model must be found. One measure which summarizes the size of the kth generation is the <u>average</u> or <u>expected size</u> of the kth generation. Its significance will be discussed later.

To find the expected size of the kth generation requires the expected value of Z_k, denoted EZ_k. By definition [1],

$$EZ_k = 0\cdot p(Z_k = 0) + 1\cdot p(Z_k = 1) + 2\cdot p(Z_k = 2) + \ldots$$

$$= \sum_{i=0}^{\infty} i\, p(Z_k = i)$$

As an example,

$$EZ_1 = 0\cdot p(Z_1 = 0) + 1\cdot p(Z = 1) + 2\cdot p(Z = 2)$$

$$= 0\cdot p_0 + 1\cdot p_1 + 2\cdot p_2 = \mu$$

according to (4). The calculation of EZ_2 requires some algebra. From equations (5) and the fact that $p_0 = 1-p_1-p_2$

$$EZ_2 = 1\cdot(p_1^2 + 2p_2p_0p_1) + 2\,(p_1p_2 + 2\,p_2^2p_o + p_2p_1^2) + 3\,(2p_1p_2^2) + 4p_2^3$$

$$= p_1^2 + 2p_1p_2(1-p_1-p_2) + 2p_1p_2 + 4p_2^2\,(1-p_1-p_2) + 2\,p_1^2p_2 + 6p_1p_2^2 + 4p_2^3$$

$$= p_1^2 + 4p_1p_2 + 4p_2^3 = (p_1 + 2p_2)^2 = \mu^2$$

The direct calculation of EZ_k for large values of k is difficult. Once again, generating functions prove useful. By Theorem 8.1 of Eisen and Eisen [2]

$$EZ_k = f_k^1(1) = \mu^k \qquad k = 1,2,3,\ldots \tag{8}$$

The reader should verify (8) for $k = 1$ and 2.

Now suppose there are n_0 cells initially. Since, by assumption, each cell acts independently of any other cell and has the same reproduction law, the number of cells in the kth generation is just

$$N_k = n_0Z_k \tag{9}$$

Hence, using the fact that E is a linear operator [2],

$$EN_k = n_0\ EZ_k = n_0\mu^k \tag{10}$$

Note that the expected size of the kth generation in the stochastic model is equal to the size of the kth generation in the deterministic model (c.f. (3)). Yet, even if the average size is used, the behavior of the two models is not exactly the same. The reason is that there is a positive probability that the size of the nth generation is zero. More precisely, define extinction (See Definition 6.1 of Chapter 1 in [7].) to be the event that the randon sequence $\{N_k\}$ consists of zero for all but a finite number of values of k. Let q be the probability of extinction. It follows from Theorem 6.1 of Chapter 1 in [7] that if $\mu \leq 1$ then q = 1; if $\mu > 1$ then q is the unique nonnegative solution less than 1 of the equation $s = f(s)$. Hence, if $\mu = 1$ equation (10) shows that the expected size of the population does not change - its size remains n_o. However, if one observed a tissue or tumor grown from n_o cells its size would nearly always be zero after a sufficiently long time since q = 1. On the other hand, if the deterministic model were valid the tissues size would always be n_o. Now consider the stochastic model when $\mu > 1$. There is still a positive probability that the size of the nth generation will be zero if n is sufficiently large. For example, if $p_o = \frac{1}{4}$ and $p_2 = \frac{3}{4}$ then

$$\mu = 0\cdot\frac{1}{4} + 2\cdot\frac{3}{4}$$

and q satisfies the equation $f(s) = s$ or

$$\frac{1}{4} + \frac{3}{4}s^2 = s$$

It is easy to verify that $q = \frac{1}{3}$. This means that if a large number of tumors were grown from n_o cells the sizes of the tumors would be zero in about $\frac{1}{3}$ of the cases. Further, the mean size a_k of the kth generation obtained by only considering non zero populations would overestimate the true mean. Using conditional expectation [2] it follows that

$$EN_k = 0\cdot q + a_k(1-q)$$

Therefore,

$$a_k = \frac{EN_k}{1-q} = \frac{3}{2} n_o\mu^k.$$

when $q = \frac{1}{3}$. This problem does not arise in the deterministic cases.

Assume that the growth of a tumor is governed by the stochastic model. What would be the size of a single tumor at the kth generation? Theorem 8.1 of Chapter 1 in [7] gives the answer to this question - namely,

$$N_k \sim W\mu^k n_o$$

where W is a random variable that is 0 when and only when Z_n goes to 0; otherwise $EW = 1$[3]. The randomness of W is significant. It arises because the size of the tumor may undergo proportionally large random fluctuations in the early generations. These fluctuations act as multiplicative factors whose effects persist in later generations. Therefore formula (3) or (10) could not be used to predict the size of a tumor. However, if a large number of tumors were observed and their sizes were averaged then these formulas could be used to predict the average size. Hence, averaging serves two purposes. It not only reduces the effects of experimental errors but also allows deterministic models to be used predictively in the face of randomness.

Note that an expectation of a product of random variables never appeared in the calculations of the mean size of the kth generation in the stochastic model. As a consequence of this linearity the predictions of the stochastic model (using the mean) and corresponding deterministic models agreed. If products of random variables are involved the results predicted by both types of models generally do not agree, as is well known from the theories of epidemics.

(b) <u>continuous models</u>

(i) deterministic model

The earliest model describing the growth of populations[4] was due to Thomas Malthus in 1798. He assumed that the birth (or growth) rates γ and the death rate δ per unit of population were constant. Let $N(t)$ be the size of the population at time t; then

$$N(t+\Delta t) = N(t) + \gamma N(t)\Delta t - \delta N(t)\Delta t$$

or

$$\text{(11)} \qquad \frac{N(t+\Delta t) - N(t)}{\Delta t} = (\gamma-\delta) N(t)$$

---

[3] The symbol $\sim$ is read "is asymptotic to" ; $f(k) \sim g(k)$ means that $\lim_{k\to\infty} \frac{f(k)}{g(k)} = 1$.

[4] The populations were people.

Letting Δt approach zero in (11) yields

$$\frac{dN}{dt} = (\gamma-\delta)\ N(t) \tag{12}$$

The solution of (12) is

$$N(t) = N(0)\, e^{\mu t} \tag{13}$$

where

$$\mu = \gamma-\delta$$

and N(0) is the size of the population at t = 0.

At this point the reader might object that a non-integer growth rate, say $\frac{1}{3}$, per unit cell is nonsensical. Of course this does not occur for a single cell. However, consider a large number of cells - for example, a tumor. If one observes the growth of the tumor with the naked eye its size appears to increase continuously - no sudden jumps in size are observed. A continuous model seems more appropriate than a discrete model. We reiterate that models should be considered on the basis of a high degree of relevance to the data. Thus, if observations are taken at a discrete set of time points the continuous model need only fit closely at these times. The reader need only compare equation (13) and (2) to see that a continuous model can be used at times 0, c, 2c,... to describe a discretely growing population.

The behavior of the cell population is completely described by (13). As t increases, N(t) decreases, remains constant, increases when $\mu < 0$, $\mu = 0$, $\mu > 0$, respectively.

(ii) stochastic model (Age-dependent Branching Process)

Assume that a cell born at time 0 has a random life length (or cycle time) C with probability distribution

$$G(t) = P(C \leq t).$$

At the end of its life it is replace by i similar cells of age 0 with probability p_i (i = 0,1,2). The probabilities p_i are assumed to be independent of absolute time, of the age of the cell at the time it is replaced and of the number of other cells present. The life length of any cell has the same distribution G and is independent of the life length of any other cell. Cell growth growth continues as long as cells are present. Note that the essential difference between this process and the discrete model is that the cell's cycle time is a random variable while it is a constant in the discrete model.

Let Z(t) be the number of cells present at time t. The generating function for the number of descendants of a single cell is

$$h(s) = p_0 + p_1 s + p_2 s^2.$$

The mean number of descendants of a single cell is

$$\mu = 1p_1 + 2p_2 = h'(1).$$

The trivial cases $p_1 = 1$ and $p_0 = 1$ are always excluded. Finally, it is assumed that $G(0+) = 0$.

The above assumptions imply that $EZ(t) < \infty$ and as a consequence $P(Z(t) < \infty) = 1$ (Theorem 13.1 in Chapter 6 of [7]). It can be shown that

$$M(t) = EZ(t) = \sum_{i=0}^{\infty} iP(Z(t) = i)$$

is monotone. In fact, $\mu < 1$, $\mu = 1$, $\mu > 1$, imply that M is decreasing, M = 1, M is increasing, respectively. (See Section 15.1 in Chapter 6 of [7].)

In general, M can be calculated only by a numerical or series method, However, the asymptotic behavior of M(t) as $t \to \infty$ can often be determined. (See Sections 16 and 17 in Chapter 6 of [7]). In both situations described below, G is not a lattice distribution.[5] The statement of the results requires the equation

$$\mu \int_0^\infty e^{-\alpha t}\, dG(t) = 1 \tag{14}$$

(i) $\mu > 1$

Let α be the smallest positive root of (14); then

$$M(t) \sim n_1 e^{\alpha t} \qquad t \to \infty \tag{15}$$

where

$$n_1 = (\mu-1)/(\alpha\mu^2 \int_0^\infty te^{-\alpha t}\, dG(t) \tag{16}$$

[5] Assigns a positive probability only to points of the form a+nb where a and b are constants and n = 0, ±1, ±2, ±3,... .

(ii) $\mu < 1$

Suppose there exists a real α (necessarily negative) satisfying (14) and in addition $\int_0^\infty te^{-\alpha t}\, dG(t) < \infty$; then (15) and (16) remain valid.

If G is a lattice distribution then M(t) is constant on each interval $(0,\Delta)$, $(\Delta,2\Delta)$,... and its value on the interval $(k\Delta,(k+1)\Delta)$ is

$$M(k\Delta+0) \sim [\Delta(\mu-1)e^{\alpha k\Delta}]/[\mu^2(1-e^{-\alpha\Delta})\int_0^\infty te^{-\alpha t}\, dG(t)] \tag{17}$$

The reader should verify that (17) reduces to (8) when $G(t) = 0$ $(t < c)$ and 1 $(t \geq c)$.

As in the discrete case, it follows that for lage t, the mean size of the population increases exponentially just as in the deterministic case (c.f. (13)). However, the behavior of the multiplicative constants is different. Assume that initially there is one cell. In the deterministic process $N(0) = 1$ and this multiplicative factor in (13) does not change with time. However, in the stochastic model for large t the multiplicative factor n_1 need not be one. From (16) its value depends on the distribution G.

As in the discrete stochastic process it can be shown that if $\mu > 1$ then (See Sections 19, 20 and 21 in Chapter 6 of [7].)

$$Z(t) \sim n_1 We^{\alpha t}$$

as $t \to \infty$. W is a random variable independent of t and $EW = 1$. Once again according to this stochastic model tumors starting with the same number of cells and growing will show fluctuations in size at a given time.

The question naturally arises as to whether a stochastic or deterministic model should be used. In many problems the numbers of individuals are so large (the role played by chance fluctuations being correspondingly so small) that a description of the deterministic type is adequate. For some situations involving very small number of cells (e.g., the first steps of tissues regeneration or the initiation of bacteremia) the stochastic model is more appropriate. However, in some cases, the variance and higher moments of the population calculated by using sophisticated stochastic models may not reflect the true state of affairs. This is because experimentally observed fluctuations may be due to variations in the system parameter as opposed to chance effects in a system with fixed parameters. We reiterate that the final criterion is the agreement between the predictions of the theory and the observed experimental results.

All the models discussed in the section give rise asymptotically to exponential growth of the form

$$y = y_0 e^{\alpha t} \tag{17}$$

when average size is recorded. The constant α is called the Malthusian parameter. Cell populations that are described by (17) are said to be in "log phase". This terminology is unfortunate since they are more correctly described as being in exponential (growth) phase. The term log phase originated from experimental considerations.

In biology a growth curve $y(t)$ is usually determined experimentally. The results are frequently plotted on semilog paper. During the early phases of cell growth a straight line can often be fitted to the data by the method of least squares [11]. The cells are said to be in log phase growth because of the type of the graph paper employed. Taking logarithms of both sides of (17) yields

$$\log y = \log y_0 + (\log e)\, \alpha t,$$

a straight line on semilog paper. It is apparent that (17) describes log phase growth.

The Malthusian parameter α can easily be estimated graphically. Let t_1 and t_2 be the values of the abscissa at any two points on the line and let y_1 and y_2 be the corresponding values of y; then

$$\alpha = \ell n \left(\frac{\log y_2 - \log y_1}{t_2 - t_1} \right) \tag{18}$$

The parameter α is closely related to the doubling time of a population, t_d, defined by

$$e^{\alpha t_d} = 2.$$

Hence,

$$t_d = \frac{\ln 2}{\alpha} \quad .$$

If there were no cell deaths and C had zero variance then t_d would equal the expected cycle time $t_c = EC$. Many biologists have attempted to estimate the "cycle time" by equating it with t_d. Assuming that the population grows according to the continuous stochastic model, (14) can be rewritten as

$$1 = mE(e^{-\alpha C}) \geq me^{-\alpha t_c}$$

The last inequality is a consequence of Jensen's inequality[5]. Thus

$$e^{\alpha t_c} \geq m$$

and so

$$t_c \geq t_d\left(\frac{\ln m}{\ln 2}\right)$$

In practice Var C > 0 and the inequality is strict. Hence, $t_c - t_d$ might be significantly large.

As observed in Chapter 1 a population of cells does not grow indefinitely for many possible reasons. In the next section we will examine some other more realistic growth laws.

2.3 Growth Laws

The notion of negative feedback on population growth rate due to competition for and depletion of resources was considered by Paul Verhulst [15]. He assumed that birthrates should decrease and deathrates increase as the population size, y, increases. A linear law is the simplest. The birthrate at any time t is

$$b(t) = b_0 - b_1 y(t)$$

---

[5] Suppose f is a continuous, convex function defined on an interval. Then $f(E(X)) \leq E(f(X))$. At each point $(s, f(s))$ on the graph of f there is, by definition, a line L lying below the graph - i.e., $f(s) + m(x-s) \leq f(x)$, where m is the slope of L. Jensen's inequality follows by choosing $s = E(X)$ and taking expectations.

and the death rate

$$d(t) = d_o + d_1 y(t)$$

where all the constants are positive and $b_o > d_o$. Then

(20) $$\frac{dy}{dt} = b(t)-d(t) = (b_1+d_1)(\frac{b_o-d_o}{b_1+d_1} - y)$$

Setting $b_1 + d_1 = k$ and $\frac{b_o-d_o}{b_1+d_1} = a$, equation (20) reduces to the Verhulst equation

(21) $$\frac{dy}{dt} = k(a-y)$$

Note that $a-y < 0$ when $y > a$. Hence, the population cannot grow indefinitely since $\frac{dy}{dt} < 0$ when $y > a$.

Valuable information about the solution of (21) can be obtained directly from the differential equation. We shall see in the next section that such techniques are extremely valuable, especially when an analytic solution cannot be found. These concepts first arose in mechanics and it seems appropriate to introduce them by considering the behavior of balls located at A, B and C in Fig. 2.2.

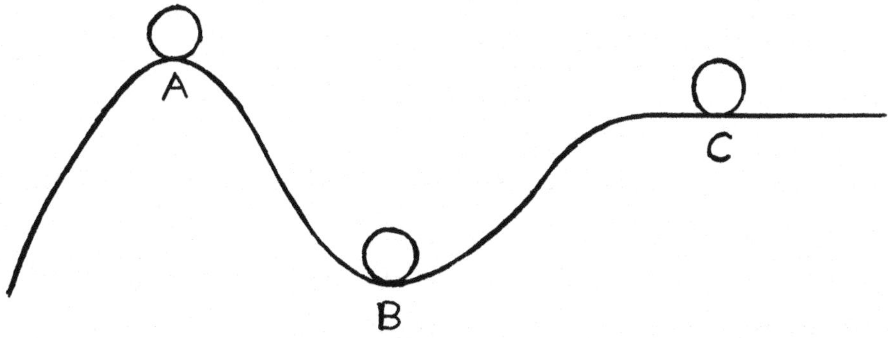

Figure 2.2 Types of stability.

Carefully balanced balls at A, B and C in Fig. 2.2 will remain in those positions. These states of the mechanical system are called <u>equilibrium</u> or <u>stationary states</u> since they do not change with time. The equilibrium states can be further characterized by the response of a ball to a small perturbation. A is a position of <u>unstable equilibrium</u> since the ball will not return to A after a small displacement. On the other hand, B is a state of stable equilibrium since the ball returns to B after a small displacement. Position C is frequently called a position of neutral equilibrium (or stability); the ball remains at the position to which it was displaced.

These intuitive notions can be made precise by considering the planar equations of motion of a ball. The equations will be of the form

$$\frac{dy_1}{dt} = f_1(y_1,y_2)$$

(22)

$$\frac{dy_2}{dt} = f_2(y_1,y_2)$$

The stationary states, being time independent, are obtained by solving the equations $\frac{dy_1}{dt} = \frac{dy_2}{dt} = 0$ or $f_1(y_1,y_2) = 0$ and $f_2(y_1,y_2) = 0$. It is convenient to assume that the stationary state is the origin $y_1 = 0$ and $y_2 = 0$ or $f_i(0,0) = 0$ for $i = 1,2$. There is no loss in generality in this assumption since any equilibrium state $y_1 = a_1, y_2 = a_2$ can be translated to the origin by the coordinate change $y_1^* = y_1 - a$, and $y_2^* = y_2 - a_2$.

The following notation will be used. Let $y = (y_1,y_2)$; then

$$|y| = \sqrt{y_1^2 + y_2^2}$$

The open circular region $|y| < R$ will be denoted by D(R) and the circle itself $|y| = R$ by C(R).

Assume that there is a unique path p of the system (22) through each point y of some open spherical region D(A). The path of p described by y(t) when $t \geq 0$ will be designated by p^+.

The origin is said to be (see Fig. 2.3)

(a) stable (in the sense of Lyapunov) whenever for each $R < A$ there exists an $r \leq R$ such that if a path p^+ originates at a point y_o of the circular region D(r) then it remains in the circular region D(R) ever after. In other words under any sufficiently small initial perturbation from the equilibrium point the resulting path (or motion) remains close to the equilibrium point.

(b) asymptotically stable whenever it is stable and further every path p^+ starting inside some $D(R_0)$, $R_0 > 0$ tends to the origin as time increases indefinitely - i.e., any sufficiently small initial perturbation from the equilibrium point results in a path which eventually approaches the origin.

(c) unstable whenever for any circle C(R) and any r, no matter how small, there is always a point y_0 in D(r) such that the path through y_0 reaches the boundary circle C(R). This means that given any R there is always some very small initial perturbation which results in a path that is eventually further than R from the equilibrium point.

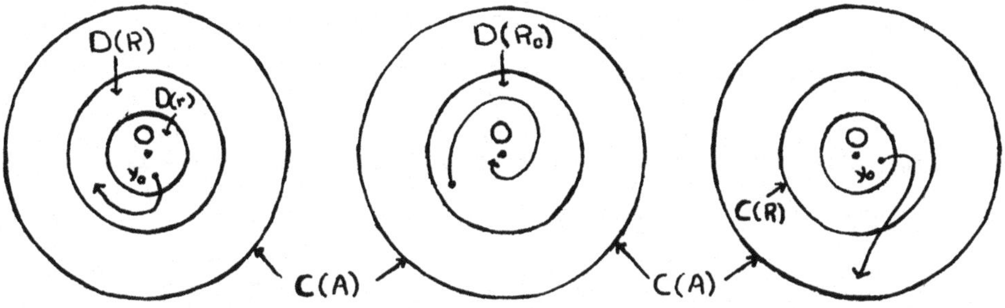

stable asymptotically stable unstable

Figure 2.3 Analytic definitions of stability.

These definitions can easily be generalized to higher dimensional systems. Further details can be found in [9].

If the motion of each ball in Fig. 2.2 were described by a system of differential equations of the type (22), then point A, B and C would be unstable, asymptotically stable and stable points, respectively.

The equilibrium state of (21), when $\dot{y} = 0$, is y = a. Setting $y^* = y-a$ transforms the equilibrium state to the origin. Equation (21) becomes

$$\frac{dy^*}{dt} = -ky^*$$

and has the solution $y^* = y^*(0)\, e^{-kt}$. It follows that the origin $y^* = 0$ and hence y = a is asymptotically stable. Moreover the solution of the system (21)is

$$y-a = (y(0)-a)\ e^{-kt}$$

Usually, $y(0) < a$ and the curve increases as shown in Fig. 2.4 but if $y(0) > a$ the curve decreases.

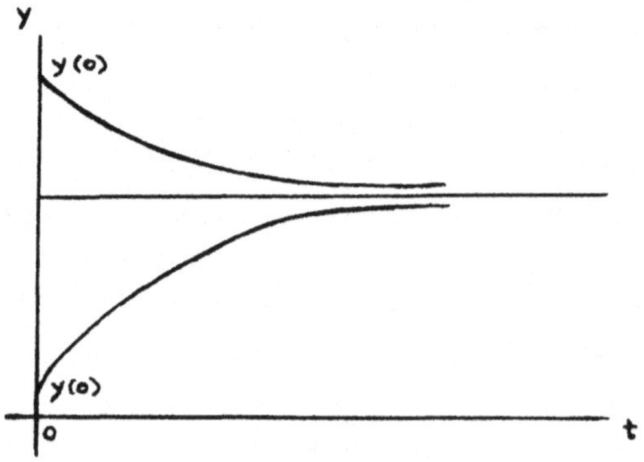

Figure 2.4   Verhulst law of growth.

Another attempt to represent cell growth in a more realistic way is by means of the Logistic differential equation

(23) $$\frac{dy}{dt} = ky - ay^2 \qquad k > 0,\ a > 0$$

The term ky represents the excess of birth rate over death rate. If the limiting factor is the appearance of a toxic or growth limiting substance, a plausibility argument for its representation by the term $-ay^2$ can be given. Suppose the toxic material diffuses freely throughout the intercellular medium. A given cell detects the cumulative toxic effect of all y cells. Thus, the toxic effect on a given cell is proportional to y. The toxic effect on all y cells is y times the effect on a single cell and hence proportional to y^2. However, the representation of the growth of cells by equation (23) is a pure assumption whose validity is tested by comparing the predictions of the theory and experience.

Stationary states of equation (23) satisfy

$$ky-ay^2 = ay\ (\frac{k}{a} -y) = 0$$

and so are $y = 0$ and $y = \frac{k}{a}$. In a small neighborhood of $y = 0$ equation (23) can be approximated by

$$\frac{dy}{dt} = ky$$

The solution is $y = y(0)e^{kt}$, which grows exponentially away from $y = 0$. Hence, the equilibrium value $y = 0$ is unstable. The rigorous justification of this method of small perturbations can be found in [9].

In the neighborhood of $y = k/a$

$$ky-ay^2 = -k(y-\frac{k}{a})-a(y-\frac{k}{a})^2 \approx -k(y-\frac{k}{a})$$

Equation (23) can be approximated by

$$\frac{dy}{dt} = -k(y-\frac{k}{a})$$

whose solution is

$$y = \frac{k}{a} + (y(0-\frac{k}{a})e^{-kt}$$

This solution approaches $y = \frac{k}{a}$. Hence, the equilibrium value $\frac{k}{a}$ is asymptotically stable.

Equation (23) can be solved by the method of separation of variables by rewriting it in the form

$$\frac{y_e\, dy}{y(y_e-y)} = kdt \tag{24}$$

where $y_e = \frac{k}{a}$. Integrating both sides of (24) as time varies from $t = 0$ to an arbitrary time t and y varies from an initial y_o to a final value y yields

$$\int_{y_o}^{y} \frac{y_e\, dy}{y(y_e-y)} = \int_{t_o}^{t} kdt \tag{25}$$

The integration on the left side of (25) is facilitated by rewriting the integrand as a sum of partial fractions, namely,

$$\frac{y_e}{y(y_e-y)} = \frac{1}{y} + \frac{1}{y_e-y} \tag{26}$$

Substituting (26) into (25) and integrating yields

$$\ln \frac{y}{y_e-y} - \ln \frac{y_o}{y_e-y_o} = kt$$

$$\frac{y}{y_e-y} = \frac{y_o}{y_e-y_o}\, e^{kt}$$

or

$$(27) \qquad y = \frac{y_0\, y_e}{y_0 + (y_e - y_0)e^{-kt}}$$

This equation is called the logistic law of growth.

If $y_0 < y_e$, then $y_0 < y_0 + (y_e-y_0)e^{-kt}$. The denominator of the fraction in (27) is less than y_0 and so

$$y(t) < \frac{y_0\, y_e}{y_0} = y_e.$$

Thus y never exceeds y_e but approaches y_e as time increases indefinitely. Further if $y_e >> y_0$, then for small t, $y_e\, e^{-kt} >> y_0$ and the term $y_e\, e^{-kt}$ is much larger than the other terms in the denominator in (27). Thus (27) can be approximated by

$$y = \frac{y_0\, y_e}{y_e\, e^{-kt}} = y_0\, e^{kt}$$

the Malthusian growth law. However, after a sufficiently long time, $y_e\, e^{-kt} << y_0$. Then the term y_0 in the denominator is dominant, and the solution can be expanded as

$$(28) \qquad y = \frac{y_e}{1+[(y_e-y_0)/y_0]e^{-kt}} = y_e - \frac{y_e}{y_0}(y_e-y_0)e^{-kt}+\ldots$$

This equation indicates that the equilibrium value is approached exponentially. Thus the curve has the characteristic S-shaped appearance shown in Fig. 2.5.

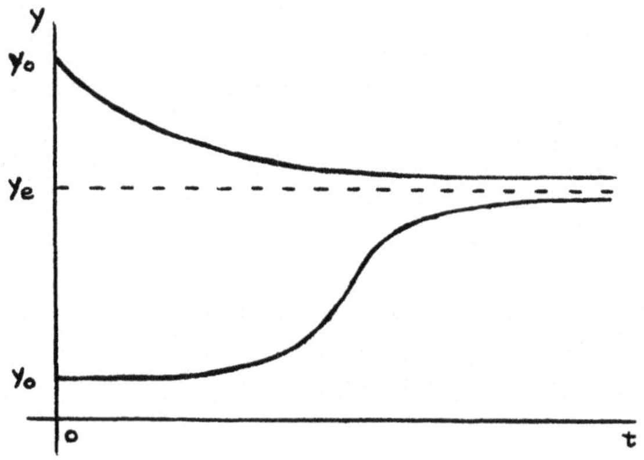

Figure 2.5 The logistic law of growth.

If $y_o > y_e$, then $y_o > y_e + (y_e - y_o)e^{-kt}$ and it follows from (27) that y is never less than y_e. However, y eventually approaches y_e and (28) indicates the approach is approximately exponential. (See Fig. 2.5.)

A bacterial colony grown in a nutrient medium usually grow according to the logistic law [10]. The growth of the infusorian[6] Parmecium caudatum also follow this law [4]. However, if bacteria are transferred to a different nutrient medium there may be an initial phase of growth, called the lag phase, during which time the growth of the population is retarded. The lag phase is believed to be due to the formation of new enzyme systems required for growth using the new nutrients.

A population very close to its stationary value, y_e, is said to be in its stationary phase of growth. After a long time, the population usually decays to zero from its equilibrium value y_e. New growth and death conditions have arisen, such as depletion of a necessary nutrient or the accumulation of a new toxic metabolic product, which are not accounted for by equations (23).

It has been noted by many investigators that the growth rate of a tumor decreases with time [14]. This suggests the system of equations

$$\frac{dy}{dt} = \gamma y \tag{29}$$

$$\frac{dr}{dt} = -a\gamma \tag{30}$$

where γ is the time dependent specific growth rate and $a > 0$ is the retardation constant. Initially, $y(0) = y_o$ and $\gamma(0) = k$, the log phase growth rate. From equation (29)

$$\frac{d(a \ln y)}{dt} = a\gamma$$

and so substituting for $a\gamma$ from (30) yields

$$\frac{d(a \ln y)}{dt} = -\frac{d\gamma}{dt}$$

It follows that

$$\gamma = k - a \ln \frac{y}{y_o} \tag{31}$$

---

[6] Infusoria are a class of protozoans having cilia. Protozoa are single celled animals (eucaryotes), the lowest division of the animal kingdom. They are more complex and generally larger than bacteria.

Hence, the system of equations (29) and (30) reduces to the single differential equation

$$\frac{dy}{dt} = y(k-a \ln \frac{y}{y_o}) \tag{32}$$

when (31) is substituted in (29).

Setting $\ln y = z$ in equation (32) reduces it to the linear equation

$$\frac{dz}{dt} = -az + (k+a \ln y_o)$$

whose solution is

$$z = \frac{k}{a} e^{-at} + \frac{k}{a} + \ln y_o$$

Hence, the solution of (32) is

$$y = y_o \exp[\frac{k}{a}(1-e^{-at})] \tag{33}$$

the so called Gompertz function [6].

From (32) if y is close to y_o the population growth is close to exponential since the equation $\dot{y} = ky$ holds approximately. As time increases the size of the population increases and approaches the <u>saturation level</u> $y = y_o e^{k/a}$. Note this is a stationary state of (32). These considerations suggest the curve defined by (33) has the shape shown in Fig. 2.6.

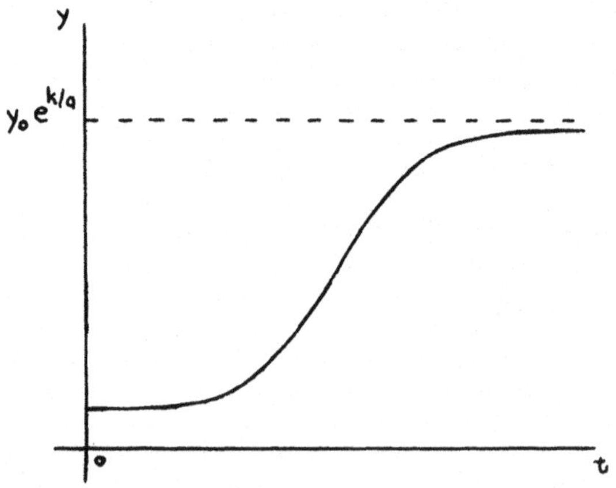

Figure 2.6 The Gompertzian law of growth.

The use of the Gompertzian function for growth curves of tumors was first suggested by Laird [8]. More recently it has been used by Simpson. Herren and Lloyd [13] to systematically represent data for experimental tumor systems. There is no profound underlying significance to the convential use of the Gompertz function. Its general acceptance and use make it a convenient representation. However, other forms of basic growth curves with saturation behavior could also be satisfactory. For example, the curve defined by the differential equation

$$\frac{dy}{dt} = (a-by^k)y$$

was introduced by Ayala et al [1] for analyzing fruit fly data. If $0 < k < 1$ the corresponding curve is growing rapidly, which may be useful in categorizing rapidly growing tumors. If $k > 1$ a slowly growing curve is generated. This curve may be useful for describing very slowly growing tumors.

Each of the differential equations introduced for describing growth can be written in the form

$$\frac{dy}{dt} = \gamma(y)y$$

where $\gamma(y)$ is nonnegative on some interval $0 < y < c$ and

$$\frac{d\gamma}{dy} < 0 \tag{34}$$

If (34) holds, then either $\gamma(M) = 0$ for some finite, positive value M in which case y increases to M as a limiting value, or $\gamma(y) > 0$ for all value of y in which case y increases with bound. In the first case γ is called a bounded <u>(specific) growth rate</u> and in the second case, an <u>unbounded (specific)growth rate</u> [3]. This generalized growth rate will be used in the next section. Goel, Maitra and Montroll [5] have also defined a generalized specific growth rate in a study of population growth in interacting populations.

Some investigators have speculated that the specific growth rate under normal growth conditions is only a function of population size.. Further, it is uniquely[7]

[7] Allowing for experimental artifacts and biological variability. Unique is not used in the mathematical sense in this paragraph.

determined for a given type of tumor. The specific growth rate is unaltered by transplant, removal or cycle-nonspecific therapy. The distribution of cells in the active cycle, resting, dying or leaving the tumor may also be a unique function of tumor size.

The hypotheses of the preceding paragraph are debatable. However, it is universally agreed that growth rates would not be determined solely by population size after more general perturbations, such as cycle-sensitive drugs. These drugs can alter the lengths of various phases of the cell cycle. Further, their effectiveness may depend on whether they act early or late in a given phase. Hence, the study of the growth of a tumor treated with cycle-specific drugs may require the cell-age distribution, the distribution of phase times and the duration of the "window" of drug effect. Hence, more complicated models of the cell are required. These will be presented in the next and later chapters.

2.4 Two Compartment Growth [3]

Many models for tissue growth have the following characteristics [11]:

(i) The system contains proliferating and nonproliferating compartments

(ii) There is a flow or transfer of cells between some or all of the compartments.

(iii) Cell death occurs in some or all compartments

For a system consisting of one proliferating (P) and one nonproliferating (Q) compartment these properties suggest the following system of differential equations:

$$\frac{dy_p}{dt} = (\gamma-\delta-\alpha)y_p + \beta y_q \qquad (35)$$

$$\frac{dy_q}{dt} = \alpha y_p - (\lambda+\beta)y_q$$

Here y_p and y_q equal the total number of cells in the proliferating and nonproliferating compartments respectively; γ is the mean specific growth rate for P; α and β are respective average transfer coefficients for P and Q; δ and λ are respective mean death rates for P and Q (Fig. 2.7)

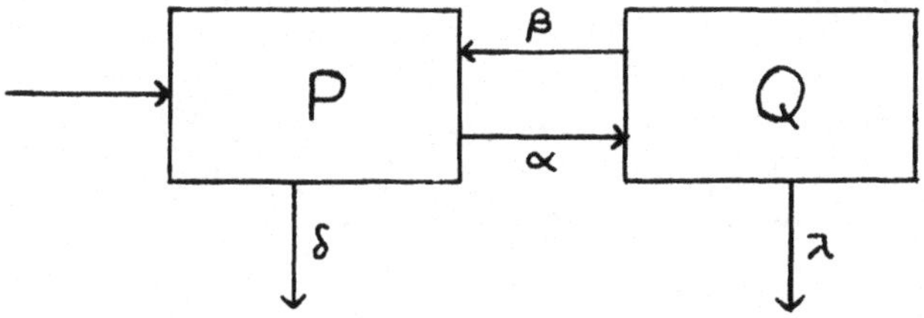

Figure 2.7 Two compartment growth.

The specific growth rate $\gamma = \gamma(y_p)$ satisfies (34) with respect to y_p. The stability of (35) will first be studied for constant transfer rates α, β and constant loss rates δ, λ. This may be a reasonable simulation of a normally growing tissue. Then the effects of a perturbation in these rates is considered.

Stability. If the limiting values of y_p and y_q are finite, they occur among the critical points of (35), i.e., among those points in the y_p, y_q plane at which the curves

$$(\gamma(y_p)-\delta-\alpha)y_p + \beta y_q = 0 \qquad \text{(P)}$$

$$\alpha y_p-(y+\beta)y_q = 0 \qquad \text{(Q)}$$

intersect. The curves intersect at some non-zero point precisely when the equation

$$\gamma(y_p) = \delta + \alpha - \frac{\alpha\beta}{\lambda+\beta} = \delta + \frac{\alpha\lambda}{\lambda+\beta}$$

has a non-zero solution for y_p. Since $\gamma(y_p)$ is a decreasing function, this equation has a (unique) solution for y_p if and only if γ satisfies

(36) $$\gamma(\infty) < \delta + \frac{\alpha\lambda}{\lambda+\beta} < \gamma(0^+).$$

where $\gamma(\infty)$ denotes the value of γ as y_p approaches infinity. For stability purposes it is important to determine the nature of the crossing of two curves. We note that Q-curve is a straight line which goes through the origin and has positive slope.

The P-curve satisfies

$$y_q(0) = \frac{1}{\beta} \lim_{y_p \to o^+} (\delta+\alpha-\gamma(y_p))y_p = -\frac{1}{\beta} \lim_{y_p \to o^+} \gamma(y_p)y_p \leq 0$$

Hence, the P-curve is initially not above the Q-curve. If the P-curve does rise above the Q-curve for some value of y_p, then

$$\frac{\delta+\alpha-\gamma(y_p)}{\beta} y_p > \frac{\alpha}{\lambda+\beta} y_p$$

Since $\gamma(y_p)$ is decreasing, this equality continues to hold y_p increases. Hence, if the two curves intersect at some non-zero point, then the P-curve is below the Q-curve to the left of the intersection and above it to the right. If inequality (36) is not satisfied, than either

$$\gamma(o) \leq \delta + \alpha\lambda/(\lambda+\beta) \text{ or } \gamma(\infty) \geq \delta + \alpha\lambda/(\lambda+\beta).$$

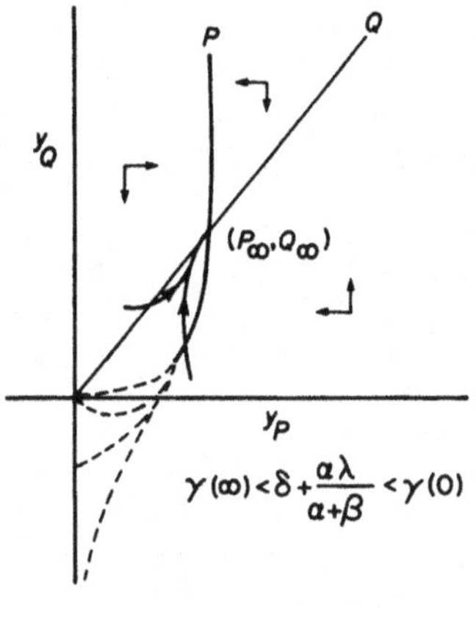

Figure 2.8 Bounded growth.

In the first case, the P-curve is above the Q-curve for all $y_p > 0$, and in the second case, the P-curve lies below the Q-curve for all positive y_p.

Figures 2.8, 2.9 and 2.10 illustrate the essential features of all the possibilities. Above the Q-curves dy_q/dt is negative which implies y_q is decreasing. Analogously, below the Q-curve y_q is increasing. This is indicated in the figure by the vertical vectors. The horizontal vectors illustrate that y_p is increasing above the P-curve and decreasing below it. In the bounded growth case (Figure 2.8), the vector field about the non-zero critical (P_∞, Q_∞) indicates that this point is stable.

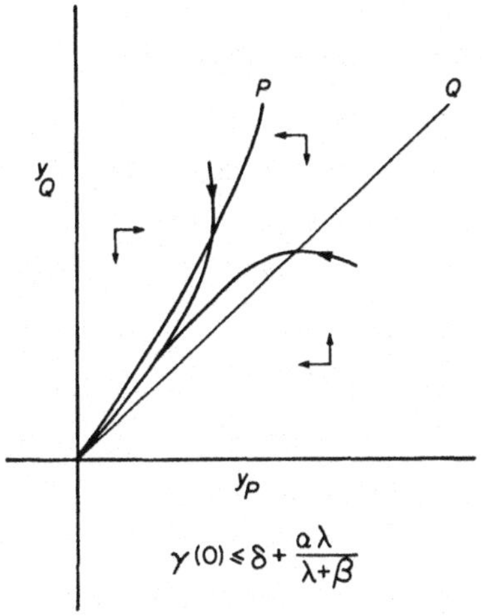

2.9 Decay.

It is not difficult to show that the point is in fact asymptotically stable. Hence, the time directed paths in Figure 2.8 represent typical solutions of (35). They flow toward (P_∞, Q_∞) as time approaches ∞. Similarly, in Figures 2.9 (decay) and 2.10 (unbounded growth) typical solution paths flow toward the origin and infinity respectively.

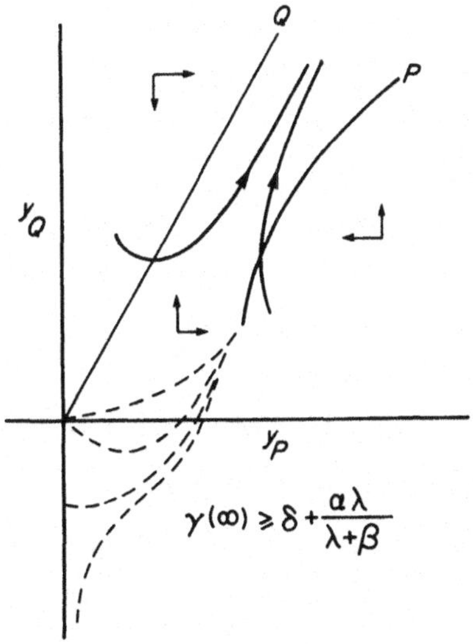

Figure 2.10 Unbounded growth.

In summary, if γ is a decreasing function of y_p and α,β,δ and λ are constant, then

(i) y_p and y_q approach finite stable values P_∞ and Q_∞ respectively if and only if

$$\gamma(\infty) < \delta + \frac{\alpha\lambda}{\lambda+\beta} < \gamma(0);$$

(ii) y_p and y_q decay to zero if and only if

$$\gamma(0) \leq \delta + \frac{\alpha\lambda}{\lambda+\beta} ;$$

(iii) $y_p + y_q$ increases without bound if and only if

$$\gamma(\infty) \geq \delta + \frac{\alpha\lambda}{\lambda+\beta}$$

In case (i) P_∞ satisfies

$$\gamma(P_\infty) = \delta + \frac{\alpha\lambda}{\lambda+\beta} \tag{37}$$

which represents the limiting value of the rate function γ.

Q_∞ satisfies

$$Q_\infty = \frac{\alpha}{\lambda+\beta} P_\infty$$

from which the limiting value of the growth fraction[8], g_∞, can be computed as

$$g_\infty = \frac{\lambda+\beta}{\lambda+\beta+\alpha}$$

Also,

$$g_\infty \gamma(P_\infty) = g_\infty\delta + (1-g_\infty)\lambda,$$

which indicates that a steady state of renewal has been reached in which proliferation is just enough to overcome cell loss.

We remark that the above results were derived under the assumption that the transfer rate β from Q to P was not zero. However, the same results hold even if β is equal to zero.

---

[8] proliferating cells ÷ total cells.

Perturbations. A change in the controlling parameters of a normally growing system may result in neoplastic growth. Equation (37) is the key expression in the foregoing stability analysis. If the right side of (37) decreases, then P_∞ will increase since $\gamma(y_p)$ is a decreasing function. Hence, any change in the parameters $\alpha,\beta,\delta,\lambda$ which results in a smaller value of

$$\delta + \frac{\alpha\lambda}{\lambda+\beta} \tag{38}$$

will cause the proliferating compartment to grow. Quantity (38) decreases when β increases or when any one of α,δ, and λ decreases (it is interesting to note that if λ is zero, then P_∞ is independent of α and β, and if either α or β is zero, then P_∞ is independent of λ). Assuming that some of these parameters are altered, then the degree of change in the tissue size will depend on the function $\gamma(y_p)$. There are two possibilities:

(I) Unbounded-growth function. In the previous section we defined an unbounded-growth function $\gamma(y_p)$ as one for which $\gamma(y_p)$ is positive for all values of y_p. For this type, if (38) decreases to a value less than or equal to $\gamma(\infty)$, then the tissue will grow without bound.

(II) Bounded-growth function. If $\gamma(y_p)$ is a bounded-growth function, then $\gamma(M) = 0$ for some finite value M, and it can be shown that P_∞ is less than M for all values of α,β,δ and λ. Hence, in this case, no change in these parameters can result in unbounded growth. However, a sufficiently large decrease (38) could increase the new equilibrium point to a fatal size. In either of the two previous cases, a sufficiently large increase in (38) would result in decay.

REFERENCES

1. Ayala, F.J., Gilpin, M.E. and Ehrenfield, J.G., Competition between species: theoretical models and experimental tests, Theo. Pop. Biol., 4, 331-356, 1973.

2. Eisen, M. and Eisen, C. Probability and Its Applications, Quantum Publishers Inc., New York, 1975.

3. Eisen, M. and Schiller, J. Stability analysis of normal and neoplastic growth, Bul. Math. Bio., 66, 799-809, 1977.

4. Gause, G.F. The Struggle for Existence, Hafner, New York, 1969.

5. Goel, N.S., Maitra, C. and Montroll, E.W. On the volterra and other nonlinear models of interacting populations, Rev. Mod. Phys, 43, 231-376, 1971.

6. Gompertz, B. Phil. Trans. Roy. Soc. London, 115, 513, 1825.

7. Harris, T. The Theory of Branching Processes, Springer-Verlag, Berlin, 1963.

8. Laird, A.K. Dynamics of tumor growth, Brit. J. Cancer, 18, 490-502, 1964.

9. Lasalle, J.P. and Lefshetz, S. Stability by Liapunov's Direct Method with Applications, Academic Press, New York, 1961.

10. Monod, J. Recherches sur la croissance des cultures bacteriennes, Hermann and Cie, Paris, 1942.

11. Quastler, H. and Sherman, F.G. Cell population kinetics in the intestinal epithelium of the mouse, Exp. Cell Res., 17, 420-438, 1959.

12. Rubinow, S.I. Introduction to Mathematical Biology, Wiley, New York, 1975.

13. Simpson-Herren, L. and Lloyd, H.H. Kinetic parameters and growth curves for experimental tumor systems, Cancer Chemother. Rep. Part I, 54, 143-174, 1970.

14. Skipper, H.F. Kinetics of mammary tumor cell growth and implications for therapy, Cancer, 28, 1479-1499, 1971.

15. Verhulst, P.F. Nuox. Mem. Acad. Roy. Bruxelles 18, 1, 1847. See also, 20, 1, 1847.

III. SOME KINETIC CELL MODELS

3.1 Introduction

This chapter summarizes several models which will be used in cell cycle analysis. These models are included in the chapter because of their use of radioactive-labeling experiments and chemotherapy. A common feature of these models is that they are based on a conservation principle of hydrodynamics.

In section 1 a discrete model is described by a system of ordinary differential equations. Then three different continuous versions of the discrete model are obtained by limiting operations. Section 4 describes how to solve the partial differential equations representing the continuous models. An alternative viewpoint which stresses the conceptual differences between the continuous models appears in section 5. Finally an integral equation approach is presented

3.2 A Discrete Differential Model

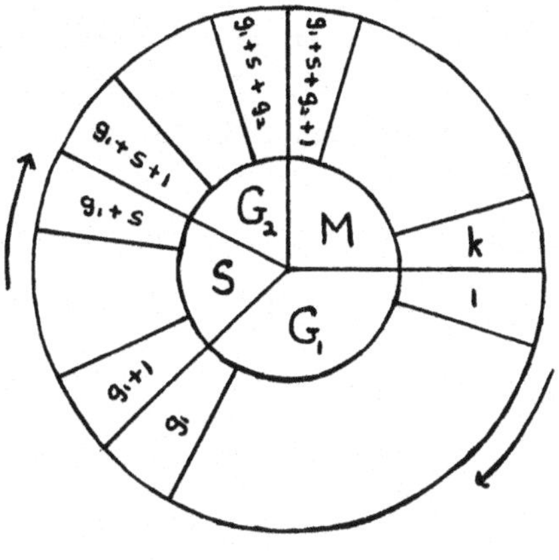

Figure 3.1 The Takahashi model.

The basic model of Takahashi [35,36] is illustrated in Fig. 3.1. It is a generalization of Kendall's earlier work [15]. The Takahashi model is specified by the following postulates.

(i) Each cell passes through k hypothetical (mathematical) compartments as it proceeds in its cycle. A cell which has just been born is located in the first compartment. A cell which is about to divide in the kth compartment.

(ii) The G_1,S,G_2 and M phases of the cell cycle are decomposed into g_1,s,g_2 and m compartments respectively, where

$$g_1 + s + g_2 + m = k$$

(iii) Each cell transits from the ith compartment to the (i+1)th compartment (i=1,2,...,k-1) with probability $\lambda_i \Delta t$[1].

(iv) Each cell in the kth compartment gives birth to β daughter cells with probability $\lambda_k \Delta t$. These daughter cells pass into the first compartment.

(v) The probability of death in any compartment is $\mu_i \Delta t$ (c.f. footnote 1).

(vi) Each cell behaves independently of every other cell with regard to transit times, births and death.

Let $\eta_i(t)$ be the [redacted] number of cells in the ith compartment. The change $\Delta\eta_i(t)$ in $\eta_i(t)$ during a time Δt is

$$\Delta\eta_i(t) = \text{influx of cells into ith compartment - efflux of cells out of ith compartment}$$

by the principal of cellular conservation. The influx into the ith compartment ($i \neq 1$) in Δt is

$$\lambda_{i-1}\, \eta_{i-1}\, \Delta t$$

and the efflux from it in Δt is

$$(\lambda_i + \mu_i)\eta_i \Delta t$$

(See Fig. 3.2.) Hence,

$$\Delta\eta_i(t) = \lambda_{i-1}\eta_{i-1}\Delta t - (\lambda_i+\mu_i)\eta_i\Delta t \qquad i = 2,3,\ldots,k \tag{1}$$

Since each cell leaving the last compartment of mitoses (i=k) results in β cells in G_1 (i=1) the conservation principle yields

---

[1] Alternatively the transit times are exponentially distributed and the probability is $\lambda_i \Delta t + O((\Delta t)^2)$.

(2) $$\Delta n_1(t) = \beta\lambda_k n_k(t) - (\lambda_1+\mu_1)n_1\ \Delta t$$

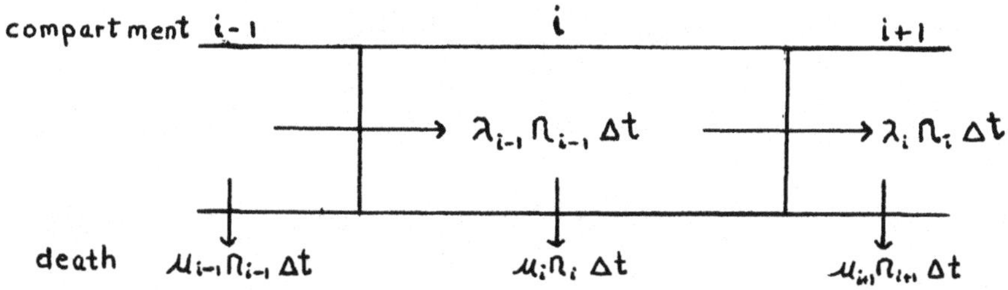

Figure 3.2 Efflux and influx in the ith compartment.

Dividing both sides of equations (1) and (2) by $\Delta t \to 0$ yields

(3) $$\frac{dn_i}{dt} = \lambda_{i-1}n_{i-1} - (\lambda_i+\mu_i)n_i \qquad i = 2,3,\ldots,k$$

(4) $$\frac{dn_1}{dt} = \beta\ \lambda_k n_k - (\lambda_1+\mu_1)n_1$$

The parameters λ_i, μ_i and β can depend on the population size, time etc.

If all the λ_i's are constant = λ, cell death is negligible and $\beta = 2$ equations (3) and (4) reduce to

(5) $$\frac{dn_i}{dt} = \lambda(n_{i-1}-n_i) \qquad 1 = 2,3,\ldots,k$$

(6) $$\frac{dn_1}{dt} = \lambda(2n_k-n_1).$$

A consequence of this particular model is that the density function for the duration of each phase and the cycle time is a gamma distribution[2]. This result does not follow from biology but rather from the model. If the general model described by (3) and (4) is used, other distributions can be obtained.

Next suppose λ_i is a constant for each phase - that is, $\lambda_i = \lambda_{G_1},\lambda_S,\lambda_{G_2},\lambda_M$ during the corresponding phases. For simplicity assume $\mu_i = 0$ and $\beta = 2$. Then the density function for the duration of each phase is also a gamma density. For

---

[2] This follows from the fact that a sum of independent exponential random variable is a gamma random variable.

example, the density function for the transit time T_2 of the S phase is

$$\delta_S(t) = \frac{\lambda_S^s}{(s-1)!} \; t^{s-1} e^{-\lambda_S t} \qquad (t \geq 0).$$

Accordingly,

$$t_2 = E\,T_2 = \frac{s}{\lambda_S}$$

$$\sigma_2^2 = \mathrm{Var}\,T_2 = s/\lambda_S^2$$

These equations can be solved for the model parameters yielding

$$\lambda_S = \frac{t_2}{\sigma_2^2} \qquad s = \frac{t_2^2}{\sigma_2^2}.$$

Similar equations can be derived for the other states. It follows that all the model parameters can be expressed in terms of the mean and variance of the sojourn time in each phase. The results for $\mu_i \neq 0$ can be found in [36].

Numerical studies of (3) and (4) show that the fraction of cells in each compartment

$$F_i = \frac{\eta_i}{\eta} \qquad (\eta = \eta_1 + \eta_2 + \ldots + \eta_k)$$

approaches a constant as $t \to \infty$. This constant is independent of the initial conditions. To prove these statements rigorously equations for the F_i must be derived. For simplicity, the equations for the F_i will be found using (5) and (6). The derivation for the system (3) and (4) is similar.

Adding both sides of equations (5) and (6) for $i = 1,\ldots,k$ yields

$$\frac{d\eta}{dt} = \lambda\eta_k.$$

Now, from the last equation and (6),

$$\frac{dF_1}{dt} = \frac{\dot{\eta}_1}{\eta} - \frac{\eta_1}{\eta} \cdot \frac{\dot{\eta}}{\eta} = 2\lambda F_k - \lambda F_1 - \lambda F_1 F_k$$

or

$$\frac{dF_i}{dt} = \lambda(2F_k - F_1 - F_1 F_k) .$$

Similarly,

$$\frac{dF_i}{dt} = \lambda(F_{i-1}-F_i-F_iF_k) \qquad i = 2,\dots,k.$$

To find the stationary state, set $\frac{dF_i}{dt} = 0$ and solve the resulting system of of equations. It can be shown that a unique solution exists. Hence, the compartment fractions F_i approach a constant which is independent of the initial conditions. Although the last result has not been proven for the more general system (3) and (4), it is assumed valid because of physical considerations.

To actually find the limiting fractions Laplace transforms of (3) and (4) are used. Since the steady state is independent of the initial conditions set $\eta_1(o) = 1$ and $\eta_i(o) = 0$ $(i \neq 1)$. This gives

$$(\lambda_1+\mu_1+p)\bar{\eta}_1-\beta\lambda_k\,\bar{\eta}_k = 1$$

$$-\lambda_{i-1}\bar{\eta}_{i-1} + (\lambda_i+\mu_i+p)\bar{\eta}_i = 0 \qquad i = 2,\dots,k$$

where $\bar{\eta}_i = \bar{\eta}_i(p)$ is the Laplace transform of η_i. Solving these equations

$$\bar{\eta}_i = \frac{f_i(p)}{g(p)}$$

where

$$f_i(p) = \prod_{i=0}^{i-1} \lambda_j \prod_{j=i+1}^{k} (\lambda_j+\mu_j+p) \qquad (\lambda_0=1;\ i=1,\dots,k-1)$$

$$f_k(p) = \prod_{j=0}^{k-1} \lambda_j$$

$$g(p) = \prod_{j=1}^{k} (\lambda_j+\mu_j+p) - \beta \prod_{j=1}^{k} \lambda_j.$$

Taking inverse transforms of the $\bar{\eta}_i(p)$ leads to the solution

$$\eta_i(t) = \sum_{j=1}^{k} \frac{f_i(e_j)}{g'(e_j)} \exp(e_jt)$$

of (3) and (4), where $e_1,\dots,e_k$ are the roots of the characteristic equation $g(p) = 0$. Let E be the largest real root of the characteristic equation. Then as $t \to \infty$

$$n_i(t) \sim \frac{f_i(E)}{g'(E)} \; e^{Et}.$$

Consequently,

$$F_i = f_i(E) \;/\; \sum_{j=1}^{k} f_j(E) \;.$$

Takahashi and Inouye [37] assumed that $\lambda = \lambda(t)$ in equations (5) and (6). They obtained an asymptotic relation between $\lambda(t)$ and the actual observed population growth $N(t)$ of the form

$$\lambda(t) \sim \frac{1}{2^{1/k}-1} \; \frac{N'(t)}{N(t)} \; .$$

This enabled them to propose two different explanations for the variable growth rate of certain tumors in agreement with experimental observations that a tumor's volume increases as a cubic function of time (cube root growth). In one explanation, λ is variable while the <u>growth fraction</u>

$$g = \frac{\text{number of proliferating cells}}{\text{total number of cells}}$$

is fixed. In the other explanation, λ is a constant while the growth fraction decreases with time.

The two-compartment model in Chapter 2 can also be used to study tumor size. If equations (2.35) are added, the following system equation is obtained [8]:

$$\frac{dy}{dt} = (g\gamma - \bar{\delta})y, \tag{7}$$

where

$$y = y_p + y_q \; = \text{the total number of cells in P and Q}$$

$$g = \frac{y_p}{y} = \text{the growth fraction}$$

$$\bar{\delta} = g\delta + (1-g)\lambda = \text{the system loss rate}$$

Assuming that $\bar{\delta} = 0$ and $y = (1+st)^3$, it follows from (7) that

$$\gamma g = \frac{3s}{1+st} \tag{8}$$

Equation (8) shows that γg decreases as time increases in agreement with the conclusions of [37]. Instead of (8) it may be more realistic[3] to assume that $\gamma g = 3sy^{-1/3}$ to obtain cube root growth.

Tannock [38] in a study of tumor proliferation parameters showed that the distribution of cycle times changes as the ascites tumor ages. The basic growth curves for this tumor show that tumor size increases mainly monotonically with age in unperturbed growth. Thus, it should be possible to obtain an explicit representation of cycle parameters in terms of tumor population size. This was done by Jansson [14] using Takahashi's compartmental analysis.

The representations are in terms of the log-phase cell kinetic parameters (subscripted 0). The subscript "D" refers to the dynamic variables when growth is retarded from its log-phase value. Jansson's results suggest that:

(a) The transition probability λ decreases when the number of cells increases. Explicitly,

$$\lambda_D = \frac{\lambda_o}{1+\Delta N},$$

where λ_o is the log-phase value of λ, λ_D is the dynamic value of λ, Δ is a parameter (determined by fitting to Tannock's data) and N is the total population size.

(b) The number of compartments in S, G_2 and M is invariant with population size

(c) The number of compartments in G_1 increases as the population increases - in fact, $\lambda_D g_{1D} = \lambda_{g_1}$.

The model parameters are expressible in terms of observable quantities of the cell population as shown in Table 1 of [36].

Note that (c) is not consistent with the hypothesis in [37] used to prove the cubic root law. Takahaski and Inouye assumed that the total number of compartments compartments (k) remains constant. This discrepancy should be investigated.

---

3 Since the total number of cells and not time may be the pertinent parameter.

A difference equation analogue of the Takahashi equations was developed by Steward and Hahn [33] for describing the effects of drugs on cells. This model was based on the earlier work of Hahn [12]. The cycle is divided into discrete compartments corresponding to divisions of the range of some variable often called a maturity variable. This variable is a function of any cellular variable that changes monotonically through the cell cycle, for example, cell volume or DNA content. As time is increased incrementally, cells are advanced to other compartments as shown in Fig. 3.3.

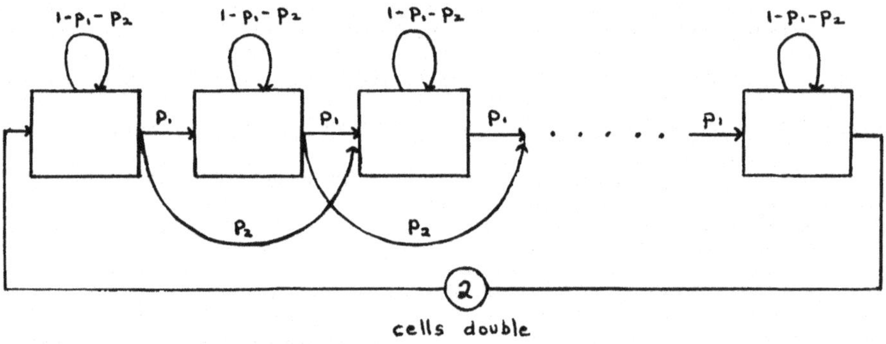

Figure 3.3 Discrete age-time model of Hahn.

The practical difficulty with the Hahn model is that values for the transition probabilities are required. Thames and White [39] show how to determine the parameters in the Hahn model from continuous labelling experiments. (See Chapter 4). For their data the cell gain in mitosis is less than 2. Previous uses of the Hahn model assumed each cell doubles during mitosis.

3.3 Continuous Versions of the Takahashi-Kendall Equations

Continuous models of the cell cycle can be obtained from Takahashi's discrete model by considering a very large number of compartments. This approach produces unified but nonrigorous derivations of these apparently diverse models. These continuous models were originally derived separately by other techniques. They have proved useful in theoretical studies of chemotherapy. However, they are not detailed enough to analyze the effects of cycle-specific drugs nor to find distribution of the lengths of the various phases of the cell cycle.

(a) The Stuart-Merkle Model [34]

Let α be a continuous variable which describes the progress of the cell through the cycle. For convenience, let $\alpha = o$ at the beginning of G_1 and $\alpha = 1$ at the end of mitosis. The length, $\Delta\alpha = k$, of each compartment is not necessarily related to the time spent in each compartment, since the time of transit across the cycle from $\alpha = o$ to $\alpha = 1$ is not necessarily uniform. Hence, α does not correspond to the chronological age of the cell but to a physiological age.

Let $\eta(t,\alpha)$ be a continuous population density function so that

(9) $N(t',\alpha) = \int_o^\alpha \eta(t,s)ds$ = number of cells of age at most α.

The population of the ith compartment

(10) $$n_i = N_i - N_{i-1}$$

where

$$N_i = N(t, i\Delta\alpha)$$

If N_i is sufficiently well behaved it can be expanded in a Taylor's series

(11) $$N_{i-1} = N_1 - \eta_i\Delta\alpha + \eta_i' \frac{(\Delta\alpha)^2}{2} - \eta_i'' \frac{(\Delta\alpha)^3}{6} + \ldots$$

where

$$\eta_i = \frac{\partial N_i}{\partial\alpha} = \eta(i\Delta\alpha, t)$$

and primes denote differentiation with respect to α. It follows from (11) that (10) can be rewritten as

(12) $$\eta_i \Delta\alpha + 0(\Delta\alpha)^2$$

and consequently

(13) $$n_{i-1} - n_i = (\eta_{i-1} - \eta_i)\Delta\alpha + 0(\Delta\alpha)^2$$

Substituting (12) and (13) in equation (6) and dividing by $(\Delta\alpha)$ yields

$$\frac{d\eta_i}{dt} = \lambda(\eta_{i-1}-\eta_i) + O(\Delta\alpha) \tag{14}$$

Expanding η_{i-1} in a Taylor's series (c.f. (11)

$$\eta_{i-1}-\eta_i = -\eta_i'\Delta\alpha + \eta_i''\frac{(\Delta\alpha)^2}{2} + O(\Delta\alpha)^3 \tag{15}$$

Substituting (15) into (14) yields

$$\frac{\partial\eta_i}{\partial t} = \lambda(-\eta_i'\frac{1}{k} + \eta_i''\frac{1}{2}\frac{1}{k^2})$$

Assuming both sides of the last equation are equal and setting $\eta_2 = n$ le ds to the translation-diffusion equation

$$\frac{\partial n}{\partial t} + v_o\frac{\partial n}{\partial\alpha} = D\frac{\partial^2 n}{\partial\alpha^2} \tag{16}$$

where

$v_o = \frac{\lambda}{k}$ = a characteristic velocity or rate of change of cell compartment;

$D = \frac{1}{2}\frac{v_o}{k}$ = a diffusion coefficient.

Takahashi's discrete model is not a good approximation to (16). This equation was used by Stuart and Merkle [34] in their pioneering study of mathematical models for cancer chemotherapy. They proposed (16) by analogy with a particle transport problem in physics.

The solution of (16), if c_o cells commence their cycles (α=0) at t=0, is [27]

$$n(t,\alpha) = \frac{c_o}{\sqrt{4\pi Dt}}\exp -\left[\frac{(\alpha-tv_o)^2}{4Dt}\right] \tag{17}$$

The density function for the cycle time c is

$$n(c,1)/\int_0^\infty n(c,1)dc.$$

As in the Takahashi-Kendall model, the cycle-time distribution is not arbitrary but is fixed by the model. More complex models, obtained by a limiting process from (3), lead to the translation-diffusion equation with variable v_o and D. These give rise to a variety of cycle distribution functions.

This model was presented mainly for historical interest and has not been used recently for studying chemotherapy. It has the conceptual disadvantage that a certain percentage of cells necessarily grow younger.

(b) Rubinow's Maturity-Time Model [29]

Let α be a continuous variable which describes the cell's progress through its cycle in terms of maturity. Maturity has different interpretations. Two possibilities are cell volume and amount of DNA in a cell. Hence, α will vary between two finite limits which need not be 0 and 1 as in (a). The same notation and expansions that were used in (a) will be used in the section except that age is now interpreted as maturity. The more elaborate compartmental equation (3), with $\mu_i = 0$, will be used instead of (5).

Assume that $\lambda = \lambda(\alpha)$ and that λ can be expanded in a Taylor's series. Let $\lambda_i = \lambda(i\Delta\alpha)$; then

(18) $$\lambda_{i-1}n_{i-1} = \lambda_i n_i - \lambda_i n_i' \Delta\alpha - \lambda_i' n_i \Delta\alpha + O(\Delta\alpha)^2$$

Using (12) equation (3), with $\mu_i = 0$, reduces to

(19) $$\frac{dn_i}{dt}\Delta\alpha = (\lambda_{i-1}n_{i-1} - \lambda_i n_i)\Delta\alpha + O(\Delta\alpha)^2$$

Substituting $\lambda_{i-1}n_{i-1}$ from (18) into (19) yields

(20) $$\frac{dn_i}{dt} = (-\lambda_i n_i' \Delta\alpha - \lambda_i' \Delta\alpha) + O(\Delta\alpha)$$

Setting $n_i = n$ and

$$\lambda_i \Delta\alpha = v(\alpha), \qquad \lambda_i' \Delta\alpha = \frac{\partial v}{\partial\alpha}$$

and neglecting $O(\Delta\alpha)$, equation (20) assumes the form

(21) $$\frac{\partial n}{\partial t} + \frac{\partial}{\partial\alpha}(vn) = 0.$$

A slightly generalized version of (21)

(22) $$\frac{\partial n}{\partial t} + \frac{\partial}{\partial \alpha}(vn) = -\lambda n \qquad \alpha_o < \alpha \leq \alpha_1$$

was derived by a different method by Rubinow [29]. The prescribed function v is called the maturation velocity. The function λ represents cell disappearance due to death or other means, but not by cell division. The birth process is represented represented by the boundary condition

(23) $$n(t,\alpha_o)\, v(\alpha_o) = 2n(t,\alpha_1)\, v(\alpha_1)$$

The factor 2 is present since each cell is assumed to produce two cells. An initial condition

(24) $$n(o,\alpha) = g(\alpha)$$

is also required. Here g is a given maturation density.

A more intuitive and rigorous derivation of (22) is based on the following cellular conservation law for the change in the number of cells at a given maturity level in a time Δt.

$$\begin{pmatrix}\text{change in}\\ \text{no. of cells}\end{pmatrix} = \begin{pmatrix}\text{no. of entering}\\ \text{cells from next}\\ \text{lower level}\end{pmatrix} - \begin{pmatrix}\text{no. of departing}\\ \text{cells from}\\ \text{given level}\end{pmatrix} - \begin{pmatrix}\text{no. of}\\ \text{cells lost}\end{pmatrix}$$

To formulate this conservation law analytically, the cell density function $n(t,\alpha)$ is introduced. Note that $n(t,\alpha)\Delta\alpha$ represents the number of cells in the maturity level α to $\alpha + d\alpha$. Assume that the rate of loss of cells from the maturity level α to $\alpha + d\alpha$ is proportional to the number of cells $n(t,\alpha)\Delta\alpha$, where the proportionality loss function λ may depend on t,α,n etc. Let

$$v = \frac{d\alpha}{dt}$$

be the maturation velocity; hence $\Delta\alpha \approx v\Delta t$. Then the conservation law can be written

$$(n(t+\Delta t,\alpha+\Delta\alpha) - n(t,\alpha+\Delta\alpha))\Delta\alpha = n(t,\alpha)\, v(t,\alpha)\Delta t$$
$$- n(t,\alpha+\Delta\alpha)\, v(t,\alpha+\Delta\alpha)\,\Delta\alpha - \lambda n(t,\alpha+\Delta\alpha)\Delta\alpha\, \Delta t$$

up to terms of the second order in $\Delta\alpha$ and Δt. Dividing both sides of this last equation by $\Delta\alpha\ \Delta t$ and letting Δt and so $\Delta\alpha$ approach zero, yields (22) upon using the definition of the partial derivative.

The other partial differential equations representing the continuous models can also be derived in an analogous direct manner without using Takahashi's equations.

(c) The Von Foerster-Scherbaum-Rasch equation [45].

The cell density function n satisfies

(25) $$\frac{\partial n}{\partial t} + \frac{\partial n}{\partial a} = -\lambda n.$$

Here a is the chronological age of a cell, that is, the time elapsed since its birth. Once again λ is a loss function and may be divided into two parts

(26) $$\lambda = \lambda_m + \lambda_d$$

Here λ_m is the fractional probability per unit time that cell division takes place. The subscript m stands for mitosis. Upon division the cell is considered lost. The fractional probability per unit time for cell death or disappearance is denoted by λ_d. The boundary condition at a = 0 is

(27) $$n(t,o) = 2\int_o^\infty \lambda_m(s)\, n(t,s)ds.$$

The factor 2 is a consequence of the assumption that each cell gives birth to two daughter cells. The initial condition is

(28) $$n(o,a) = f(a)$$

where f(a) is a given age density function.

Equation (25), with $\lambda = o$, was proposed earlier by Scherbaum and Rasch [31].

The age-time and maturity-time representations have certain formal similarities. In the next section their differences will be studied. However, one weakness of the von Foerster model should be stated at this point. All cells born at the same time behave in an identical manner throughout their life cycle even though they divide at different times. In actuality cells born at the same time spend different times in each of the phases of the cycle. They disperse in "maturity".

To remedy the above deficiency the cell cycle can be decomposed into several phases (usually G_1,S,G_2,M). The cells are assumed to behave identically in each of these phases and so can be described by separate von Foerster equations. However, the cells can leave each phase at a different time. The boundary conditions for each of the phases must, of course, be matched. This approach was used by Chuang [4] to describe chemotherapy of L1210, a murine leukemia and also in a cell model proposed by Trucco [40].

(d) Generalizations

Frequently more than 1 trait is required to describe the progress of a cell through its cycle. Hence, a density function n(t,a,x) will be considered. Here a is the chronological age and

$$x = (x_1,\ldots,x_k)$$

is a vector of the revelant phenotyptic traits. Using the conservation principle the equation for the density function can be written as

$$\frac{\partial n}{\partial t} = -\nabla\cdot J + B-D+M \tag{29}$$

where B,D and M are the birth, death and migration rates and the operator

$$\nabla = (\frac{\partial}{\partial a}, \frac{\partial}{\partial x_1},\ldots,\frac{\partial}{\partial x_k}).$$

The migration rate is required if differention of cells is allowed. The birth death and migration terms may all be functions of n, a and x. The flux

$$J = vn$$

is a connective flow with velocity

$$v = (1,v_1,\ldots,v_k)^T$$

where $v_i = \frac{dx_i}{dt}$ is the growth rate of trait i.

As an example, consider a single trait x = V, the volume of a cell. Usually the birthrate B is accounted for by a boundary condition. It is the boundary conditions that typifies the cell population equation. Equation (29) can then be written as

$$(30)\qquad \frac{\partial n}{\partial t} + \frac{\partial n}{\partial a} + \frac{\partial}{\partial V}(Fn) = (P+D)n$$

where F(a,V) is the rate of change in volume of a cell of age a and volume V, D(a,V)dt is the fraction of cells of age a and volume V that die in time dt and P(a,V)dt is the probability that a cell of age a and volume V divides, in the time dt = da. The boundary condition

$$(31)\qquad n(t,o,V) = b\int_0^\infty P(a,2V)n(t,a,2V)da$$

is a consequence of the fact that dividing mothers of age a and volume 2V give rise on the average to b live offspring of age 0 and volume V. To completely specify the solution the initial distribution of a cells

$$(32)\qquad n(o,a,V) = n_o(a,V)$$

must be given. These equations were proposed by Anderson and Bell [3] in their studies ([1], [2], [3]) of volume growth of mammalian cell cultures.

In several prokaryotes and lower eukaryrtic cells volume seems to be a major factor in controlling cell division. Several mathematical models based on this fact have been proposed ([31], [17], [6] for exponential growth; [22] for non-exponential growth). These models assume that volume growth rate and all division depend on cell volume.

The dynamic behavior of the volume distribution of a population of cells that are not in steady-state growth has also been studied. Kim and Woo [16], formulated a difference equation model for cellular growth which is analogous to Hahn's model [12]. They assume that the volume of the cells in each age class is normally distributed. The mean and variance of this distribution depend on the age compartment. This assumption is valid for steady-state growth and certain non-steady state populations. However, it is not valid for populations of cells treated with agents which interfere with a specific metabolic pathway. Different distributions of cell volumes can be obtained at the same age! The reader should be able to explain this after reading the next two paragraphs.

Ross ([25], [26], [27]) treated cultured mammalian cells (Chinese Hamster V79 fibroblasts and murine lymphoma L5178Y) with drugs that inhibit a specific reaction of the DNA synthesis pathway. The cells were observed to increase their volume at normal or nearly normal rates in the presence of colcenid, vinblastine, excess thymidine and hydroxyurea and at slightly reduced rates with Ara-C, 5-Fu, actinomycin-D and bleomycin. This can be explained by the fact that many metabolic functions not related to the DNA synthetic pathway continue at their normal rates [20]. These processes determine the volume of the cell. Consistent with this last assumption is the following discovery. Drugs which have a more general effect like the inhibition of RNA or protein synthesis (puromycin and cycloheximide) inhibit the volume growth rate.

On the basis of the above experiments it has been hypothesized that these DNA inhibitors decouple two components of the cell cycle - DNA division and volume growth.[4] Other experimental findings ([20], [44]) have also been explained by the decoupling of the DNA division cycle from other cellular processes. Further support for the uncoupling hypothesis is supplied by experiments with drugs whose effects can be reversed. When the drug effects wear off the cells return to their original volume distribution and generation time (to a first approximation).

The observation that the cycle times do not change after a perturbation is not consistent with the models proposed by Fantes et al. [9]. The basic premise of these models is that a cell can sense its volume. Under a perturbation it changes its cycle time to maintain a constant average cell size. Evidence for this feedback type of control is cited for several cell systems. However, it is not true for some mammalian systems.

---

[4] The resulting unbalanced growth is thought to be the primary reason for cell death after exposure to such lethal agents.

A model for mammalian systems was introduced by Zietz [47] based on equations (30)-(32). He assumed that the generation time distribution is independent of volume. The probability P(a,v)da that a cell of age a and volume V divides in the interval a to a+da is just the probability that a cell of volume V divides in the interval a to a+da given that its age is at least a. Using the definition of conditional probability and the independence of generation time and volume

$$P(a,V) = \frac{f(a)}{1-F(a)}$$

where f(a) is the density function of the generation time and $F(a) = \int_0^\infty f(s)ds$. The number of nonviable cells was assumed to be negligible; therefore

$$D(a,V) = 0.$$

Cell growth was assumed to be proportional to age and so

$$F(a,V) = r$$

(or more generally, F(a,V) = g(a)). This last assumption is consistent with the earlier experimental work of Sinclair and Ross [32] on the volume growth rate of Chinese Hamster V79 fibroblasts after synchronization by mitotic selection. They observed that both exponential and linear growth rates were consistent with their simple model. A linear growth rate was also observed after the application of DNA inhibiting drugs. Exponential growth was rejected on theoretical grounds. Trucco and Bell [43] showed that if the volume growth rate was exponential there had to be a strong coupling between cell division and cell volume if the system was to exhibit an asymptotically stable volume distribution.

Under the above assumptions Zietz showed that asymptotically

(i) the number of cells must grow exponentially;

(ii) there is a stable time independent age distribution;

(iii) the volume distribution must be time independent.

The first two results are proved by integrating equations (30)-(32) with respect to volume. This leads to the von Foerster equation from which (i) and (ii) follow (see §4 and Nooney [21]). The third result is obtained by computing the moments of the volume distribution. The first two moments were found earlier by Trucco

[40] by using a renewal type equation of volume growth.

Zietz then applied his model to the experimental data on thymidine treated cells. Since the effects of thymidine were reversible, the assumption of negligible cell death (D = 0) was retained. To account for the drug action four von Foerster like equations were required - one for each of the phases G_1, S_1, G_2 and M. The equations for the phases in which the drug did not act were

$$\frac{\partial n_i}{\partial t} + \frac{\partial n_i}{\partial a} + r \frac{\partial n_i}{\partial V} = - \frac{f_i}{1-F_i} n_i \qquad i = G_1, G_2, M$$

where $f_i(a)$ is the probability density function of leaving compartment i between a and a+da. They required no modification.

To simulate the effect of thymidine the S equation had to be modified. It was assumed that when the drug is present all cells in the S phase are stopped from maturing but continue to increase in volume at their normal rate. This gives the equation

$$\frac{\partial n_s}{\partial t} + \mu \frac{\partial n_s}{\partial a} + r \frac{\partial n_s}{\partial V} = - \frac{\mu f_s}{1-F_s} n_s$$

where

$$\mu = \begin{cases} 1 & \text{if the drug is not present;} \\ 0 & \text{if the drug is present.} \end{cases}$$

Further all cells completing G_1 during the application of the drug do not enter S. They enter a compartment R but continue to increase their volume at a normal rate. This leads to the equation

$$\frac{\partial n_R(t,V)}{\partial t} + r \frac{\partial n_R(t,V)}{\partial V} = 0$$

and the boundary condition

$$n_R = (1-\mu) \int_0^\infty \frac{f_{G_1}}{1-F_{G_1}} n_{G_1}(t,a,V)da.$$

The last two equations are not used before the drug is applied. After removal of the drug, all cells in R enter S at age 0. This last assumption along with the fact that cells completing G_1 enter S gives

$$n_S(t,0,V) = \mu \int_0^\infty \frac{f_{G_1}(a)}{1-F_{G_1}(a)} n_{G_1}(t,a,V)da + n_R(t,V)I$$

where I(t) = 1 at the time of drug removal and 0 otherwise. Since cells enter G_2 only when they complete S (and so do not enter G_2 when the drug is present) the following boundary condition is also required

$$n_{G_2}(t,0,V) = \mu \int_0^\infty \frac{f_S(a)}{1-F_S(a)} n_S(t,a,V)da.$$

The remaining two boundary conditions are not directly influenced by the drug. Since a cell enters M after completing G_2

$$n_M(t,0,V) = \int_0^\infty \frac{f_{G_2}(a)}{1-F_{G_2}(a)} n_{G_2}(t,a,V)da.$$

Finally, a mother cell of volume 2V gives birth to 2 daughter cells of volume V. Since both of these cells enter G_1,

$$n_{G_1}(t,0,V) = 2 \int_0^\infty \frac{f_M(a)}{1-F_M(a)} n_M(t,a,2V)da.$$

The initial distribution of cells in each phase (c.f. (32)) were also specified.

The model predicted a damped oscillatory decay of the mean volume to its equilibrium value (its original volume in exponential growth) after removal of the drug. This agreed with the experimental observation.

Zietz also modelled the lethal effects of a drug on cells. His results show that this can be done by monitoring cell volume increase after administration of the drug. The results, because of possible clinical application, are detailed below.

Cultured CHV79 cells were treated with various concentrations of vinblastine. This drugs inhibits cells from completing mitosis by interfering with microtubule formation [10]. It is assumed that a certain fraction, f, of cells in mitosis are unable to complete division. However, all cells, whether blocked or not, continue to increase their volume at a constant rate of r(mm^3/hr.)

The cells at time $t = 0$ are assumed to be in exponential growth, in the absence of the drug. Hence, the total population

(33) $$n(t) = n(o)e^{\lambda t}$$

where λ is the exponential growth rate. The assumption of exponential growth implies that any subpopulation of cells grows at the same rate. Hence, the number of cells in mitosis

(34) $$m(t) = m(o)e^{\lambda t}.$$

The total number of cells dividing at time t is

(35) $$\dot{n}(t)dt = \lambda n(o)\ e^{\lambda t}dt$$

Suppose that the drug is administered at time $t = 0$. The shortest possible generation time for CHV79 is about 8 hours. Hence, for $t < 8$ only unaffected cells can reach their next mitosis. Therefore the <u>flow</u> of cells into mitosis is unaltered for $t < 8$. Using this last observation the number of cells observed in mitosis ($t < 8$) after the drug block is equal to the number of cells that would be in mitosis if there was no drug treatment (since $t < 8$) plus the number of cells still in mitosis because they have been blocked from dividing by the drug. Hence,

$$m(t) = m(o)e^{\lambda t}+f \int_o^t \dot{n}(t)dt = m(o)e^{\lambda t}+f[n(t)-n(o)]$$

Simplifying the last equality by using (33) yields

(36) $$m(t) = m(o)e^{\lambda t}+fn(o)[e^{\lambda t}-1]$$

In the remaining derivations it is still assumed that $t < 8$. Actually measurements on the cells were made 5 hours after the drug was administered. Hence, the derived formulas can be used to check the agreement between the model and the experimental data.

The total number of cells t hours after the drug is applied is

(37) $$n(t) = n(o) + (1-f) \int_o^t \dot{n}(t)dt = n(o)[1+(1-f)(e^{\lambda t}-1)]$$

where the second term represents the number of cells that have divided after the administration of the drug.

Suppose a cell is blocked at least until time t. Its volume at time t is

$$V(t) = V(o)+rt. \tag{38}$$

The additional volume over that given in (38) added to the culture by a cell which divides at time $\tau < t$ is required to find the total volume of the culture. Just before division such a cell's volume is $V(o) + r\tau$. The mother divides into two daughter cells of volume V_1 and V_2. In the remaining $t-\tau$ hours each daughter cell increases its birth volume by $r(t-\tau)$. Therefore, the total volume of the progeny at time t is

$$V_1+V_2+2r(t-\tau) = V(o)+rt+r(t-\tau) \tag{39}$$

Comparing (38) and (39) the additional volume added to the culture by each cell which divides at $\tau < t$ is $r(t-\tau)$.

The foregoing considerations show that the volume of the culture is

$$V(t) = V(o)+n(o)rt+(1-f) \int_0^t r(t-z)\dot{n}(z)dz$$

After substituting (35) and simplifying the last equation reduces to

$$V(t) = V(o)+n(o)rt + \frac{r(1-f)n(o)}{\lambda} [e^{\lambda t}-\lambda t-1]. \tag{40}$$

The mean volume of the culture at time t after drug administration is (from (37) and (40))

$$\bar{V}(t) = \frac{V(t)}{n(t)} = \frac{\bar{V}(o) + rt + \frac{r(1-f)}{\lambda} [e^{\lambda t}-\lambda t-1]}{1+(1-f)[e^{\lambda t}-1]} \tag{41}$$

where $\bar{V}(o)$ is the average volume of the culture at t = o.

The fraction of the population in mitosis, the mitotic index, is (from (36) and (37))

(42) $$MI(t) = \frac{m(t)}{n(t)} = \frac{MI(o)e^{\lambda t}+f(e^{\lambda t}-1)}{1+(1-f)(e^{\lambda t}-1)} .$$

Solving (41) for the fraction of cells in mitosis affected by the drug gives

(43) $$f = \frac{[\bar{V}(t)-\bar{V}(o)]e^{\lambda t}}{rt+[\bar{V}(t)-\bar{V}(o)][e^{\lambda t}-1]}$$

since $V(o) = \frac{r}{\lambda}$. A rigorous proof can be found in [3]. Intuitively, the total volume increase in the population during the time interval dt is either

(44) $$nr\, dt = (\text{number of cells}) \times (\text{volume increase of one cell}).$$

or

(45) $$\bar{V}dn = \bar{V}\,\lambda n\, dt = (\text{mean volume}) \times (\text{increase in cell number}).$$

The result follows by equating (44) and (45). Let g be the fraction of the population affected by the drug. Since the effects of vinblastine are mitotic arrest it seems reasonable to assume that $fm = gn$ or $f(\frac{m}{n}) = g$. Hence the fraction of the total population affected by the drug is

(46) $$g = f\, MI(t).$$

The predictions using (41), (42), and (43) agreed with the experimental data.

A possible clinical application of the above work is in determining which of several drugs is the most effective for treating a particular tumor. The effectiveness of a drug could be measured by monitoring cell volumes from biopsies of the tumor. Alternatively, cells from the tumor might be cultured and the cultured cells subjected to the drugs.

In acute leukemia, Clarkson et al. [5], Sarna et al. [30] and Lampkin et al. [18] have shown that the proliferative activity of leukemia cells in the blood and bone marrow are related to cell size. In addition, the kinetics of leukemia cells in the bone marrow is reflected by the kinetics of leukemia cells in the blood.

Studying the effects of drugs by using blood samples instead of bone marrow samples offers the advantage of constant monitoring in a relatively painless and easy manner. Dr. Dennis Ross is now investigating whether the above methods can predict the clinical response of patients undergoing chemotherapy for acute leukemia.

Note on Volume Measurement

The volume of cells can be measured quite accurately by means of a Coulter counter or a fluorescent activated cell sorter (FACS). A Coulter counter determines the volume of a cell by forcing an electrically conducting saline solution containing suspended cells to flow through a tiny aperature and measuring the change in the electrical resistance of the fluid in the aperature as each cell passes through it.

FACS separates cells according to the amount of fluorescent dye bound to each cell. A liquid suspension of cells is forced under pressure through a micronozzle which vibrates rapidly breaking the jet stream into uniform droplets. Before the droplets form, the jet stream is illuminated by a focused blue or green light of an argon-ion laser, operating at a wavelength selected to excite fluorescence in cells tagged with the fluorescent dye. Some of the fluorescent light, filtered to remove the exciting wavelength, is focused onto a photomultiplier tube, which generates an electrical signal proportional to the number of fluorescent molecules on each cell. The forward scattered light from the jet stream generates a second signal related to the volume of the passing cell. The signals are then processed, delayed and combined in order to produce electric pulses for changing the liquid stream at the exact time that the droplets containing the cell is forming. The charged droplets are deflected appropriately by an electrostatic field established by two charged plates. Uncharged droplets continue in their original course.

3.4 Solutions of the Continuous Models

Equation (29) is an example of a quasi-linear equation of the first order - that is, it is of the form

$$b^1 \frac{\partial z}{\partial x^1} + b^2 \frac{\partial z}{\partial x^2} + \ldots + b^n \frac{\partial z}{\partial x^n} = b \tag{47}$$

Here b^i and b are differentiable functions of x_i and of the dependent variable z which do not vanish simultaneously.

A general solution of (47) can be found by finding solutions, called <u>first integrals</u>, of the auxiliary equations

$$\frac{dx^1}{b^1} = \frac{dx^2}{b^2} = \ldots = \frac{dx^n}{b^n} = \frac{dz}{b} \tag{48}$$

It can be shown that if $f_\alpha(x^i, z) = c_\alpha$ are N functionally independent first integrals of (48), then every function z implicitly defined by an equation

$$\pi(f_\alpha(x^i, z)) = 0$$

where π is an arbitrary differentiable function is a solution of (47) (see Duff [7]).

Let z be a solution of (47) defined implicitly by $u(x^1, \ldots, x^n, z) = 0$. The set of points $(x^1, \ldots, x^n, z)$ satisfying $u = 0$ is called an <u>integral surface</u> S. Through each point of S passes a unique <u>characteristic curve</u> which is determined by adjoining a parameter t to (48) - namely,

$$\frac{dx^i}{dt} = b^i \; ; \quad \frac{dz}{dt} = b \qquad (i = 1,2,\ldots,n) \tag{49}$$

Therefore, S is composed of an (n-1)-parameter family of characteristic curves. It should thus be possible to find an (n-1) - dimensional hypersurface H_{n-1} such that each of the characteristic curves meets the hypersurface in precisely one point (see Fig. 3.4).

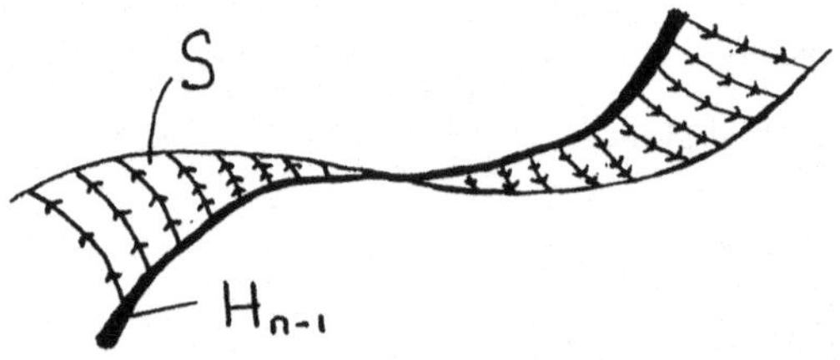

Figure 3.4 Integral surface generated by characteristic curves.

Conversely, given a hypersurface H_{n-1} construct an integral surface $S = H_n$ containing H_{n-1}. This problem has a unique solution provided that the vector b^i $(i = 1,\dots,n)$ is not tangent to the projection of H_{n-1} on the plane $z = o$, the space of the variables $x^1,\dots,x^n$. Then it can be shown [7] that the integral surface is generated by the characteristic curves through H_{n-1}. Every point on S is determined by a unique characteristic through some initial point (when $t = o$) on H_{n-1}. The integral surface and hence z is uniquely determined.

To illustrate the uniqueness result consider Rubinow's equation (22) rewritten in the following form

(50)
$$\frac{\partial n}{\partial t} + v\frac{\partial n}{\partial \alpha} = -(\lambda + \frac{dv}{d\alpha})n$$

Here $\alpha_o = 0$, $\alpha_1 = 1$; $v = v(\alpha)$ and $\lambda = \lambda(\alpha)$ are only functions of α. The initial surface H_1 is obtained by writing boundary condition (24) in the following parametric form[4]

(51)
$$t = 0 \qquad \alpha = u \qquad n = g(u)$$

The characteristic equations are:

(52)
$$\frac{dt}{ds} = 1 \qquad \frac{d\alpha}{ds} = v \qquad \frac{dn}{ds} = -(\lambda + \frac{dv}{d\alpha})n$$

The solution of the first two equations are

(53)
$$t = s$$
$$s = \int_u^\alpha \frac{d\xi}{v(\xi)}$$

since the solutions must satisfy the initial conditions (51) when $s = 0$. The last equation in (52) can be written as

$$\frac{dn}{n} = -(\lambda + \frac{dv}{d\alpha})\frac{d\alpha}{v}$$

since $ds = \frac{d\alpha}{v}$. Its solution is

$$\ell n \frac{n}{n(o,u)} = -\int_u^\alpha \frac{\lambda(\xi)}{v(\xi)} d\xi - \ell n \frac{v(\alpha)}{v(u)}$$

4 The general proof uses parametric form and is similar to the method used in this example.

or

$$n = n(o,u)\,\frac{v(u)}{v(\alpha)}\,\exp\left(-\int_o^\alpha \frac{\lambda(\xi)}{v(\xi)}\,d\xi\right) \tag{54}$$

To obtain the final form of the solution, the parameters s and u are expressed in terms of α and t by using the equations labelled (53). These values are then substituted in (54). In order to accomplish this, the function η defined by

$$\eta(\alpha) = \int_o^\alpha \frac{d\xi}{v(\xi)} \tag{55}$$

and its inverse h defined by

$$\alpha = h(\eta) \tag{56}$$

are required. It follows from equations (53) and (55) that $t = \eta(\alpha) - \eta(u)$ or $\eta(u) = \eta(\alpha)-t$. Applying the inverse function h to this last equation gives

$$u = h(\eta(\alpha)-t) \tag{57}$$

Substituting the value of u from (57) into (54) yields

$$n(t,\alpha) = n(o,h(\eta-t))\,\frac{v(h(\eta-t))}{v(\alpha)}\,\exp\left(-\int_{h(\eta-t)}^\alpha \frac{\lambda(\xi)}{v(\xi)}\,d\xi\right) \tag{58}$$

Similarly, it can be shown that if the boundary condition $n(t,o)$ is specified the solution of (50) is

$$n(t,\alpha) = n(t-\eta,o)\,\frac{v(o)}{v(\alpha)}\,\exp\left(-\int_o^\alpha \frac{\lambda(\xi)}{v(\xi)}\,d\xi\right) \tag{59}$$

These solutions will be utilized in a biological example at the end of this section.

As another example consider the von Foerster equation (25) with $\lambda = \lambda(t,a)$. The characteristic equations are

$$\frac{dt}{ds} = 1 \qquad \frac{da}{ds} = 1 \qquad \frac{dn}{ds} = -\lambda n$$

Eliminating the parameter s from the first two equations yields $\frac{da}{dt} = 1$. Hence, the projections of the characteristics on the plane $n = 0$ are

$$a = \begin{cases} t + a_0 & a > t \\ t - t_0 & a < t \end{cases}$$

where a_0 is the initial age at $t = 0$ of a cell in the original population and t_0 is the time of birth of a cell after $t = 0$ (see Fig. 3.5) Hence, the solution to

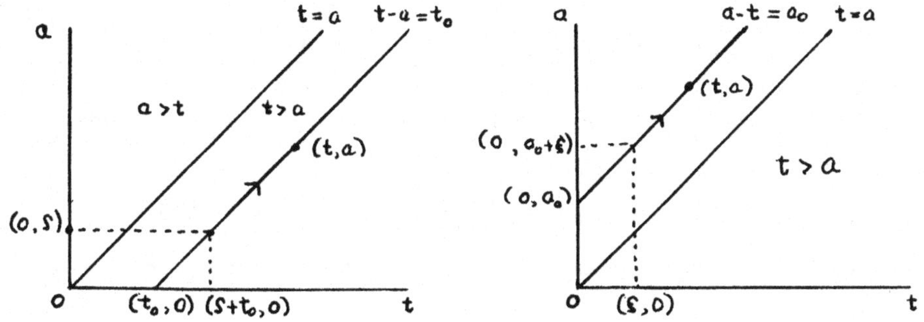

$n(t_o,0)$ determines $n(t,a)$ for $t > a$ $\qquad$ $n(0,a_0)$ determines $n(t,a)$ for $a > t$

Figure 3.5 Projection of characteristics for von Foerester's equation.

the third characteristic equation

$$\frac{dn}{da} = -\lambda n \tag{60}$$

can be written in two parts. Cells whose ages a are less than t are born at t_0 $t_0 > 0$. The density of such cells, $n\,(t_0,0)$, uniquely determines $n(t,a)$ for a point (t,a) on the projection of the characteristic through $(t_0,0)$ (See Fig. 3.5.) To verify this analytically **parametrize** the line $t-a = t_0$ by means of the equations

$$a = s, \qquad t = s + t_0$$

The parametric form of the characteristic equation (60) is

$$\frac{dn}{ds} = \lambda(s+t_0,s)\; n$$

Its solution is

$$n(s+t_o,s) = n(t_o,o)\ \exp(-\int_o^a \lambda(s+t_o,s)ds)$$

Since $t_o = t-a$, this last equation can be written as

$$n(t,a) = n(t-a,o)\ \exp(-\int_o^a \lambda(s+t-a,s)ds) \qquad (t > a) \tag{61}$$

Similar reasoning shows that for cells belonging to the original population

$$n(t,a) = n(o,a-t)\ \exp(-\int_o^t \lambda(\xi,a-t+\xi)d\xi) \qquad (a > t) \tag{62}$$

(See Fig. 3.5.). These solutions were obtained by Trucco [42] in an extensive investigation of von Foerster's equation.

Usually, $n(o,a)$ is specified and $(n(t,a)$ for $a > t$ is thus completely determined by (62). However, $n(t,o)$ is given by a boundary condition of the form (c.f. (27))

$$n(t,o) = 2\int_o^\infty b(t,a)\ n(t,a)da. \tag{63}$$

Hence, the solution is not determined for $t > a$ by (61) since the unknown function $n(t-a,o)$ is involved. To determine this unknown function

$$n(t,o) = B(t)$$

the values of $n(t,a)$ from equations (62) and (63) are substituted into (64) yielding

$$B(t) = 2\int_o^t b(t,a)\ n(t,a)da + 2\int_t^\infty b(t,a)\ n(t,a)da$$

$$= 2\int_o^t \phi(t,a)\ B(t-a)da + I(t)$$

Here

$$\phi(t,a) = b(t,a)\ \exp(-\int_o^a \lambda(s+t_o,s)ds$$

and

$$I(t) = 2\int_t^\infty b(t,a)\ n(t,a)da$$

where n(t,a) is completely determined by (62). The term I(t) gives the contributions to the births at time t from cells already in the culture at t = 0. The integral equation for the births can be rewritten in the form of a Volterra equation

$$B(t) = 2 \int_0^t B(s)\ \phi(t,t-s)ds + I(t). \tag{64}$$

by letting s = t-a. In this form the von Foerster model is called the Lotka equation. It can be shown that if I and ϕ are continuous and bounded there exists a unique solution on the finite interval [o,T].

Most work has been done when ϕ is only a function of age - for example, if b = b(a) and $\lambda = \lambda(a)$. In this situation, equation (64) is of the form

$$B(t) = 2 \int_0^t B(t-s)\ f(s)ds + I(t). \tag{65}$$

This equation for the number of cells B(t) born at time t was derived by Hirsh and Engelberg [13] by considering the cell cycle as a renewal process. The time between divisions was considered to be a random variable with a given stationary distribution f. The density f can be measured by photographing a culture of cells sequentially in time and observing the length of time between mitosis for individual cells. It is assumed that there is no correlation between the division time of the mother and daughter cells. Using this model it was found that the generation time distribution is not symmetric about a mean value. It is skewed toward longer cycle times for the investigated cultures. Towards the end of this section, it will be seen that for certain cell lines the division times of mother and daughter cells are correlated for a few generations.

An analysis of the renewal equation (64) by Nooney [21] shows that the number of newly born cells eventually increase at a constant exponential rate. Moreover there is an asymptotically stable age distribution. This limiting distribution is independent of the initial age distribution. From the fact that the births increase exponentially, it follows that the total number of cells increase at the same rate. Once again exponential growth is predicted! These results will be derived by the method of Laplace transforms.

Denote the Laplace transform of a function by the subscript L - for example,

$$B_L(s) = \int_0^\infty e^{-st} B(t)dt.$$

Taking Laplace transforms of both sides of (65) yields

$$B_L = 2 B_L f_L + I_L$$

upon using the convolution theorem. Hence,

$$B_L = \frac{I_L}{1-2 f_L} .$$

The solution obtained by inverting the equation for B_L, can be written as

$$B(t) = \sum a_k e^{r_k t}$$

where r_k are the roots of the characteristic equation

$$1-2 f_L(s) = 0.$$

It can be shown that the characteristic equation has one real root, r_0, larger in amplitude than any of the other roots, r_i. Thus the birthrate is asymptotically exponential - i.e.,

(66) $$B(t) \sim a_0 e^{r_0 t}$$

as $t \to \infty$. Recalling that $n(t-a,o) = B(t-a)$ and that λ was assumed to be a function only of the age, it follows from equation (65) and (62) that

(67) $$n(t,a) \sim \eta(a) e^{r_0 t}$$

The explicit form of the function η can easily be determined by the reader. From (67) the asymptotic formula for the total number of cells

$$N(t) = \int_0^\infty n(t,a)da$$

is

(68) $$N(t) \sim (\int_0^\infty \eta(a)da)e^{r_0 t}$$

Thus the age density function, d, asymptotically approaches a stationary shape -

(69) $$d(t,a) = \frac{n(a,t)}{N(t)} \sim \eta(a) / \int_0^\infty \eta(a)da$$

Further details can be found in [19] and [23]. In general, these results do not hold for (64) if ϕ is time dependent. The effect of the initial conditions may not decay.

These results have been generalized by Gurtin and MacCarny [11]. They allow the birth and death rates to be functionals of the total population size N - that is, $b = b(a,N)$ and $\lambda = \lambda(a,N)$. The resulting model consists of two coupled integral equations and has a unique stationary age distribution under certain conditions.

A steady state maturity distribution can also be found for Rubinow's model. Assume a solution of the form

(70) $$n(t,\alpha) \sim \eta(\alpha)e^{\gamma t}$$

where γ is a constant. Substituting (70) into (22) yields

$$\eta' = -\frac{(\lambda+\gamma+v')}{v}\eta$$

where differentiation is with respect to α. The solution is

(71) $$\eta(\alpha) = \eta_0 \frac{v(\alpha_0)}{v(\alpha)} \exp\left(-\int_{\alpha_0}^{\alpha}\left[\frac{\gamma+\lambda(\xi)}{v(\xi)}\right]d\xi\right.$$

where η_0 is a constant. By substituting (70) into the boundary condition (23) the following relation for determining γ is obtained

(72) $$\exp\left(\int_{\alpha_0}^{\alpha}\frac{\gamma+\lambda(\xi)}{v(\xi)}d\xi\right) = 2.$$

In particular, if α represents cell volume then $\alpha_1 = 2\alpha_0$. Assume that $v = c\alpha$, where c is a constant and that cell death is negligible - that is, $\lambda = 0$. Then (71) reduces to

$$\eta(\alpha) = \eta_o \frac{\alpha_o}{\alpha}^2 .$$

Hence, the steady state volume distribution for a cell population in steady state exponential growth follows a characteristic inverse square law ([21], [17]).

Rubinow's paper [29] shows how the solutions of the continuous models in this section can be applied to actual data. The experimental results of Prescott [24] were used. Prescott measured the generation times of a population of Tetrahymena geleii HS cells. The cells were grown under uniform conditions and all the cells were of age zero at time zero. The number of cells observed in a minute with a given generation time were plotted. It was found that the data points could be fitted by a gamma distribution, $u = u(T)$. The generation times were skewed to the right. Note that they are not symmetric about the mean value as required by the Stuart-Merkle theory.

In a related experiment, Prescott carefully recorded the total number of cells $N_e(t)$ grown from an initial population of 50 cells of the same strain. The cells were initially synchronized at $t = 0$ to be newly born. These results were used to test the continuous models for the birth process.

A theoretical growth curve was obtained by Stuart and Merkle [34] by using (16). The constants v_o and D were determined by least-square fitting to Prescott's data on the generation times. The agreement between the theoretical curve and Prescott's experimental growth curve was not good (as suspected from the skewness of the generation time distribution).

Next, von Foerster's equation (25) will be used to fit the data. It is assumed that there is no cell death and so $\lambda = \lambda_m(a)$. A solution will be found by Lagrange's method in which the population is decomposed into generations. Let

$$(73) \qquad n(t,a) = \sum_{j=1}^{\infty} n_j(t,a)$$

where j represents the jth generation of cells. For the first generation only the initial condition (28) will be required and satisfied, since no cells have divided. In fact,

$$n(o,a) = f(a) = N_o\delta(a)$$

where $\delta(a)$ is the Dirac delta function and N_o is the total number of cells in the initial population. This initial condition expresses the fact that there are N_o cells of age 0 at $t = 0$, in agreement with the data. Hence, from (62)

$$n_1(t,a) = N_o\delta(a-t)\exp(-\int_{a-t}^{a}\lambda_m(s)ds) \tag{74}$$

The first generation gives birth to a second generation according to the boundary condition (27). Hence,

$$n_2(t,o) = 2\int_o^\infty \lambda_m(s)\, n_1(t,s)ds\ . \tag{75}$$

Substituting (75) into (61) yields

$$\begin{aligned} n_2(t,a) &= 2\int_o^\infty \lambda_m(s)\, n_1(t-a,s)ds\, \exp(-\int_o^a \lambda_m(\varsigma)d\varsigma) \\ &= 2\int_o^{t-a} \lambda_m(s)\, n_1(t-a,s)ds\, \exp(-\int_o^a \lambda_m(\varsigma)d\varsigma) \end{aligned} \tag{76}$$

since $s \le t-a$ or $n_1(t-a,s) = 0$. By using (61) it can be shown that

$$n_1(t-a,s)\exp(-\int_o^a \lambda_m(\varsigma)d\varsigma) = n_1(t-s,a)\exp(-\int_o^s \lambda\xi)d\xi \tag{77}$$

Substituting the right side of (77) in (76) gives

$$\begin{aligned} n_2(t,a) &= 2\int_o^{t-a} \lambda_m(s)\, n_1(t-s,a)\exp(-\int_o^s \lambda_m(\xi)d\xi)ds \\ &= 2\int_o^t y(s)\, n_1(t-s,a)ds \end{aligned}$$

where

$$y(s) = \lambda_m(s)\exp(-\int_o^s \lambda_m(\xi)\, d\xi).$$

The general result

$$n_j(t,a) = 2\int_o^t y(s)\, n_{j-1}(t-s,a)ds \qquad (j=2,3,4,\ldots) \tag{78}$$

can be proved by similar reasoning.

The number of cells of the initial generation entering mitosis per unit time is

$$M(t) = \int_0^\infty \lambda_m(a) n_1(t,a)da$$

which reduces to

$$M(t) = N_0 \lambda_m(t) \exp(-\int_0^t \lambda_m(\xi)d\xi)$$

upon using (74). Integrating this last equation yields

$$\int_0^t M(s)ds = N_0 - N_0 \exp(-\int_0^t \lambda_m(\xi)d\xi).$$

Hence,

$$\lambda_m(t) = \frac{M(t)}{N_0 - \int_0^t M(s)ds} \tag{79}$$

Substituting the distribution u = u(T), obtained from Prescott's experiment, for M in (79) determines λ_m. Equations (74) and (78) then completely specify the functions $n_j(t,a)$.

By integrating (73) for all ages, the total population at any time t is

$$N(t) = \int_0^\infty n(t,a)da = \sum_{j=1}^\infty N_j(t) \tag{80}$$

where

$$N_1(t) = N_0 \exp(-\int_0^t \lambda_m(\xi)d\xi)$$

$$N_j(t) = 2 \int_0^t N_{j-1}(t-\xi)y(\xi)d\xi \qquad j=2,3,4,\ldots$$

The last two equations follow from (74) and (78).

Rubinow compared N(t) with $N_e(t)$, the experimental growth curve obtained by Prescott. The agreement was not too good. Therefore, the concept that an individual cell has no memory of the generation time of its parent, but is subject to the same probability of mitosis as a function of age as every other cell, is not supported.

The same problem will now be studied from the viewpoint of Rubinow's maturity-time formalism. This approach postulates that there is a heterogeneous distribution of velocities of maturation, with some memory from one generation to the next generation. The extreme form of this hypothesis will be used. It will be assumed that each generation imparts to its daughters the same generation time, that is, perfect memory.

In the "perfect memory" model let T be the generation time. The conditional density for the subpopulation whose generation time lies between T and T + dT will be denoted by $n(t,\alpha|T)$. Then

(81) $$n(t,\alpha) = \int_o^\infty n(t,\alpha|T)dT$$

The function $n(t,\alpha|T)$ satisfies (21) with $v = \frac{1}{T}$. The initial and boundary conditions are:

(82) $$n(o,\alpha|T) = N_o g(T)\delta(\alpha)$$

(83) $$n(t,o|T) = 2\, n(t,1|T)$$

where g is the distribution of the generation times. Since g is a density

(84) $$\int_o^\infty g(T)dT = 1$$

The function g will be determined below from Prescott's data.

A solution to the system (21), (82) and (83) will be found, as previously, by the method of generations. Let

(85) $$n(t,\alpha|T) = \sum_{j=1}^{\infty} n_j(t,\alpha|T)$$

where j represents the jth generation of cells. The first generation of cells have not divided and so n_1 satisfies (21) and (82). Hence, from (58),

(86) $$n_1(t,\alpha|T) = N_o\, g(T)\, \delta(\alpha-t/T)$$

since $\lambda = 0$, $v = \frac{1}{T}$ is a constant and

(87) $$h(\eta) = \eta/T$$

being the inverse function of

$$\eta = \int_o^\alpha \frac{d\xi}{v(\xi)} = T \int_o^\alpha d\xi = \alpha T$$

The first generation gives birth to the second generation according to the boundary condition (83). Hence

(88) $$n_2(t,o|T) = 2\ n_1(t,1|T) = 2\ N_o\ g(T)\delta(1-t/T)$$

Substituting (88) into (59) yields

$$n_2(t,\alpha|T) = n_2(t-\eta,o) = 2\ N_o\ \alpha(T)\delta(1-\frac{(t-\eta)}{T})$$

or

(89) $$n_2(t,\alpha|T) = 2\ N_o\ g(T)\ \delta(1+\alpha-t/T)$$

More generally, the (j-1)th generation gives birth to the jth generation and so from boundary condition (83)

(90) $$n_j(t,o|T) = 2\ n_{j-1}(t,1|T)$$

Substituting (90) into (60) yields

(91) $$n_j(t,\alpha|T) = n_j(t-\alpha T,o|T) = 2\ n_{j-1}(t-\alpha T,1|T)$$

Hence, (89) and (91) with j = 3 yield

(92) $$n_3(t,\alpha|T) = 2^2\ N_o\ g(T)\ \delta(2+\alpha-t/T)$$

It can be shown by induction that

(93) $$n_j(t,\alpha|T) = 2^{j-1}\ N_o\ g(T)\ \delta(j-1+\alpha-t/T) \qquad j=1,2,\ldots$$

The number of first generation cells completing mitosis per unit time is

$$M(t) = \int_o^\infty \frac{1}{T} n_1(t,1|T)dT = \int_o^\infty \frac{1}{T} N\ g(T)\ \delta(1 - t/T)\ dT$$

Utilizing the relation

$$\delta(1-t/T) = T^2\delta(T-t)$$

M(t) reduces to

$$M(t) = N_0 \int_0^\infty g(T)T\delta(T-t)dT = N_o tg(t) \tag{94}$$

Substituting the experimentally determined function u for M in (94) yields

$$g(t) = \frac{u(t)}{tN_0} .$$

Hence, the functions n_j and so $n(t,\alpha|T)$ are completely specified by (93) and (86), respectively.

From equations (81), (86) and (93)

$$n(t,\alpha) = N_0 \sum_{j=1}^{\infty} 2^{j-1} \int_0^\infty g(T)\ \delta(j-1+\alpha - \tfrac{t}{T})dT \tag{95}$$

$$= N_0 \sum_{j=1}^{\infty} 2^{j-1} \frac{t}{(j-1+\alpha)^2}\ g(\tfrac{t}{j-1+\alpha})$$

and

$$N(t) = \int_0^1 n(t,\alpha)d\alpha = N_0 \sum_{j-1}^{\infty} 2^{j-1} \int_{t/j}^{t/(j-1)} g(\xi)d\xi \tag{96}$$

Note that if all cells have the same generation time T_0 (i.e. $g(T) = \delta(T-T_0)$) then (96) reduces to exponential growth.

Rubinow plotted N(t), given by (96), as a function of time and compared the resulting curve with $N_e(t)$ obtained from Prescott's data. The predictions of the theory agreed well with the data. Hence, these Tetrahymena geleii cells have a memory of the generation time of their parents for several generations. Fast (slow) growing cells arise from fast (slow) growing parents.

However, such a model cannot be correct for large times because fast growing cells would eventually swamp out the more slowly growing cells. For example, consider an initial population of 10^3 cells with generation time T_o and 1 cell with generation time $T_o/2$. Then

$$N_o g(t) = \delta(t-T_o/2) + 10^3\ \delta(t-T_o)$$

where $N_o = 1001$. The entries in the following table were computed from (96). Here $\varepsilon > 0$ is a very small number.

time	no. of cells with generation time		mean generation time
	T_o	$T_o/2$	
0	10^3	1	$1.00T_o$
$10T_o-\varepsilon$	5.12×10^5	5.24×10^5	$.75T_o$
$20T_o-\varepsilon$	5.24×10^8	5.50×10^{11}	$.50T_o$

Note that after about $10T_o$ there are almost the same number of both types of cells. After $20T_o$ the faster growing cells are predominant and the average generation time is about $T_o/2$.

The assumption that maturation velocities persist for a given cell line leads to the conclusion that cells with the shortest generation times will be predominant. The question naturally arises how the initial population of cells with many different maturation velocities could exist. One possible answer to this question is that a real cell population has a variety of maturation velocities which are the result of random variation on a time scale which is large compared to a generation time. Thus, over a few generations there would be a persistence of memory of parental generation time. However, this memory would gradually disappear on a longer time scale. The model may be considered to be an approximation which is only valid over a few generations. Another possibility for real cell populations is that the offspring of cells with a given generation time have a random generation time. These times have a unimodal continuous distribution about a mean generation time which is the same as that of their parent. Here the model is also a simple approximation to reality which is valid over a time span of a few generations

3.5 Another General Approach to Continuous Models

The differences between the physical processes underlying the von Foerster and Rubinow models are not apparent from the differential equations and boundary conditions. Their behavior can be classified by studying the life history of a cell and its descendants. Only 1 descendant of each cell is followed in time.

Figures 3.6 illustrates the time history of a cell according to Rubinow's model. Note that each cell of a cell line has the same generation time τ which is

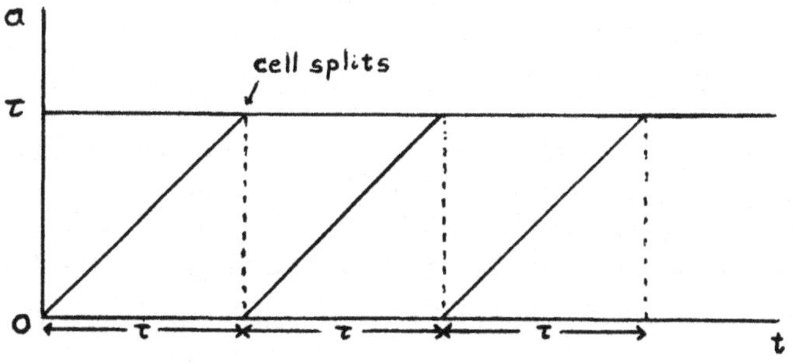

Figure 3.6 Life history of a cell. Perfect memory model.

just the generation time of its parent. All cells with the same generation time constitute a subpopulation. Cells from two different subpopulations have different generation times. The total population is composed of many different subpopulations. This leads to the observed variability of generation times. This model seems different than the perfect memory model of Section 4 since the maturity velocity $da/dt = 1$. Their equivalence will be shown below. The new model is presented for constrast with the von Foerster model whose new description follows.

The time history of a cell governed by the age-time model can be depicted as in Figure 3.7 as will be shown below. Note that each cell has the same maturation velocity $v = 1$ and that the cells divide at different ages. The generation time of a cell is a random variable which is independent of the generation time of any other cell. Consequently the generation times $\tau_1, \tau_2, \ldots$ vary.

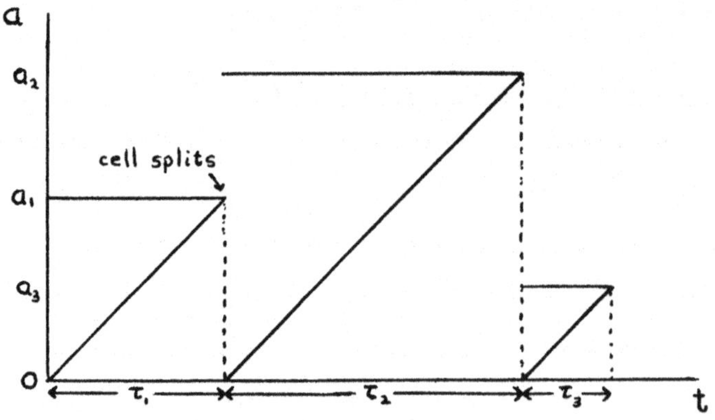

Figure 3.7 Life history of a cell. Age-time model.

To show the equivalence of the two new models with the old models an extension of the age-time formalism will be introduced. Assume that the total population consists of different subpopulations. Each subpopulation has its own characteristic generation time τ as proposed in Rubinow [29]. The density function associated with the subpopulation having generation time τ will be denoted by $n = n(t,a|\tau)$. Then

(97) $$\frac{\partial n}{\partial t} + \frac{\partial n}{\partial a} = -\lambda_d(a)n \qquad o < a < \tau.$$

The initial condition is

(98) $$n(o,a|\tau) = \phi(a,\tau)$$

where ϕ is a given function. The boundary condition is

(99) $$n(t,o|\tau) = 2 \int_0^\infty K(\tau|s)n(t,s|s)ds$$

Here, $K(\tau|s)$ is the probability of cells of generation time s giving birth to cells having a generation time in the interval τ to $\tau+d\tau$. Since K is a conditional probability density function

(100) $$\int_0^\infty K(t|s)dt = 1.$$

To obtain the maturity-time or perfect memory model let

(101) $$K(\tau|s) = \delta(\tau-s),$$

$\tau = T$ and $\alpha = \frac{a}{T}$. Then

$$n(t,\alpha T|T) = \bar{n}(t,\alpha|T)$$

satisfies Rubinow's equation (22) with $v = \frac{1}{T}$. Utilizing (101) the boundary equation (99) reduces to

$$\bar{n}(t,o|T) = 2\ \bar{n}(t,1|T)$$

which is just boundary condition (83) for $\bar{n}$. Boundary condition (82) is also satisfied by n by a proper choice of ϕ in (98). Hence, $\bar{n}$ satisfies all the conditions of the perfect memory model.

The model reduces to the age-time model by choosing

(102) $$K(\tau|s) = P(\tau).$$

To verify this let

$$n(t,a) = \int_o^\infty n(t,a|\tau)d\tau.$$

Integrating with respect to τ reduces equation (97) to the form of the von Foerster Foerster's equation (25). By a similar process (98) is transformed into (28) where

$$f(a) = \int_o^\infty \phi(a,\tau)d\tau$$

Substituting (102) into (99) gives

(103) $$n(t,o)\tau) = 2\ P(\tau) \quad \int_o^\infty n(t,s|s)ds$$

The probability of a cell having generation time s is

$$\lambda_m(s) = \frac{n(t,s|s)}{\int_o^\infty n(t,s|\tau)d\tau} = \frac{n(t,s|s)}{n(t,s)}$$

Hence,

$$n(t,s|s) = \lambda_m(s)\ n(t,s) \tag{104}$$

Note that (100) implies $\int_0^\infty P(\tau)d\tau = 1$. Therefore, integrating both sides of (103) with respect to τ and utilizing (104) gives (27). Hence, this model is equivalent to von Foerster's model.

A model in which some of the cells have a perfect memory while others have no memory can usually be constructed by setting

$$K(\tau|s) = p\ \delta(\tau-s) + (1-p)\ P(\tau).$$

In this model a fraction p, of the cells maintain the mother's cycle time and the remaining fraction 1-p resume the cycle randomly. Note that p = 0,1 yields (102) and (101), respectively.

The kernel K must be determined experimentally. Only one theoretical constraint exists. This is a consequence of the long-time behavior. It can be shown that

$$n(t,a|\tau) \sim A\ u(\tau)e^{\gamma(t-a)}$$

It then follows from (99) that

$$u(\tau) = 2\int_0^\infty K(\tau|s)e^{-\gamma s}\ u(s)ds.$$

The generation time distribution function u is an eigenfunction of the operator associated with the kernel K.

3.6 Trucco's Model ([40],[44])

In this section a model of the cell cycle based on integral equations will be introduced. The model will be applied in Chapters 4 and 5 to show how the duration of each phase of the cell cycle and the proportion of cells in each phase can be experimentally determined.

The cellular population is assumed to be in a steady-state of exponential growth. The expected number of cells in each phase of the cell cycle increases as

$e^{\gamma t}$, where γ, the specific growth rate of the population is a constant. This is equivalent to using the asymptotic solution of the von Foerster equation. A different equation is used for each phase.

Trucco's model is represented schematically in Figure 3.8.

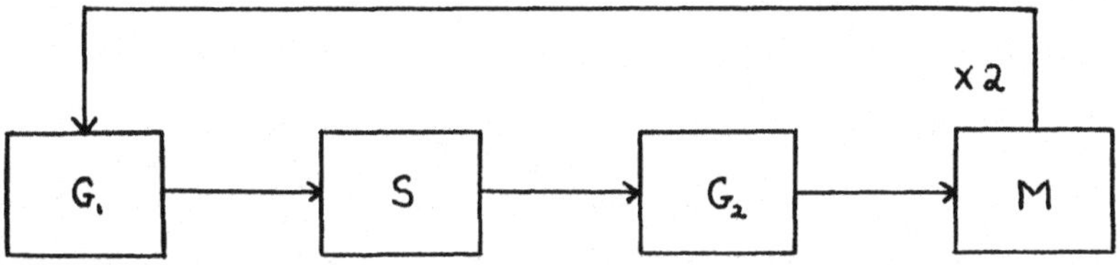

Figure 3.8 Trucco's model.

Binary fission occurs at the end of the mitotic period and newborn cells begin the cycle in G_1. The sojourn times of cells in the jth phase are randomly distributed according to the probability density functions $\delta_j (j = 1,2,3,4)$. The probability that a cell will leave the jth phase with age between a and a+da is $\delta_j(a)da$. The sojourn time of a cell in any phase is independent of the time spent in other phases and of the sojourn time of other cells.[5] The probability that the sojourn time of a cell in phase j exceed a is

$$\phi_j(a) = 1 - \int_0^n \delta_j(x)dx.$$

The expected number of cells in phase j at time t with ages in the small interval (a,a+da) is $n_j(t,a)da$, where the functions n_j is called the <u>age density function</u>. It will be shown that the age density function n_j can be expressed in terms of the <u>generalized birth rate</u>.

$$\alpha_j(t) = n_j(t,o)$$

and the <u>initial age density</u>

---

[5] This assumption is not true for all cells - for example, the ones studied by Prescott.

$$\beta_j(a) = n_j(o,a).$$

Consider a cell at time t of age $a < t$ in phase j. This cell must have entered phase j at time t-a. At that time its age was zero. The expected number of such cells is $\alpha_j(t-a)$. In order to reach age a, the sojourn time of a cell of age 0 must exceed a. The probability of the occurence of this event is $\phi(a)$. Hence,

$$n_j(t,a) = \alpha_j(t-a)\ \phi_j(a) \qquad (a < t) \tag{105}$$

Next consider a cell in phase j at time t whose age $a > t$. Such a cell must have been in phase j with age a-t at time zero. The expected number of such cells is $\beta_j(a-t)$. The probability that a cell in phase j of age a-t at time zero reaches age a is $\phi_j(a)/\phi_j(a-t)$. Therefore

$$n_j(t,a) = \beta_j(a-t)\ \frac{\phi_j(a)}{\phi_j(a-t)} \qquad (a > t). \tag{106}$$

The function $\beta_j(a)$ is usually prescribed as an initial condition. The value of $n_j(t,a)$ is given by (105) and (106) once α_j is determined.

Integral equations for determining α_j can be found as follows. Consider a cell whose age is zero at time t and which has just entered phase j+1 (j = 1,2,3). This event can occur in two distinct ways:

(a) A cell of age zero could have entered phase j at time t-a ($a \leq t$). The total birth rate of such cells is $\alpha_j(t-a)$. In order for these cells to enter, phase j+1 at time t, they must leave phase j when their age is a. The probability of this last event is $\delta_j(a)da$. Hence, the total birth rate of cells exhibiting this behavior is

$$\int_o^t \alpha_j(t-a)\ \delta_j(a)da$$

(b) A cell could have been in phase j with age $y (o < y < \infty)$ at time zero. The number of such cells is $\beta_j(y)dy$. Since such cells are in phase j+1 at time t, they must depart with age y+t. The probability per unit time of this last event is $\delta_j(y+t)/\phi_j(y)$. The total birth rate of such cells is

$$\int_0^\infty \beta_j(y)\ \delta_j(t+y)[\phi_j(y)]^{-1}dy$$

It follows that for j = 1,2 or 3

$$\alpha_{j+1}(t) = \int_0^t \alpha_j(t-a)\delta_j(a)da + \int_0^\infty \beta_j(y)\delta_j(t+y)[\phi_j(y)]^{-1}dy \tag{107}$$

Setting x = t+y, the last integral in (107) can be written as

$$\int_t^\infty \beta_j(x-t)\ \delta_j(x)[\phi_j(x-t)]dx$$

Finally, by utilizing (105) and (106), equation (107) can be written as

$$\alpha_{j+1}(t) = \int_0^\infty n_j(t,a)\delta_j(a)[\phi_j(a)]^{-1}da. \ (j = 1,2,3) \tag{108}$$

In phase 4, cell division must be taken into account. Let m/2 be the probability that a mitotic cell undergoes binary fission and 1-m/2 be the probability that it disappears from the population by death or emigration. The average number of cells entering state 1 is m. Taking cell multiplication into account and using analogous reasoning t can be shown that

$$\begin{aligned} \alpha_1(t) &= m\int_0^t \alpha_4(t-a)\ \delta_4(a)da + m\int_0^\infty \beta_4(y)\ \delta_4(t+y)[\phi_4(y)]^{-1}dy \\ &= m\int_0^\infty n_4(t,a)\ \delta_4(a)[\phi_4(a)]^{-1}da. \end{aligned} \tag{109}$$

For stationary state exponential growth

$$n_j(t,a) = \beta_j(a)e^{\gamma t}$$

so that

$$\alpha_j(t) = n_j(t,o) = \beta(o)e^{\gamma t} = K_j e^{\gamma t} \tag{110}$$

where K_1 is a constant depending on the size of the initial population; K_2, K_3 and K_4 are related constants (see (117) below). Since $\beta_j(a) = n_j(t,a)e^{-\gamma t}$, it follows from (105) that

$$\beta_j(a) = K_j e^{-\gamma a}\ \phi_j(a) \tag{111}$$

for a < t. However, $\beta_j(a)$ is independent of t and so (111) holds for all a. This implies that

(112) $$n_j(t,a) = K_j e^{\gamma(t-a)} \phi_j(a) \qquad (j = 1,2,3,4)$$

From (108) and (112) it follows that

(113) $$\alpha_{j+1}(o) = K_j \int_o^\infty e^{-\gamma a} \delta_j(a)\, da \qquad (j = 1,2,3)$$

Denoting the Laplace transform of δ_j by

$$\bar{\delta}_j(\gamma) = \int_o^\infty e^{-\gamma a} \delta_j(a) da$$

and using the fact that $\alpha_j(o) = K_j$ (see (110)), equation (113) can be rewritten as

(114) $$\alpha_{j+1}(o) = \alpha_j(o)\ \bar{\delta}_j(\gamma) \qquad (j = 1,2,3)$$

Similarly, from (109),

(115) $$\alpha_1(o) = m\ \alpha_4(o)\ \bar{\delta}_4(o)$$

By substituting $\alpha_4(o)$ from (114) into (115) and repeating this process on the resulting equation leads to

(116) $$m\ \bar{\delta}_1(\gamma)\bar{\delta}_2(\gamma)\bar{\delta}_3(\gamma)\bar{\delta}_4(\gamma) = 1$$

Note that $\bar{\delta}_j(\gamma)$ is a decreasing function of γ, $\bar{\delta}_j(o) = 1$ and $\lim_{\gamma\to\infty} \bar{\delta}_j(\gamma) = 0$. In particular, $\gamma = 0$ if $m = 1$ and $o < \gamma < 1$ if $m > 1$. If $m < 1$, then $\gamma < 0$ which corresponds to an exponentially decaying population.

By using (110) equations (114) and (115) can be rewritten as

(117) $$K_{j+1} = K_j\ \bar{\delta}_j(\gamma) \qquad j = 1,2,3$$

(118) $$K_1 = m\bar{\delta}_4(\gamma)\ K_4$$

Finally, in stationary state exponential growth, the expected number of cells in each phase

(119) $$N_j(t) = \int_0^\infty n_j(t,a)da = N_j(o)e^{\gamma t}$$

by using (112). Here

(120) $$N_j(o) = K_j \int_0^\infty e^{-\gamma a} \phi(a)da = K_j[1-\bar{\delta}_j(\gamma)]/\gamma \ .$$

Techniques similar to those presented in this chapter can be used to model the resting state (G_0) of cells. Some of these refined models will be presented later when they are required.

REFERENCES

1. Anderson, E.C., Bell, G.I. Petersen, D.F. and Tobey, R.A., Cell growth and division, IV. Determination of volume growth rate and division probability, Biophys. J., 9, 246-263, 1969.

2. Bell, G.I. Cell growth and division, III. Conditions for balanced exponential growth in a mathematical model, Ibid., 8, 431-444, 1968.

3. Bell, G.I. and Anderson, E.C. Cell growth and division, I. A mathematical model with applications to cell volume distributions in mammalian suspensions, Ibid., 7, 329-351, 1967.

4. Chaung, S.N. Mathematical models for cancer chemotherapy - pharmacokinetic and cell kinetic considerations, Cancer Chemo. Reports, 59, 827-842, 1975.

5. Clarkson, B. et al. Studies of cellular proliferation III, Cancer, 25, 1237-1260, 1970.

6. Collins, J.F. and Richmond, M.H. Rate of growth of Bacillus cereus between divisions, J. Gen. Microbiol., 28, 15-33, 1962.

7. Duff, G.F.D. Partial Differential Equations, University of Toronto Press, Toronto, 1956.

8. Eisen, M. and Schiller, J. Stability analysis of normal and neoplastic growth, Bul. Math. Bio., 66, 799-809, 1977.

9. Fantes, P. et al. The regulation of cell size and the control of mitosis, J. Theor. Biol., 50, 213-244, 1975.

10. Greenwald, E.S. Cancer Chemotherapy, 2nd edition. Medical Outline Series, Medical Examination Publishing Co., New York, 274, 1973.

11. Gurtin, M. and MacCarny, R. Nonlinear age-dependent population dynamics, Arch. Rat. Mech. Anal., 54, 281-300, 1974.

12. Hahn, G.M. State vector descriptions of the proliferation of mammalian cells in tissue culture. Biophys. J., 6, 275-290, 1966.

13. Hirsch, H. and Engelberg, J. Determination of the cell doubling-time distribution from culture growth rate data, J. Theo. Biol., 9, 297-302, 1965.

14. Jansson, B. A Dynamic Description of the Cell Cycle and Its Phases, FOAP Rapport, C8299-M6, Forsvarets Forsknings Planeringsbyran, Stockholm, 1971.

15. Kendall, D.G. On the role of variable generation time in the development of a stochastic birth process, Biometrika, 35, 316, 1948.

16. Kim, M. and Woo, K. Kinetic analysis of cell size and DNA content distributions during tumor cell proliferation: Ehrlich Ascites tumor study, Cell Tissue Kinetics, 8, 197-218, 1975.

17. Koch, A.L. and Staechter, M. A model for statistics of the cell division process, J. Gen. Microb., 29, 453-54, 1962.

18. Lampkin, B. et al. Synchronization and recruitment in acute leukemia, J. Clin. Invest. 50, 2204, 1971.

19. Lopez, A. Asymptotic properties of a human age distribution under a continuous net fertility function, Demography 4, 680-687, 1967.

20. Mitchison, J.M. The Biology of the Cell Cycle, Cambridge University Press, Cambridge, 1971.

21. Nooney, G. Age distributions in dividing populations, Biophysical J., 7, 69-76, 1967.

22. Painter, P. and Marr, A. Mathematics of microbial populations, Ann. Rev. Microbiol., 22, 519, 1968.

23. Pollard, J.H. Mathematical Models for the Growth of Human Populations, Cambridge University Press, Cambridge, 1973.

24. Prescott, D.M. Variations in the individual generation times of Tetrahymena geleii HS, Exp. Cell Res. 16, 279-284, 1959.

25. Ross, D.W. and Sinclair, W.K. Cell cycle compartment analysis of Chinese hamster cells in stationary phase cultures, Cell and Tissue Kinetics, 5, 1-14, 1972.

26. Ross, D.W. Cell volume growth after cell cycle block with chemotherapeutic agents. Cell and Tissue Kinetics, 9, 379-387, 1976.

27. Rubinow, S.I. Introduction to Mathematical Biology, Wiley, New York, 1975.

28. Rubinow, S.I. Mathematical Problems in the Biological Sciences, S.I.A.M., Philadelphia, 1973.

29. Rubinow, S.I. A maturity-time representation for cell populations, Biophys, J., 8, 1055-1073, 1968.

30. Sarna, G., Omine, M. and Perry, S. Cytokinetics of human acute leukemia before and after chemotherapy, Europ. J. Cancer, 11, 483-492, 1975.

31. Scherbaum, O. and Rasch, G. Cell size distribution and single cell growth in tetrahymena pyrformis GL, Acta Path. Microbiol. Scand., 41, 161-82, 1957.

32. Sinclair, W. and Ross, D.W. Modes of growth in mammalian cells. Biophys. J., 9, 1056-70, 1969.

33. Steward, P. and Hahn, G. The application of age response functions to the optimization of treatment schedules, Cell Tissue Kinetics, 4, 279-291, 1971.

34. Stuart, R.M. and Merkle, T.C. The Calculation of Treatment Schedules for Cancer Chemotherapy, Part II, UCRL-14505, University of California, Lawrence Radiation Laboratory, Livermore, California, 1965.

35. Takahashi, M. Theoretical basis for cell cycle analysis, I. labelled mitosis wave method, J. Theor. Biol., 13, 203-211, 1966.

36. Takahashi, M. Theoretical basis for cell cycle analysis, II, further studies on labelled mitosis wave method, Ibid., 195-209, 1968.

37. Takahashi, M. and Inouye, K. Characteristics of cube root growth of transplantable tumors, Ibid, 14, 275-283, 1967.

38. Tannock, I.F. A comparison of cell proliferation parameters in solid and Ascites Ehrlich tumors, Cancer Res. 29, 1527-1534, 1953.

39. Thames, H.D., Jr. and White, R.A. State-vector models of the cell cycle I. Parameterization and fits to labelled mitoses data, J. Theor. Biol. 67, 733-756, 1977.

40. Trucco, E. Collection functions for non-equivivant cell populations, J. Theoret. Biol., 15, 180-189, 1967.

41. Trucco, E. On the average cellular volume in synchronized cell populations, Bull. Math. Biophys., 32, 459-473, 1970.

42. Trucco, E. Mathematical models for cellular systems the von Foerster equation, Part I and II, Ibid., 27, 285-304, 449-471, 1965.

43. Trucco, E. and Bell, G. A note on the dispersionless growth laws for single cells, Ibid, 32, 475-483, 1970.

44. Trucco, E. and Brockwell, P.J. Percentage labelled mitoses curves in exponentially growing cell populations, J. Theor. Biol., 20, 321-337, 1968.

45. Tyson, J. and Kauffman, S. Oscillator control of mitosis, J. Math Biol., 1, 289-310, 1975.

46. Von Foerster, J. Some Remarks on Changing Populations, in The Kinetics of Cell Proliferation (F. Stohlman, Ed.), Greene and Stratton, New York, 382-407, 1959.

47. Zietz, S. Mathematical Modelling of Cellular Kinetics and Optimal Control Theory in the Service of Cancer Chemotherapy, Ph.D. Thesis, Dept. of Math., Univ. of California, Berkeley, 1977.

IV. AUTORADIOGRAPHY

4.1 Introduction

In autoradiography, localized sites of radioactivity are produced within a biological specimen. This causes tissue sections prepared from the specimen to photograph themselves.

One of the earliest investigations of the cell cycle by the use of radioactive labelled molecules (H^3 - leucine) is described in the classic paper by Howard and Pelc [17]. The use of tritiated thymidine was introduced by Verley and Hunebelle [40] and Taylor et al. [37] in 1957. About two years later the quantitative study of cell kinetics based on autoradiography was introduced by Quastler and Sherman [25]. These two investigators employed tritiated thymidine (H^3 - TdR) to obtain the percentage labelled mitosis (PLM) curves. This curve gives, at each instant of time, the percentage of cells in mitosis which were labelled with the radioactive molecule. Quastler and Sherman showed how the mean length of each phase of the cell cell cycle could be obtained from the PLM curve.

Refinements of the analytical methods of Quastler and Sherman for studying PLM curves will be discussed in this chapter. Mathematical and biological problems still inherent in PLM analysis will be described.

The importance of analyzing the kinetics of cellular proliferation will be seen in the later chapters. The three most important reasons for such analysis are:

(a) learning about the normal processes of cell division, growth and growth control;

(b) understanding the dynamics of a disease process;

(c) designing optimal schedules for chemotherapy and radiation therapy.

4.2 Fractional Labelled Mitosis Curve

The most frequently used radioactive molecule in determining all growth kinetics has been ^{3}H-thymidine.[1] In tritiated thymidine several of the hydrogen

---

[1] Thymidine = thymine with sugar (see Appendix A)

atoms of the thymidine molecule have been replaced by tritium, a radioactive isotope of hydrogen. Thymidine is the precursor of one of the deoxyribonucleoside - triphosphates which are used during replication of DNA (see Appendix A). Since ^{3}H - TdR reacts similarly to Tdr the synthesized DNA will contain radioactive molecules. Thus, only cells that are synthesizing DNA when exposed to H^3 - Tdr - that is, cells which are in their S phase, become radioactively tagged. Slides for microscopic study are then prepared from the cells. In a darkroom, the slides are dipped into a photographic emulsion. β-particles are emitted from the sites of radioactivity. Some of these particles pass through the emulsion, strike the silver halide crystals in the emulsion and so produce an image. Finally, the unexposed crystals are dissolved out as silver thiosulfate complexes by the photographic fixer. Further details on autoradiography can be found in [5] and [9].

There are three main techniques for labelling proliferating cells. These techniques depend on the number of different radioactive labels used and the length of time in which the label or labels are available. A pulse labelling experiment is one in which a single label is available for a short period of time (frequently assumed to be zero). The condition for pulse labelling can be realized for cells growing in culture. For example, by washing or dilution with untritiated thymidine the uptake of radioactive labels can be made practically instantaneous. However, in vivo (in rats), it is only known that the unincorporated ^{3}H-TdR is rapidly (within one hour [19]) catabolized to neglible concentrations. A continuous labelling experiment is one where a single label is available for an extended period of time, usually until a sample is taken. The condition for this experiment is easy to achieve. However, there is a greater danger that the cells will be injured than in pulse labelling, since the cells are exposed to the radiation for a longer period of time. A double labelling experiment is one where two different radioactive labels are given at two points in time. Usually the label is available at each point for only a brief period [6, 14, 42].

During the last two decades the most popular method of studying the cell cycle has been the method of fraction of labelled mitosis (FLM). Experimentally it starts by briefly introducing H^3-TdR into the cell population. Then specimens of tissue[2] or cell culture are taken at fixed intervals of time and microautoradiographs are prepared.

A slide is then placed under a microscope. The labels which are silver particles appear as black grains. Due to background radiation (γ-rays) and the delicacy of the experimental procedure, grains may also appear over a non-radioactive cell. The statistical problems of distinguishing between radioactive and non-radioactive labelled cells will not be discussed here [4,10]. The total number of cells in mitosis[3] and the number of labelled mitotic cells is counted. The fraction

$$FLM(t) = \frac{\text{no. of labelled mitotic cells at time t after the } H^3\text{-Tdr pulse}}{\text{total no. of mitotic cells at the same time}}$$

is called the <u>fraction labelled mitoses</u> (FLM) curve. (Capital letters will be used in this way when referring to empiric entities and small letters for the corresponding functions in mathematical models.) The earlier name for this curve was the percentage labelled mitoses (PLM) curve.

Some general properties of the FLM curve can be deduced from the biology of the cell cycle. The curves should start at zero since there are no labelled mitotic cells when the H^3-TdR pulse was applied. It should grow at a rate dependent upon the distribution of S and G_2 (the time from H^3-TdR uptake to mitosis) and approach one. The FLM curve should decrease again after a period mainly influenced by S and increase again when daughters of initially labelled cells enter mitosis. Theoretically, the wave form should continue, though levelling off due to the variance in cycle times. Practically, only a few cycles can be detected because of dilution of the label by cell division. An experimental FLM curve is shown in Fig. 4.1.

The solid line is the best-fitting synthetic FLM curve as judged by the least squares criterion.

---

[2] Usually by sacrificing an animal.

[3] Mitosis is the only phase of the cell cycle which can ordinarily be distinguished with the light microscopes.

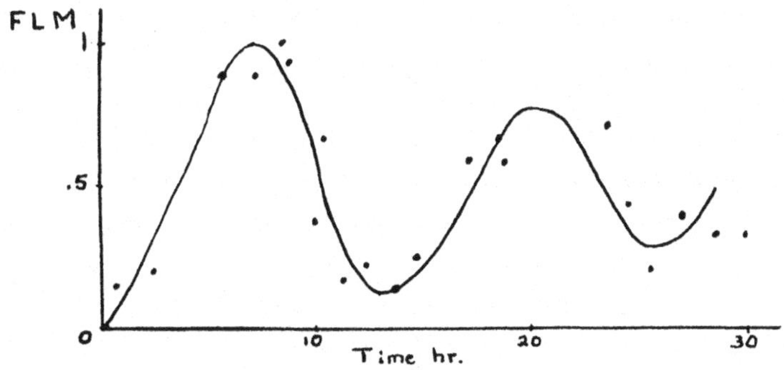

Figure 4.1 Experimental FLM curve.

At present, all methods of FLM analysis do not explicitly consider nonproliferating cells. Suppose nonproliferating cells are considered to be in the G_0 phases and there is a steady rate of flux through G_0. Then a G_0 cell is simply regarded as a cell which happens to spend a long time in G_1. Hence, the mean cycle time will appear to be longer. On the other hand, nonproliferating cells can be thought of as forming a separate population Q. If the transitions from Q to P are neglible the growth rate will be smaller. This will affect the FLM curve indirectly through the age distribution. The experimental FLM curve, being based on samples of mitotic cells, will apply to P alone as long as all mitotic cells may be assumed to be P cells.

The purpose of the FLM method is to deduce the distribution of the cycle time and its phases from the observed FLM curve. This presupposes some model. To illustrate the basic ideas consider the following highly simplified one. It is assumed that

(i) the length of each phase is a constant;

(ii) each mother cell gives birth on the average to one daughter;

(iii) the population has a stable age distribution at time zero, when the thymidine pulse is administered;

(iv) all cells in S are instantaneously labelled at time zero.

It follows from these assumptions, using either the von Foerster model ($\lambda_d = 0$; λ_m = a delta function) or Trucco's model ($p = 1/2$; δ_j = a delta function), that cell age is a uniformly distributed over each phase and also the whole cycle.

Furthermore, the total number of cells in a given time interval is proportional to the length of that interval. This last fact will enable us to deduce the form of the FLM curve. The length of time spent in each phase will be denoted by the same letter as the phase itself. Let t be the time after administration of the label. If $0 \leq t \leq G_2$, then flm(t) = 0 since cells which were in S at t = 0 have not reached M. Next suppose $G_2 \leq t \leq G_2+M$. The oldest labelled cells in the M phase at time t were obviously the oldest cells (age S) in the S phase at time zero. These cells are of age $t-G_2$ in M at time t. Labelled cells of age 0 in M at time t are $t-G_2$ units younger than the oldest labelled cells in M and so must have been of age $S-(t-G_2)$ at time zero. Therefore, all cells in the S-phase at time zero whose age a satisfies

$$S-(t-G_2) \leq a \leq S$$

will be in the M phase at time t. Recalling that the number of cells in an interval is proportional to the length of that interval it follows that

$$\mathrm{flm}(t) = \frac{t-G_2}{M} \qquad G_2 \leq t \leq G_2+M$$

By analogous reasoning, it can be shown that

$$\mathrm{flm}(t) = \begin{cases} 0 & 0 \leq t \leq G_2 \\ (t-G_2)/M & G_2 \leq t \leq G_2+M \\ 1 & G_2+M \leq t \leq G_2+S \\ (G_2+S+M-t)/M & G_2+S \leq t \leq G_2+S+M \\ 0 & G_2+S+M \leq t \leq C \end{cases}$$

where C is the cycle time. Since similar events occur in the following cycles

$$flm(nC+t) = flm(t) \qquad n = 1,2,3,\ldots$$

(see Fig. 4.2).

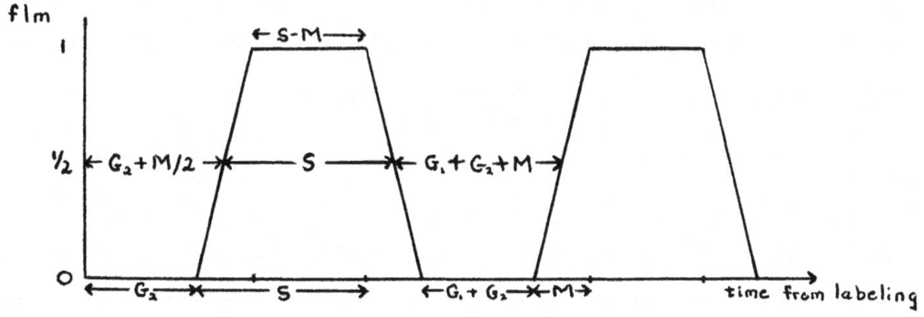

Figure 4.2 FLM curve for a uniform age distribution.

Clearly the theoretical FLM curve does not resemble an actual FLM curve (c.f. Fig. 4.1 and Fig. 4.2). The above model is too simple. More sophisticated models will be studied in the next section. However, this simple model has led to some approximate methods which have been used in the analysis of actual curves.

(a) Boundary or .5 Level Method. [25]

The expected lengths of G_2 + M/2, S, G_1+G_2+M are estimated from the distances between successive intersections of the FLM curve with the level 1/2 (see Fig. 4.2). Although this method gives the exact answer for the simple model studied above, it may break down. In actuality, the second wave may not reach 1/2. To avoid this a lower level can be chosen or the level 1/2 might be replaced by half the height of the first peak [21].

(b) Area (First Wave) Method.

The expected duration of the S phase is estimated by the area under the first wave. The reader should verify that for the simple model studied above this procedure provides the exact answer. This method was hinted at in [25] and suggested in [21].

(c) Plateau Method

Some actual FLM curves rapidly converge to a plateau value. It has been suggested [2,21] on the basis of a stationary model (each cell has exactly one daughter) with random independent phases, that the asymptotic value estimates the ratio between the expected duration of S and the expected cycle time.

It has been shown by simulation of special cases that the errors involved in applying these estimators can be large [22]. In the next section an explanation for the errors arising from using the area and plateau estimates will be given.

4.3 Mathematical Models for FLM Curves: Pulse Labelling

Certain physical assumptions are required in order to carry out the mathematical analysis. It is assumed that there is an instantaneous labelling of all cells in DNA synthesis and so a cell will be labelled if and only if it or one of its ancestors was in the S phase during the administration of the label. The effects of dilution of the label [8] due to cell division are assumed negligible. This assumption is valid for a few cycles. The delayed labelling of cells by labelled thymidine released from dying cells [35] is assumed insignificant. A key assumption is that the presence of radioactive label does not influence the growth fraction or the passage of proliferating cells through the cycle. Only proliferating cells will be considered.

The following model [20] is a theoretical description of the situation where a steady state population is subject to a pulse label of duration τ. The phase G_1,S,G_2 and M will be denoted by the phase numbers 1,2,3 and 4, respectively. It will be assumed that:

(i) At a point in the distant past a cell population consists of a large number of cells. This population can be described by an age-dependent branching process (See Chapter 2, Section 2(b), part (ii))

(ii) A cell begins the cell cycle in phase 1 and successively passes through phases 2,3 and 4.

(iii)[4] The mean number of daughters per mitosis is $m > 1$ (Note that cell loss through death or removal during the cycle is permitted as long as the probability of loss is the same for every cell in the population at any

[4] $0 < m < 1$ corresponds to a population decreasing to extinction. It will not be considered; the concept of a limiting age distribution is not meaningful. Nooney [24] includes this case.

given time)[5].

(iv) There is a joint probability density function, $\delta_{1,2,3,4}(t_1,t_2,t_3,t_4)$ $0 < t_i < \infty$, associated with the sojourn times of the phases of the cell cycle for a cell just beginning phase.

(v) The transit times of a cell through its phases are independent of those of the mother cell or any other cell in the population.

(iv) Labelling begins at time 0 and continues until time τ. The cell population has developed under stable conditions for a long enough time for a limiting age distribution to be reached. Further, τ is sufficiently small relative to the density of cycle times so that the daughter of a labelled cell does not take up a new label.

(vii) At time $t \geq 0$, a cell is considered labelled if and only if it or one of its ancestors was in phase 2 during the interval of labelling[6].

(viii) The presence of a label does not alter the progression of a proliferation cell through the cycle.

(ix) Random samples of cells are obtained after labelling

Let

$$flm_M(t) = P(\text{cell is labelled} \mid \text{it is just entering M at time } t),$$

and the steady state probability.

$$g_4(y) = P(\text{age of a cell in mitosis lies between } y \text{ and } y+dy).$$

Then, by definition,

$$flm(t) = P(\text{cell is labelled and it is in M at time } t)$$

and so by the total probability theorem

$$(1) \qquad flm(t) = \int_0^t flm_M(t-y)g_4(y)dy.$$

---

[5] See [4,24,35] for age-dependent rates of loss.

[6] There is little information on the rate of uptake of H^3TdR at various times in S.

Hence, the theoretical FLM function will be determined once the functions flm_M and g_4 are specified.

To find flm_M the following notation is introduced. The random variables denoting the length of the phases of the cell cycle will be denoted by the same letters as the phases themselves. Let C be the random variable representing the length of the cell cycle. The superscript 0 refers to the present generation of cells which are just entering M at time t and the superscript i denotes the generation of cells i cycles back from the present generation (see Fig. 4.3).

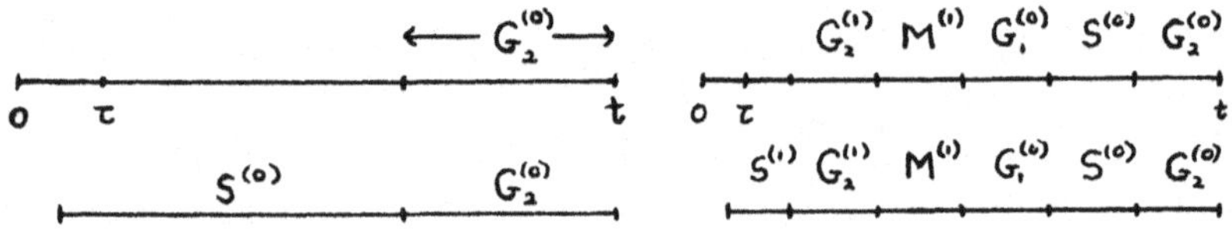

cell itself is labelled (a)

first generation labelled cell (b)

Figure 4.3 A cell is labelled given it is just entering M at time t.

The event a cell is labelled given that it is just entering M at time t is the union of the following disjoint events:

E_0: $G_2^{(o)} \le t$ and $S^{(o)} + G_2^{(o)} > t-\tau$ (see Fig. 4.3(a))

E_1: $G_2^{(1)}+M^{(1)}+G_1^{(o)} + G_2^{(o)} \le t$ and $S^{(1)} + G_2^{(1)} +M^{(1)}+G_1^{(o)}+S^{(o)}+G_2^{(o)} > t-\tau$ (see Fig. 4.3(b))

E_k: $G_2^{(k)}+M^{(k)} + \sum_{i=1}^{k-1} C^{(i)}+G_1^{(o)}+S^{(o)}+G_2^{(o)} \le t$ and $S^{(k)}+G_2^{(k)}+M^{(k)}+ \sum_{i=1}^{k-1} C^{(i)}+G_1^{(o)}$ $+S^{(o)}+G_2^{(o)} > t-\tau$

since a cell is labelled if and only if it or one of its ancestors was in the S phase during the administration of the label. Hence,

(2) $$flm_M(t) = P(E_0) + P(E_1)+\ldots+ P(E_k)+\ldots$$

The probabilistic appearing in the right side of (2) can be written as the difference of distribution functions by using the identity

$$P(A \cap B') = P(A-B) = P(A)-P(B) \qquad (A \supset B)$$

For example,

$$P(E_o) = P[(G_2^{(o)} \leq t) \cap (S^{(o)}+G_2^{(o)} \leq t-\tau)'] = P(G_2^{(o)} \leq t) - P(S^{(o)}+G_2^{(o)} \leq t-\tau)$$

$$= F_3(t) - F_{23}(t-\tau)$$

Hence, (2) can be rewritten as

$$(3) \qquad flm_M(t) = F_3(t)-F_{23}(t-\tau) + \sum_{\nu=0}^{\infty} (F_{34;\nu;123}(t)-F_{234;\nu;123}(t-\tau))$$

where the subscript 234;ν;123, for example, denotes the age distribution of cells just entering M, with age measured from the start of S ν+1 cycles earlier. The numerical subscripts indicate that consecutive phases are combined and considered as a single phase. For example, 234 denotes that phase 2,3 and 4 have been combined and its length is considered to be the single random variable $S^{(k)}+G_2^{(k)}+M^{(k)}$. Note the distinction between the subscript 1,2,3 which refers to the phases 1 to 3 jointly and the subscript 123 which refers to the sum of the lengths of the phases 1 to 3. Quantities pertaining to the whole cycle time will generally left without a subscript. Substituting (3) into (1) yields

$$(4) \qquad flm(t) = \int_o^t F_3(t-y)g_4(y)dy - \int_o^{t-\tau} F_{23}(t-\tau-y)g_4(y)dy$$

$$+ \sum_{\nu=0}^{\infty} \{ \int_o^t F_{34;\nu;123}(t-y)g_4(y)dy-\int_o^{t-\tau} F_{234;\nu;123}(t-\tau-y)g_4(y)dy\}$$

If successive cycles are independent[7], the Laplace transform (denoted by an overbar) of (4) is

---

[7] Parts of a cycle and the next are also assumed independent.

(5) $$\overline{flm}(s) = s^{-1}\bar{g}_4\{\bar{f}_3-e^{-\tau s}\bar{f}_{23}) + \sum_{\nu=0}^{\infty} (\bar{f}_{34}\bar{f}^{\nu}\bar{f}_{123}-e^{-\tau s}\bar{f}_{234}\bar{f}^{\nu}\bar{f}_{123})\}$$

$$= s^{-1}\bar{g}_4\{\bar{f}_3-e^{-\tau s}\bar{f}_{23}) + \bar{f}_{123}(\bar{f}_{34}-e^{-\tau s}\bar{f}_{234})/(1-\bar{f})\}$$

where f_3, f_{23}, f_{123} ... are the corresponding density functions of the distributions $F_3, F_{23}, F_{123}, \ldots$. In particular, after the limiting age distribution has been reached

f_3 = conditional age density for cell just completing phase 3, with age measured from the start of phase 3;

f_{123} = conditional age density for cells just completing phases, with age measured from the start of phase 1.

An equation similar to (3) was used by Gilbert [15] with independent phases described by gamma distributions.

To calculate these densities the joint density function

$$\delta_{1,2,3,4}(u_1,u_2,u_3,u_4)$$

of the phase times $U_i (i = 1,2,3,4)$ for a cell just beginning phase 1 is required. Its Laplace transform is

$$L_{1,2,3,4}(s_1,s_2,s_3,s_4) = \int_0^{\infty} \cdots \int_0^{\infty} \exp(-\textstyle\sum s_i u_i)\delta_{1,2,3,4}(u_1,\ldots,u_4)du_1du_2du_3du_4.$$

The marginal joint density function for the first $r (1 \le r \le 4)$ phases is

$$\delta_{1,\ldots,r}(u_1,\ldots,u_r)$$

and its Laplace transform

(6) $$L_{1,\ldots,r}(s_1,\ldots,s_r) = L_{1,2,3,4}(s_1,\ldots,s_r,o,\ldots,o).$$

The density function δ_{1234} for the cell cycle time denoted δ, by convention, and has Laplace transform

(7) $$L(s) = L_{1,2,3,4}(s,s,s,s).$$

The density function for the time to complete the phases $1,\ldots,r$ (or the density function of the random variable $u = u_1+\ldots+u_r$) will be denoted by $\delta_{1\ldots r}$ and its Laplace transform by $L_{1\ldots r}$. Note that

(8) $$\delta_{1\ldots r}(u) = \int_{u_2+\ldots+u_r\leq u} \delta_{1,2,\ldots,r}(u-\sum_{i=2}^{r} u_i,u_2,\ldots,u_r)du_2\ldots du_r$$

and its Laplace transform

(9) $$L_{1\ldots r}(s) = L_{1,\ldots,r}(s,s,\ldots,s).$$

Finally, the conditional density function of $u_2,\ldots,u_r$ given $U = u$ is

(10) $$\delta(u_2,\ldots,u_r|u) = [\delta_{1,\ldots,r}(u-\sum_{i=2}^{r} u_i,u_2,\ldots,u_r)]/\delta_{1\ldots r}(u) \quad (\sum_{i=2}^{r} u_i \leq u)$$

Some results from the theory of branching processes will also be required (c.f. (115), Chapter 3). The rate of increase of the cell population is the positive root γ of the equation

(11) $$mL(\gamma) = 1$$

where L is defined by (6) and m is the mean number of daughters per mitosis. The limiting expected density of cells between ages u and u+du in the population is

(12) $$g(u) = \left(\frac{m}{m-1}\right) \gamma e^{-\gamma u} (1-\Delta(u))$$

where

$$\Delta(u) = \int_0^u \delta(x)dx.$$

More generally [4], if phases $1\ldots r$ are regarded as a single phase

(13) $$g_{1\ldots r}(u) = Ce^{-\gamma u}(1-\Delta_{1\ldots r}(u))$$

where $C = \gamma/(1-L_{1\ldots r}(\gamma))$, applies with the combined phase, even if phases 1 to r are not independent of phases r+1 to p.

The conditional age density, $f_{1\ldots r}$, for cells just completing phase r, with age measured from the start of the cycle can now be found. Given that a cell has attained age u in phase 1...r the probability that it will complete phase r between ages y and y+dy is

$$\delta_{1\dots r}(y)dy/(1-\Delta_{1\dots r}(u)) \qquad (y \geq u)$$

Therefore, using (13), the joint probability function of age in phase, U, and age at transition, Y, is

$$C\ e^{-\gamma u}\ \delta_{1\dots r}(y)\ dy\ du \qquad (y \geq u).$$

Conditioning along the line Y = U gives the required result

$$f_{1\dots r}(u) = e^{-\gamma u}\ \delta_{1\dots r}(u)/L_{1\dots r}(\gamma). \tag{14}$$

Using (14) the joint density function, $f_{1,\dots,r}$ of $U_1,\dots,U_r$ for cells just completing phase r when the limiting age distribution has been reached can be determined. From this density all of the densities f_i appearing in (5) can be calculated. It follows from (10) and (14) that the joint probability density function of $U,U_2,\dots,U_r$ is

$$\delta(u_2,\dots,u_r|u)\ f_{1,\dots,r}(u) = e^{-\gamma u}\delta_{1,\dots,r}(u-\sum_{i=2}^{r} u_i, u_2,\dots,u_r)/L_{1\dots r}(\gamma)$$

$$(\sum_{i=2}^{r} u_i \leq u)$$

Changing variables gives

$$f_{1,\dots,r}(u_1,u_2,\dots,u_r) = \exp(-\gamma \sum_{i=1}^{r} u_i)\delta_{1,\dots,r}(u_1,\dots,u_r)/L_{1\dots r}(\gamma) \tag{15}$$

A few examples will illustrate how the $\bar{f}_i$ in (5) can be found from (15). Actually, it is more convenient to use the Laplace transform of (15)

$$\bar{f}_{1,\dots,r}(s_1,\dots,s_r) = \frac{L_{1,\dots,r}\ (s_1+\gamma, s_2+\gamma,\dots,s_r+\gamma)}{L_{1,\dots,r}(\gamma,\gamma,\dots,\gamma)} \tag{16}$$

where the denominator of (15) has been rewritten using (9). It follows immediately from the properties of Laplace transforms of densities and (16) that

(17) $$\bar{f}_{34}(s) = \bar{f}_{1,2,3,4}(o,o,s,s) = \frac{L_{1,2,3,4}(\gamma,\gamma,s+\gamma,s+\gamma)}{L_{1,2,3,4}(\gamma,\gamma,\gamma,\gamma)}$$

(18) $$\bar{f}_{234}(s) = \bar{f}_{1,2,3,4}(o,s,s,s) = L_{1,2,3,4}(\gamma,s+\gamma,s+\gamma,s+\gamma)L^{-1}_{1,2,3,4}(\gamma,\gamma,\gamma,\gamma)$$

In actuality, (5) is only of theoretical interest without additional assumptions. The joint density function $\delta_{1,2,3\ 4}$ is not known experimentally and either is g_4. As an example Macdonald [20] assumes the joint density is normal.

If the flm curve approaches an asymptotic its plateau value is given by

(19) $$\text{flm}(\infty) = \lim_{s\to o} s\ \overline{\text{flm}}(s) = \lim_{s\to o} (\bar{f}_{34}(s)-e^{-\tau s}\ \bar{f}_{234})/(1-\bar{f}(s))$$

$$= (\bar{f}'_{34}(o) - \bar{f}'_{234}(o)+\tau)/(-\bar{f}'(o)) = (\tau+\mu_{234}-\mu_{34})/\mu$$

by L'Hospitals rule. Here μ_{34} is the mean of f_{34} etc. Assuming that the phase times are independent and denoting the density of the duration of phase i by δ_i it follows that

(20) $$L_{1,2,3,4}(s_1,s_2,s_3,s_4) = \bar{\delta}_1(s_1)\ \bar{\delta}_2(s_2)\ \bar{\delta}_3(s_3)\ \bar{\delta}_4(s_4)$$

Using relation (20) it follows from (17) and (18) that

(21) $$\mu_{234} - \mu_{34} = -\bar{f}'_{234}(o)+\bar{f}'_{34}(o) = \{\textstyle\int_o^\infty e^{-\gamma t}t\delta_2(t)dt\}/\{\int_o^\infty e^{-\gamma t}\delta_2(t)dt\}$$

(22) $$\mu = \sum_{i=1}^{4} (\textstyle\int_o^\infty e^{-\gamma t}\ t\delta_i(t)dt)/(\int_o^\infty e^{-\gamma t}\ \delta_i(t)dt).$$

Hence, in a steady state of renewal, m = 1 and $\gamma = 0$ from (11),

(21) $$\text{flm}(\infty) = \frac{\tau + \text{expected duration of S}}{\text{expected duration of cycle}}$$

as can readily be verified from (19), (21), and (22). Note that if $\tau \neq 0$ a larger value than that predicted by the Plateau method will be obtained. When $\tau = 0$, the Plateau approximation improves as $m \to 1(\gamma \to o)$. It can be shown that the asymptotic value exists when $m > 1$ and the total lifetime of a cell has a non-lattice distribution.

The flm curve under the assumption of independent phase times will now be calculated. If M is independent of $G_1 + S + G_2$, then

$$G_2^{(\nu+1)} + M^{(\nu+1)} + \sum_{i=1}^{\nu} C^{(i)} + G_1^{(o)} + S^{(o)} + G_2^{(o)} = G_2^{(\nu+1)} + \sum_{i=1}^{\nu+1} C^{(i)}$$

since sequences M, G_1, S, G_2 and G_1, S, G_2, M will have the same distribution. Hence, the subscript 34;ν;123 may be written 3;ν+1. Similarly, 234;ν;123 can be written 23;ν+1. It follows from (4) that (5) can be rewritten as

$$\text{(23)} \qquad \bar{flm}(s) = s^{-1}\bar{g}_4\{(\bar{f}_3 - e^{-\tau s}\bar{f}_{23}) + \sum_{\nu=0}^{\infty} (\bar{f}_3\bar{f}^{\nu+1} - e^{-\tau s}\bar{f}_{23}\bar{f}^{\nu+1})\}$$

$$= s^{-1}\bar{g}_4(\bar{f}_3 - e^{-\tau s}\bar{f}_{23})/(1-\bar{f})$$

From equations (118) and (119) of Chapter 3, the number of cells in phase 4 at time t is

$$N_4(\gamma) = [K_4(1-\bar{\delta}_4(\gamma))e^{\gamma t}]/\gamma$$

Using equation (111) of Chapter 4[8],

$$g_4(a) = \frac{n_4(t,a)}{N_4(t)} = \gamma e^{-\gamma a}(1-\bar{\delta}_4(\gamma))^{-1}\,(1-\int_0^a \delta_4(x)dx)$$

Hence,

$$\text{(24)} \qquad \bar{g}_4(s) = [\gamma\ 1-\bar{\delta}_4(s+\gamma)\}]/[(s+\gamma)(1-\bar{\delta}_4(\gamma)]$$

Substituting (24) into (23) and using (16) to compute $\bar{f}_3$, $\bar{f}_{23}$ and $\bar{f}$ yields

$$\overline{flm}(s) = \frac{\gamma\{1-\bar{\delta}_4(s+\gamma)\}}{s(s+\gamma)(1-\bar{\delta}_4(\gamma))}\left[\frac{L_3(s+\gamma)}{L_3(\gamma)} - e^{-\tau s}\frac{L_2(s+\gamma)L_3(s+\gamma)}{L_2(\gamma)L_3(\gamma)}\right] \Big/ \left[1-\frac{L(s+\gamma)}{L(\gamma)}\right]$$

Using (11), the last formula reduces to

---

[8] The same result holds if phase 4 is independent of combined phases 1,23[4].

$$(25)\qquad \overline{f\ell m}(s) = \frac{\gamma\bar{\delta}_3(s+\gamma)(1-\bar{\delta}_4(s+\gamma)(\bar{\delta}_2(\gamma)-e^{-\tau s}\bar{\delta}_2(s+\gamma))}{s(s+\gamma)\bar{\delta}_2(\gamma)\bar{\delta}_3(\gamma)(1-\bar{\delta}_4(\gamma))(1-m\bar{\delta}(s+\gamma))}$$

where δ is the density function of the cycle time. When $\tau = 0$, equation (25) reduces to formula (24) of Trucco and Brockwell [39].

The reader should verify this last statement by using the results of section 6 in Chapter 3. Briefly, the age density functions for labelled cells, $n_j^*(t,a)$, are given by (104) and (105) of Chapter 3, with n_j, α_j and β_j replaced by n_j^*, α_j^* and β_j^*. Immediately after the thymidine pulse there are no labelled cells in G_1, G_2 and M so that $\beta_i^*(a) = 0$ for j = 1,2 and 3. However, $\beta_2^*(a) = \beta_2(a)$ the initial age distribution for cells in S. The α^* are determined from (106) and (108) of Chapter 3 with β_j replaced by β_j^*. Once $\alpha_4^*(t)$ is known

$$N_4^*(t) = \int_0^\infty n_4^*(t,a)da$$

Then from (118) and (119) of Chapter 3

$$f\ell m(t) = \frac{N_4^*(t)}{N_4(t)} = \frac{\gamma\ N_4^*(t)\ e^{-\gamma t}}{K_4(1-\bar{\delta}_4(\gamma))}\ .$$

Other investigators have also studied particular examples of (25) with $\tau = 0$. Barrett [2] assumes that the phases are lognormal, m = 1 ($\gamma = 0$) and that the mitotic time is negligibly small. When $\gamma = 0$ and $\tau = 0$ L'Hospital's rule shows that (25) reduces to

$$(26)\qquad \overline{f\ell m}(s) = \frac{(1-\bar{\delta}_4(s))(1-\bar{\delta}_2(s))}{s^2(1-\bar{\delta}(s))D_4}$$

where D_4 is the mean sojourn time in phase 4. Another study of stationary populations using renewal equations can be found in Bronk [8].

Takahashi's result [34] can also be derived from (25) under the assumptions

(i) $\tau = 0$

(ii) $\delta_j(a) = \dfrac{k_i}{D_j\Gamma(k_j)}\ x_j^{k_j-1}\ e^{-x_j}$ (j = 1,2,3,4)

where $x_j = \dfrac{k_j}{D_j}\ a$ and the k_j's are positive integers

(iii) $\frac{k_i}{D_j} = y$ is independent of j.

The details can be found in [39]

A simple but interesting application of (25) is equivant populations (i.e. all cells undergo mitosis at the same time.) Assume that all the phase densities are Dirac delta functions, $\delta_j(a-D_j)$, where D_j is the duration of phase j. The cycle density function $\delta(a-D)$ is also a delta function with D equal to the duration of the cycle. Then

$$\bar{\delta}_j(s) = \int_0^\infty e^{-st}\,\delta(s-D_j)dt = e^{-sD_j}$$

$$\bar{\delta}(s) = e^{-sD}$$

Equation (11) becomes

$$me^{-\gamma D} = 1.$$

Hence, (25), with $\tau = 0$, becomes

$$\text{(27)} \qquad f|m(s) = \frac{\gamma e^{-\gamma D_3}[1-e^{-(s+\gamma)D_4}][1-e^{-\gamma D_2}]}{(1-e^{-\gamma D_4})s(s+\gamma)(1-e^{-sD})}$$

Assume for definiteness that $D_4 < D_2$. Let $t = mD + \tau$ where $0 \le \tau < D$ and $m = 0,1,2,\ldots$ The continuous periodic inverse transform of (27) is

$$\text{(28)} \quad f\ell m(t) = \begin{cases} \dfrac{1-e^{-\gamma(\tau-D_3)}}{1-e^{-\gamma D_4}} & D_3 \le \tau \le D_3 + D_4 \\[2ex] 1 & D_3 + D_4 \le \tau < D_2 + D_3 \\[2ex] \dfrac{e^{-\gamma(t-D_2-D_3)} - e^{-\gamma D_4}}{1-e^{-\gamma D_4}} & D_2 + D_3 \le \tau \le D_2 + D_3 + D_4 \\[2ex] 0 & 0 \le \tau < D_3 \text{ or } D_2+D_3+D_4 \le \tau < D \end{cases}$$

It is interesting that for $\gamma > 0$ the ascending and descending limbs of the fℓm curve in (28) are not straight lines (c.f. Fig. 4.2). The curve represented by

(28) approaches the curve in Fig. 4.2 as $\gamma \to 0$. Hence, the two curves will coincide under the assumptions used in obtaining the curve in Fig. 4.2. Recall that when the phases are independent $m = 1$ implies $\gamma = 0$. Note that even for these simple distributions with $\gamma \neq 0$ the area under a wave is not S.

The area under the first wave for arbitrary densities exhibits the same dependence on γ. Assume that the duration of the M phase is independent of the rest of the cycle. The area under the first wave (contributions from cells which were in the S phase at the time of labelling) is

$$A_1 = \lim_{t\to\infty} \int_0^t \left(\int_0^x [F_3(x-y) - F_{23}(x-y)] g_4(y)dy \right) dx$$

(c.f. equation (4) with $\tau = 0$). Using Laplace transforms and the facts that $\int_0^t h(x)dx = s^{-1}\bar{h}(s)$ and $\lim_{t\to\infty} h(t) = \lim_{s\to 0} s\bar{h}(s)$, it follows that

$$A_1 = \lim_{s\to 0} s[s^{-1}\{s^{-1}(\bar{f}_3 - \bar{f}_{23})\bar{g}_4(s)] \qquad \text{(c.f. (5))}$$

$$= \bar{f}_3'(0) - \bar{f}_{23}'(0) = E[(G_2+S)e^{-\gamma(G_2+S)} - G_2 e^{-\gamma G_2}]$$

If the duration of the phases G_2 and S are independent and $\gamma = 0$ then $A_1 = E[S]$.

The first wave method and the other approximate methods have been studied by simulation [23] and analytically [22,27]. In [27] the authors study the effects of correlated phases. The conclusion is that approximate hand analysis is not a match for a machine oriented approach. However, the computer methods can be greatly aided by approximate starting estimates.

There are now many programs for analyzing FMF curves that give the mean and variance of the duration of each phase and of the total cell cycle. The programs listed in Table 4.1 are based on the assumption that the phases are independent. Estimates of parameters are obtained by minimizing various sums of squares. A steady state of proliferation is assumed only for the lognormal density. Correlated phases are considered by Macdonald [20]. Parameters are estimated by the method of maximum likelihood.

TABLE 4.1 SOME METHODS OF ANALYZING FLM CURVES

phase density	basis of numerical method
lognormal	Monte Carlo [2,3,31,32]
normal	analytic formula for FLM-function involving erf [16]
gamma	inversion of Laplace transform [15]
gamma	inversion of Laplace transform (Trucco-Brockwell model) [29]
gamma	3-dimensional Newton-Raphson; orthogonal search for coefficients of variation [36]
Pearson Type III	Newton's methods; Runge-Kutta-Gill [35]
arbitrary	Fast Fourier transform [1]
arbitrary	Monte Carlo [39]
arbitrary	analytic formula for FLM-function [7]

Since Takahashi's model [35] has been discussed in Section 2 of Chapter 3, a brief description of this method will be given. Suppose that a cell population is pulse labelled at t = 0. Only cells in the S phase will be labelled (i.e. cells in compartments (g_1+1) to (g_1+s)). Any labelled mitotic cells which appears later must be the progeny of these cells. The descendents $n_i^*(t)$ of these cells at time t can be found by solving the system of differential equations (3) and (4) of Chapter 3 subject to the initial conditions

$$n_i^*(o) = 0 \qquad \text{if } 1 \leq i \leq g_1 \text{ or } g_1+s < i \leq k$$

$$= n_i(o) \qquad \text{if } g_i < i \leq g_1+s$$

The $n_i(o)$ are the steady state solutions of the system under consideration. The number of labelled mitosis

$$M^*(t) = \sum_{i=k-m+1}^{k} n_i^*(t).$$

On the other hand the total number of mitoses (labelled and unlabelled) is

$$M(t) = M^*(t) + \sum_{i=k-m+1}^{k} n_i^o(t)$$

The n_i^* (t) are solution of the system (3) and (4) of Chapter 3 subject to the initial conditions

$$n_i^*(o) = 0 \qquad \text{if } g_1 < i \leq g_1+s$$

$$= n_i(o) \qquad \text{if } 1 \leq i \leq g_1 \text{ or } g_1+s < i \leq k.$$

Therefore $f\ell m(t) = M^*(t)/M(t)$. The simultaneous set of equations were solved by a Runge - Kutta - Gill method. The steady state solutions were approximated by using the largest real root of the characteristic equation arrived at by Laplace transforms.

An investigation of fℓm curves based on the discrete time Hahn model has been carried out by Thames and White [38].

4.4 FLM Curves for Continuous Labelling [20]

The assumptions for the continuous labelling model are (i) - (ix) of Section 3 except that (vi) is altered in an obvious way - the label is available until the cell is sampled. It is assumed that there is little re-utilization of H^3 which is released upon the death of labelled cells.

Let τ be the time since the start of labelling. A cell just beginning mitosis at time t is labelled provided that the duration of its G_2 phase was at most t. Hence P(cell is labelled | it is just starting M at time t) = $F_3(t)$. Let g_4 be the age density of cells in mitosis. Then

$$f\ell m_c(t) = \int_o^t F_3(t-y)g_4(y)dy \tag{29}$$

where the subscript c denotes continuous labelling. The Laplace transform of (29) is

$$\overline{flm_c}(s) = s^{-1}\,\bar{f}_3(s)\bar{g}_4(s) \tag{30}$$

If M is independent of G_1+S+G_2, then the value of $\bar{g}_4$ is given by (24). Substituting (24) into (30) yields

$$\overline{flm_c}(s) = \frac{\gamma\, L_{12,3}(\gamma,s+\gamma)\{1-L_4(s+\gamma)\}}{s(s+\gamma)L_{123}(\gamma)\{1-L_4(\gamma)\}} \tag{31}$$

by using (16). Here γ is the positive root of equation (11).

After a sufficiently long time every cell should be labelled. Hence, $\lim_{t\to\infty} flm_c(t) = 1$. This intuitive conclusion is in agreement with the mathematical model since

$$\lim_{t\to\infty} flm_c(t) = \lim_{s\to 0} s\,\overline{flm_c}(s) = 1$$

from (30).

The Laplace transform (31) can be inverted to give

$$flm_c(t) = \{D_3(t)-L_4(\gamma)D_{34}(t)\}/\{1-L_4(\gamma)\} \tag{32}$$

where

$$D_i(t) = F_i(t) - \int_0^t e^{-\gamma(t-u)} f_i(u)du. \tag{33}$$

Formula (32) can be verified by taking the Laplace transform of both sides and noting that from (33)

$$\bar{D}_i = \bar{f}_i/s - \bar{f}_i/(s+\gamma) = \frac{\gamma}{s(s+\gamma)}\,\bar{f}_i.$$

Hence the Laplace transform of (32) reduces to (31) provided that

$$\bar{f}_{34}(s) = \frac{L_{12,3}(\gamma,s+\gamma)\;\; L_4(s+\gamma)}{L_{123}(\gamma)\;\;\; L_4(\gamma)}$$

which follows from (16).

As an example consider a stationary process ($m = 1;\gamma = o$) with constant duration of phases (the densities δ_j are delta function). Using the fact that

$\bar{\delta}_j(s) = e^{-sD_j}$ and (24) it is not difficult to show that (30) reduces to

$$\overline{flm}_c(s) = \frac{e^{-sD_3}}{s^2} \; \frac{\left(1-e^{-sD_4}\right)}{D_4}$$

Inversion of this last formula gives

$$flm_c(t) = \begin{cases} 0 & 0 \leq t < G_2 \\ (t-G_2)/M & G_2 \leq t \leq G_2+M \\ 1 & t \geq G_2 + M \end{cases}$$

Hence, for constant duration of phases, the duration of the G_2 and M phases can easily be deduced from the experimental FLM_c curve. For arbitrary distribution of phases the mean duration of the G_2 and M phases can be found by using (32). This requires computer programming.

The FLM_c curve in conjection with the continuous labelling index curve (described in the next section) can be used to deduce further properties of the cell cycle. Another method used in continuous labelling is that of grain counts [9]. An established survey for DNA labelling is [41].

4.5 The Labelling Index

Experimentally, the continuous labelling index (CLI) curve is obtained by adding tritiated thymidine continuously from time t = 0 onwards. Then, at chosen times, the fraction of labelled cells or

$$\text{labelling index} = \frac{\text{number of labelled cells}}{\text{total number of cells}}\;.$$

is recorded. The labelling index increases to one if all cells proliferate and to the growth fraction otherwise.

The simple model of a stationary process ($m=1$, $\gamma=o$) with constant phases leads to the following theoretical cli curve:

$$\text{(34)} \qquad cli(t) = \begin{cases} \frac{S}{C} & t = 0 \\ (t+S)/C & 0 < t \le C-S \\ 1 & t > C-S \end{cases}$$

where C denotes the length of the cell cycle. This theoretical curve is surprisingly close to some experimental CLI curves. Hence one estimator of $E(C-S) = E(G_1) + E(G_2) + E(M)$ proposed in the literature is the time until the CLI curve reaches one or some point close to it [9,4,40]. If E(M) and $E(G_2)$ are obtained from the FLM_c curve, then EG_1 can be calculated from the estimate of E(C-S).

The general form of the theoretical cli curve will be deduced under the assumptions listed at the beginning of Section 4.3. For convenience it is also assumed that the duration of the phases are independent. Only proliferating cells will be considered. Consider a cell at time t. There are two possibilities for the age of the cell in the cycle.

(a) The age of the cell is y and $y \le t$. Such a cell was born at t-y. This cell will be labelled at t if it was labelled after it was born ($G_1 \le y$) or if one of its ancestors were labelled ($M+G_2 \le t-y$). Hence, the probability of a cell being of age at most t and labelled is

$$\int_0^t P[(G_1 \le y) \cup (M+G_2 \le t-y)]g(y)dy$$

$$= \int_0^t \{F_1(y) + F_{34}(t-y) - F_1(y)F_{34}(t-y)\}g(y)dy.$$

(b) The cell is older than t - that is, its age is a+t where a > 0. This cell was born at time a before t = 0. It will be labelled at time t if and only if part of its S phase overlaps the interval [0,t] - that is, the event $(G_1 \le t+a) \cap (G_1+S \ge a)$ occurs. Hence, the probability of a cell being older than t and labelled is

$$\int_0^\infty P(G_1 \le t+a,\ G_1+S \ge a)g(a+t)da$$

$$= \int_0^\infty \{F_1(t+a) - F_{12}(a)\}g(t+a)da.$$

Hence, by the total probability theorem

$$(35) \qquad cli(t) = \int_0^t \{F_1(y)+F_{34}(t-y)-F_1(y)F_{34}(t-y)\}g(y)dy$$

$$+ \int_0^\infty \{F_1(t+a)-F_{12}(a)\}g(t+a)da.$$

By considering values of t lying in the intervals $(\infty, C-S]$, $[G_1,G_1+G_2+M]$ and $[0,G_1]$, it can be shown that (35) reduces to (34) for a stationary process with constant phases. However, it is much easier to verify (34) directly.

Recall that for a stationary process with constant phases the time t_1 required for the cli line to reach is C-S. It is easy to verify, from (34), that

$$(36) \qquad t_1 = \{1-cli(0)\}/cli'(0).$$

The Yamada-Puck estimator t_1 [43] seems difficult to justify for a more general model. From (35)

$$(37) \qquad cli(0) = \int_0^\infty \{F_1(a)-F_{12}(a)\}g(a)da.$$

If Var G_1, Var(G_2+S) and Var C are assumed small, then by Chebychev's inequality the densities of G_1, G_2+S and C can be approximated by delta functions concentrated at their means. Further, assuming that m = 1 (or $\gamma = 0$) it follows from (12) that

$$g(u) \simeq \frac{1}{D} \qquad 0 \leq u \leq D$$

where D = E(C). Under these assumptions it follows from (37) that

$$(38) \qquad cli(0) \simeq \frac{D_2}{D} = \frac{E(S)}{E(C)}.$$

Differentiating (35) and setting t = 0 yields

$$(39) \qquad cli'(0) = \int_0^\infty \{F_1'(a)g(a)+F_1(a)g'(a)\}da \simeq \frac{1}{D}$$

under the given assumptions. Hence, from (38) and (39), the expression (36) for t_1 approximates E(C-S).

A more general model for deriving the cli function based on the theory of branching processes and allowing cell disintegration can be found in [18]. Another model based on a system of differential equations has been proposed by J. Fried [12, 13]. The system consists of a proliferating cell compartment, denoted by P and a nonproliferating compartment, denoted by Q. Cells in the P compartment pass through the usual phases of the mitotic cycle. Each cell completing mitosis has a probability $f_\ell (0 \leq f_\ell \leq .5)$ of passing directly into the Q compartment and a probability of $1-f_\ell$ of entering G_1. Random passage of cells from P to Q with probability K_{PQ} per unit time can also occur. Cells are lost randomly from the Q compartment with probability K_Q per unit time. The P compartment is growing exponentially with growth rate constant K_g. The differential equations describing the Fried model can be obtained from (35) of Chapter 2 by specialization of the constants. In particular,

$$\beta = \delta = 0; \ \lambda = K_{PQ}; \ \alpha = \frac{K_{PQ}+2f_\ell K_g}{1-2f_\ell} \quad ; \quad \gamma = \frac{K_g+K_{PQ}}{1-2f_\ell} .$$

The labelling index (LI) curve is obtained by administering a pulse of tritiated thymidine at time $t = 0$. The fraction of labelled cells at chosen times is then recorded.

Under the same assumptions as for the continuous administration of the label, the theoretical ℓ_1 curve can be determined by analogous reasoning. A cell of age $y \leq t$ at time t will be labelled if and only if it or one of its ancestors was in phase S at $t = 0$. This event can be written as

$$\bigcup_{k=0}^{\infty} \{(M+G_2+C_k \leq t-y) \cap (M+G_2+S+C_k \geq t-y) \}$$

where C_k is the time required to complete k cycles. Hence, the probability of a cell being of age at most t and labelled is

$$\sum_{k=0}^{\infty} \int_0^t P(M+G_2+C_k \leq t-y, \ M+G_2+S+C_k \geq t-y) g(y) dy$$

The probability of a cell being older than t and labelled

$$\int_0^\infty P(G_1 \leq a, \ G_1+S > a) \ g(a+t) da$$

where $a > 0$ is the time, before $t = 0$, of the birth of the cells. Hence,

$$(40) \quad li(t) = \sum_{k=0}^{\infty} \int_0^t P(M+G_2+C_k \leq t-y, \ M+G_2+S+C_k \geq t-y) g(y) dy$$

$$+ \int_0^\infty (G_1 \leq a, \ G_1+S > a) \ g(a+t) da.$$

The Laplace transform of the sum appearing on the right side of (40) is

$$\frac{\bar{f}_3\bar{f}_4\ \bar{g}\ (1-\bar{f}_2)}{s(1-\bar{f})} \tag{41}$$

If $\lim_{t\to\infty} g(t) = 0$, it follows from (40) and (41) that

$$\lim_{t\to\infty} li(t) = \frac{E(s)}{E(c)}$$

For a stationary process with constant phases

$$li(t) = \frac{E(s)}{E(c)} .$$

4.6 Discussion

Many biological sources of error (e.g. metabolic stability of the tracer, negligible reutilization of the labels, similarity of the behavior of labelled and unlabelled cells) are discussed in [10, Chapter 5]. Errors can also arise from underexposure or overexposure of autoradiographs. The effect of background radiation must also be taken into account [27]. Another source of error is protracted labelling. If the labelling is not instantaneous, formula (25) shows that the PLM curve will vary with the exposure time. Several authorities feel that the label is pooled in the cell at the time of flashing without necessarily being incorporated into DNA. However, not all of the pooled label is fixed and utilized [11,35].

The scoring of an autoradiograph is a time consuming process. Accordingly, only 200-300 cells are counted. This leads to statistical fluctuations because of the small sample size. Sparse data points, uneven distribution of. points, variability from animal to animal or from culture to culture further limit the accuracy of kinetic estimates from FLM curves.

Even if it is granted that the above errors are small, FLM analysis has two drawbacks. The first is that present methods of FLM analysis are valid only for cell populations that have attained a stable or time invariant age distribution. In vivo, cell cycles can be influenced by external factors such as Circadian rhythms or drugs. The population may not be stable and so FLM analysis is invalid .

A second serious drawback is that FLM or any autoradiographic technique is not clinically useful. Since a high concentration of ^3H in a cell can damage it only low concentrations are used. Consequently it is necessary to expose the autoradiographs for 2-5 weeks. The scoring of an autoradiograph is also a taxing and time consuming procedure. Thus, weeks can elapse before learning whether there is a satisfactory response to chemotherapy by FLM analysis. Meanwhile, the patient may die or a useless and harmful combination of drugs may continue to be used. Moreover, since the population may not be in a steady state, FLM analysis may be invalid.

The estimates of parameters depends on the model used. Steward [33] gives an example which shows that Takahashi's method will be more likely to give shorter transit times for the S phase than Steel's method. These different results could invalidate chemotherapy studies.

No assessment of the accuracy of individual parameters, such as confidence intervals, is given for most of the models. Gilbert [15] gives the standard error of the extracted parameters. Macdonald [20] uses x^2 for a goodness of fit test while Takahashi's et al [36] use analysis of variance based on mean deviation from regression.

Nevertheless, FLM analysis remains the cornerstone of a thorough study of cell kinetics. It is the only method that simultaneously gives the mean time spent in each phase of the cell cycle and the variance of these transit times.

More sophisticated and time dependent models are required. These models should include the effects of non-proliferating cells and be able to find the correlation between phases.

REFERENCES

1. Ashihara, T. Computer optimization of the fraction of labelled mitoses analysis using the fast Fourier transform, Cell Tissue Kinet. 6, 3-16, 1973.

2. Barrett, J.C. A mathematical model of the mitotic cycle and its application to the interpretation of percentage labelled mitoses data, J. Nat. Cancer Inst., 37, 443-450, 1966.

3. Barrett, J.C. Optimized parameters for the mitotic cycle, Cell Tissue Kinet., 3, 349-353, 1970.

4. Bartlett, M.S, Distributions associated with cell populations, Biometrika, 56, 391-400, 1969.

5. Baserga, R. and Malamud, D. Autoradiography: Techniques and Application, Harper and Row, New York, 1969.

6. Baserga, R. and Lisco, E. Duration of DNA synthesis in ehrlich ascites cells as estimated by double-labeling with C^{14}- and H^{3}-thymidine and autoradiography, J. Nat. Cancer Inst., 31, 1559-71, 1963.

7. Brockwell, P.J., Trucco, E. and Fry, R.M.J. The determination of cell-cycle parameters from measurements of the fraction of labeled mitoses, Bull. Math. Biophys., 34, 1-12, 1972.

8. Bronk, B.V. On radioactive labeling of proliferating cells: the graph of labeled mitosis, J. Theoret. Biol. 22, 468-92, 1969.

9. Cleaver, J.E. Thymidine Metabolism and Cell Kinetics, North Holland, Amsterdam, 1967.

10. Eisen, M. A criterion for classifying radioactive cells, Int. J. Appl. Radiat. and Isotopes, 27, 695-697, 1976.

11. Feinendegen, L.E. Tritium-labelled Molecules in Biology and Medicine, Academic Press, New York, 1967.

12. Fried, J. A mathematical model to aid in the interpretation of radioactive tracer data from proliferating cell populations, Math. Biosciences, 8, 379-396, 1970.

13. Fried, J., Friedman, H., Zietz, S. Computer analysis of tracer kinetic data from a human hematopoctic cell line during different phases of growth, Computers and Biomedical Research, 7, 333-359, 1974.

14. Garland, P., Mainguet, P., Arguello, M., Chretien, J. and Douxfils, N. In vitro autoradiographic studies of cell proliferation in the gastrointestional tract of man, J. Nuclear Med. 9, 37-39, 1968.

15. Gilbert, C.W. The labelled mitoses curve and the estimation of the parameters of the cell cycle, Cell Tissue Kinet., 5, 53-63. 1972.

16. Hartmann, N.R. and Pedersen, T. Analysis of the kinetics of the granulosa cell populations in the mouse ovary, Cell Tissue Kinet., 3, 1-11, 1970.

17. Howard, A. and Pelc, S.R. Synthesis of deoxynbonuclei acid in normal and irradiated cells and its relation to chromosome breakage. Heredity (Suppl.), 6, 261-273, 1953.

18. Jagers, P. Branching Processes with Biological Applications, John Wiley & Sons, New York, 1975.

19. Lipkin, M. The proliferative cycle of mammalian cells. In Baserga, R. (ed.). The Cell Cycle and Cancer, 6-26, Marcel Dekker, New York, 1971.

20. Macdonald, P.D.M. Statistical inferences from the fraction labelled mitoses curve; Biometrika, 57, 489-503, 1970.

21. Mendelsohn, M.L. The kinetics of tumor cell proliferation. In: Cellular Radiation Biology, 498-513. Williams & Wilkins, Baltimore, 1965.

22. Mendelsohn, M.L. and Takahashi, M., A critical evaluation of the fraction of labelled mitoses method as applied to the analysis of tumor - and other cell cycles. In: Baserga, 161 R. (ed.), The Cell Cycle and Cancer, 58-95, Marcel Dekker, New York, 1971.

23. Nachtwey, D.S. and Cameron, I.L. Cell cycle analysis. In: Prescott, D.M. (ed.), Methods in Cell Physiology 3, Academic Press, 255, New York, 1968.

24. Nooney, G.C. Age distributions in stochastically dividing populations, J. Theoret. Biol., 314-320, 1968.

25. Quastler, H. and Sherman, F.G. Cell population kinetics in the intestinal epithelium of the mouse, Exptl. Cell Res. 17, 420-438, 1959.

26. Robinson, S.H., Brecher, J., Lourie, I.S., and Haley, J.E. Leukocyte labelling in rats during and after continuous infusion of tritiated thymidine: implication for lymphocyte longevity and DNA reutilization. Blood 26, 281-295, 1965.

27. Schotz, W.E. and Zelen, M. Effect of length sampling bias on labelled mitotic index waves, J. Theor. Biol., 32, 383-404, 1971.

28. Shackney, S.E. The effect of counting threshold and emulsion exposure duration on the percent labelled mitosis curve and their implication for cell cycle analysis, Cancer Res. 331, 2726-2731.

29. Simon, R.M., Stroot, M.T. and Weiss, G.H. Numerical inversion of Laplace transforms with application to percentage labelled mitoses experiments, Comput. Biomed. Res., 5, 596-607, 1972.

30. Stanners, C.P. and Till, J.E. DNA synthesis in individual L-strain mouse cells, Biochim. Biophys. Acta., 37, 406-419, 1960.

31. Steel, G.G. and Hanes, S. The techniques of labelled mitoses: analysis by automatic curve fitting, Cell Tissue Kinet., 4, 93-105, 1971.

32. Steel, G.G., Adams, K. and Barrett, J.C. Analysis of cell population kinetics of transplanted tumors of widely differing growth rate, Brit. J. Cancer 20, 784-800, 1966.

33. Steward, G.P. The kinetics of cell-cycle transit and its application to cancer therapy with phase-specific agents. In: The Cell Cycle in Malignancy and Immunity, CONF-731005, Tech. Information Center, USERDA, Jan., 1975.

34. Takahashi, M. Theoretical basis for cell cycle analysis I. labelled mitosis wave method, J. Theor. Biol., 13, 202-11, 1966.

35. Takahashi, M. Theoretical basis for cell cycle analysis II. Further studies on labelled mitosis wave method, J. Theoret. Biol. 18, 195-209, 1968.

36. Takahashi, M, Hogg Jr., J.D. and Mendelsohn, M.L. The automatic analysis of FLM curves. Cell Tissue Kinet., 4, 505-518, 1971.

37. Taylor, J.H., Woods, P.S. and Hughes, W.L. The organization and duplication of chromosomes as revealed by autoradiographic studies using tritium labelled thymidine, Proc. Natl. Acad. Sci. (USA), 43, 122, 1957.

38. Thames, H.D. and White, R.A. State-vector models of the cell cycle. I. Parameterization and fits to the labelled mitosis data, J. Theor. Biol., 67, 733-756, 1977.

39. Trucco, E. and Brockwell, P.J. Percentage labelled mitoses curves in exponentially growing cell populations, J. Theor. Biol., 20, 321-337, 1968.

40. Verley, W.G. and Hunebill, G. Preparation of T-labelled thymidine, Bull Soc. Chim. Belg., 66, 640, 1957.

41. Wimber, D.E. Methods for studying cell proliferation with emphasis on DNA labels. In: Lamberton, L.F. and Fry, R.J.M. (eds.), Cell Proliferation, Blackwell, Oxford, 1963.

42. Wimber, D.E. and Quastler, H. A ^{14}C and ^{3}H-thymidine double labelling technique in the study of cell proliferation in Tradescantia root tips, Exptl. Cell Res. 30, 8-22, 1963.

43. Yamada, M. and Puck, T.T. Action of radiation on mammalian cells, IV. Reversible mitotic lag in the S3 Hela cell produced by low doses of x-rays, Proc. Nat. Acad. Sci. (Wash.) 47, 1181, 1961.

V. CELL SYNCHRONY

5.1 Introduction

Biologists have been interested in cell synchrony for some time. Theoretical reasons for this interest are discussed in the introduction to the book edited by Zeuthen [10] and in James [7].

A practical reason for studying cell synchrony is its applicability to cancer chemotherapy. Unfortunately, drugs used in cancer chemotherapy kill normal as well as cancerous cells. Naturally it is desirable to kill as many cancerous cells as possible while sparing as many normal cells as possible. One way of accomplishing this goal is by taking advantage of the fact that many drugs are cycle specific - that is, they only kill cells in a certain phase of their cycle. In the cell synchronization method the cancerous cells are first synchronized by one drug. When nearly all the cancerous cells reach the desirable phase, they are treated with a second cycle-specific drug. This kills the maximum number of cancer cells while sparing large numbers of normal cells. The importance of synchrony strategies in the chemotherapy of melanoma is discussed by Brown and Thompson [2].

Synchrony is defined in section 2. Various definitions of the degree of synchrony appear in section 3 and 4. Finally, the decay of synchrony is discussed in section 5. A rapid method for measuring the degree of synchrony appears in the next chapter.

5.2 Definition of Synchrony

Only cells of a single type (a homogeneous population) will be considered in this chapter. Most biologists use the term asynchronous population to denote a naturally growing cell population not synchronized by drugs, fasting, selective detachment etc. Natural growth is synonomous with exponential growth which normally occurs after a time lag in cultural cells. Hence, natural growth tacitly assumes that the cell population has been growing for some time under normal conditions. If this time period is sufficiently large, exponential growth is also predicted theoretically. For example, consider the asymptotic behavior of age-dependent binary branching processes or the solution of Von Foerster's equation.

Hence, the following mathematical definition, appearing in Engelberg [4], is consistent with popular usage:

A population of cells is growing asynchronously (or exponentially) if

(i) the size of the population is increasing exponentially in time;

(ii) cell division is perfectly randomized - that is each cell divides independently of any other cell with a certain density which is not a Dirac delta function.

If a population of cells is not growing asynchronously then it is defined to be growing synchronously. In particular, a population of cells is perfectly synchronized if all cells pass through a certain point of their reproductive cycle at the same time.

The definition of synchrony differs from the definition in Engelberg [4]. Engelberg defines synchronous as perfectly synchronous. A growth pattern intermediate to asynchronous and perfectly synchronous is termed parasynchronous. The reason for modifying the definitions is that the new definitions are consistent with the terminology of flow microfluorometry introduced in the next chapter.

5.3 Instantaneous Indices of Synchrony

In this section some indices which have been proposed as measures of the degree of synchrony at any instant of time will be discussed. It will be assumed that the cell number, n, is sufficiently large so that n can, to a good approximation, be treated as a continuous variable. Moreover, the experimental data on the growth of the population is available as a growth curve $n = n(t)$.

Corresponding definitions exist for age-dependent binary branching processes. The population size is a random variable $Z(t)$ which takes integer values and increases by jumps. The definitions cannot be applied directly to $Z(t)$ but are applied to $E[Z(t)] = n(t)$. Details can be found in Burnett-Hall and Waugh [3].

(a) Engelberg's normalized rate of cell division [4]

Engelberg described the growth of a culture without referring to the actual size by defining a normalized rate of cell division.

$$r(t) = \frac{1}{n(t)} \frac{dn}{dt} . \tag{1}$$

For a population in asynchronous growth $n(t) = n_o e^{\gamma t}$ and $r(t) = \gamma$. At the other extreme, for a perfectly synchronous population r will take the form of a series of delta functions. Between these two extremes (for a population in which synchrony is decaying) the graph of r(t) will take the form of a damped wave about a mean value of γ (see section 5.5).

An approximation to r can be found if the mitotic index I_m is known. Recall that the mitotic index, $I_m = n_m/n$, when there are n cells in the population of which n_m are in mitosis. Hence, at time t there would be $n\, I_m$ cells in mitosis. Let T_m be the mean time for mitosis. It is assumed that T_m is small compared to the mean cycle time of the cell and that I_m only varies slightly in the interval T_m. In the time interval from t to $t+I_m$ the $n\, I_m$ cells divide and produce $n\, I_m$ new cells. Hence the rate of cell division at time t is approximately

$$\frac{dn}{dt} = n\, I_m/T_m$$

from which it follows that

(2) $$r(t) = I_m/T_m.$$

Modifications of equation (2) can be found when the time for mitosis, T_m, is large compared to the mean cycle time. In such a situation an obsevable portion of the mitotic cycle, say anaphase, the duration of which is short compared to the mean cycle, time is selected. Define an anaphase index $I_a = n_a/n$, where n_a is the number of cells in anaphase at time t. Let T_a be the mean duration of anaphase; then a formula analogous to (2), $r(t) = I_a/T_a$, can be derived.

(b) Zeuthen's index of synchrony [10]

Zeuthen's instantaneous index[1] is

(3) $$Z(t) = (I_m - I_{mo}) \;/\; (1 - I_{mo})$$

where I_m is the mitotic index at time t after synchronisation and I_{mo} is the mitotic index in exponential growth.

[1] A factor of 100 is omitted. Zeuthen used percentages.

Note that for an asynchronous population this index has the value zero. If the population is perfectly synchronized the index has the value 1 if it is recorded during the mitotic interval and

$$-\frac{I_{mo}}{1-I_{mo}}$$

otherwise. This index has the disadvantage that it can be negative.

(c) Sankoff's index of synchrony [8]

Consider a large population of cells in which the behavior of a cell is described by Trucco's model. The population eventually assumes a stationary age distribution with density

$$(4) \qquad p(\alpha) = 2\gamma e^{-\gamma\alpha}\phi(\alpha) \qquad (\alpha \geq 0)$$

where $\phi(\alpha)$ is the probability that a cell divides after age α. A similar stationary age distribution is obtained from the theory of age-dependent branching processes (see Harris [5], Chapter VI).

The unit of time is chosen to be the mean cycle time of a cell $\int t d\phi = 1$. If the cells are perfectly synchronized the number of cycles completed up to time t is [t], the largest integer in t. Each cell will be of age $\alpha = t-[t]$. Thus the age density function P of the age distribution at time t is

$$(5) \qquad \rho(t,\alpha) = \delta(\alpha-t+[t])$$

where δ is the Dirac delta function. If the cells are not perfectly synchronized, $\rho(t,\tau-[t])$ the proportion of the population near age $a = t-[t]$ at time t will decrease. For an asynchronous population $\rho(t,\alpha)$ is given by (4). Hence, a measure of synchrony could be

$$(6) \qquad \rho(t,t-[t]) - \rho(t-[t])$$

Sankoff [8] normalized (6) to obtain

$$(7) \qquad s(t) = \frac{\rho(t,t-[t]) - \rho(t-[t])}{\rho(t-[t])}$$

Sankoff's index can be expressed in terms of r(t). Assume that the population is so large that the difference between the random functions population size, age distribution and birth rate, and their respective expected values, is relatively insignificant. The expected birth rate

$$B(t) = 2\frac{dn}{dt} \tag{8}$$

where n(t) is the expected population size at time t. A cell is of age α at time t $(o \leq \alpha \leq t)$ if and only if it was born at time $t-\alpha$ and has not divided by time t. Hence

$$\rho(t,\alpha) = \frac{B(t-\alpha)\phi(\alpha)}{n(t)} \tag{9}$$

Formula (9) is valid when the operations of forming a quotient and taking expectations can be interchanged. According to our assumptions it is a valid approximation. Using equations (9), (8), (4) and (1) equation (7) can be reduced to

$$s(t) = \frac{r([t]) \quad n([t])e^{-\rho[t]}}{\gamma\, n(t)e^{-\rho(t)}} - 1 \tag{10}$$

Note that the index s gives the same information as r at integral values of t. A method for determining n appears in section 5.

Sankoff also calculates the duration of time for which the synchrony index of an initially synchronized population is at least α. These values are tabulated in [8] for a Pearson Type III generation time distribution.

5.4 Time-interval Indices of Synchrony

Several indices have been proposed to describe the degree of synchrony during an interval in which the cells divide. The descriptions appearing in the literature do not uniquely define the interval. This leads in some situations to considerable changes in the value of the indices.

(a) Engelberg's index of proportional synchronisation [4].

Consider a time interval (t_1,t_2) and let $n(t_1) = n_1$, $n(t_2) = n_2$. The mean value of r(t) is

The area between the normalized rate of cell division curve of a parasynchronous population and its mean value r is shaded in Fig. 5.1. This overlap area

(12) $$A = \int_{t_1'}^{t_2'} (r(t)-\bar{r})dt = \ell n(n_2'/n_1')-\bar{r}(t_2'-t_1').$$

The <u>proportion of synchrony</u>, e, is the ratio of this area to the total area; hence,

(13) $$e = \frac{\int_{t_1'}^{t_2'} (r(t)-\bar{r})dt}{\int_{t_1}^{t_2'} r(t)dt} .$$

Using (11) and (12) reduces (13) to

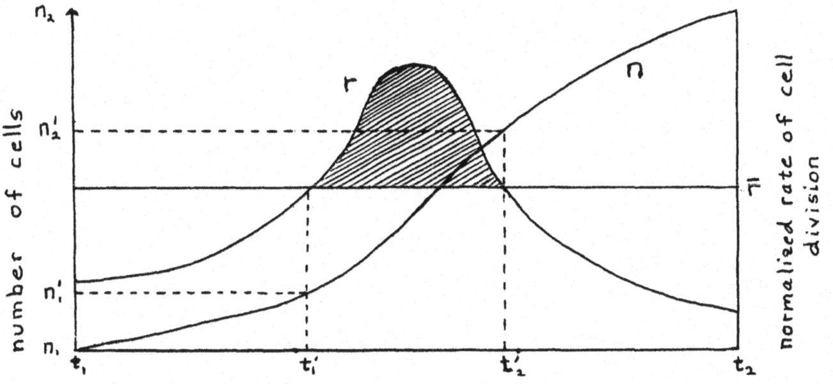

Figure 5.1 Growth curve n and normalized rate of cell division r.

(11) $$r = \frac{\int_{t_1}^{t_2} r(t)dt}{t_2-t_1} = \frac{\ell n(n_2/n_1)}{t_2-t_1} .$$

In asynchronous growth $r(t) = \bar{r} = \gamma$. Assume there is a single peak in the normalized rate of cell division between t_1 and t_2. The time interval will be chosen so that $r(t_1) < \bar{r}$ and $r(t_2) < \bar{r}$. The equation $r(t) = \bar{r}$ will therefore have two roots t_1' and t_2' (see Fig. 5.1). The population size at the start and finish of this sub-interval will be denoted by n_1' and n_2'.

$$e = \frac{\ln(n_2'/n_1') - \bar{r}(t_2'-t_1')}{r(t_2-t_1)} \tag{14}$$

Engelberg [4] defined the basic interval (t_1,t_2) as one in which $n_2 = 2n_1$ - that is, the population doubles in size. Hence, from (11), $\bar{r} = \ln 2/(t_2-t_1)$. Burnett-Hall and Waugh [3] allow the use of other time intervals, for example, one mean cycle time.

In an asynchronous population $r(t) \equiv \bar{r}$ and so $e = 0$. For a perfectly synchronous population $r(t)$ will be a delta function in a time interval containing the burst of fission. It can be shown that $e = 1$.

Displacement of the growth pattern from exponential growth displaces r below $\bar{r}$ for part of the doubling period and above r for the rest of the period. Hence there is an overlap area and so partial synchronization. Therefore, any deviation from exponential growth partially synchronizes a culture.

Care must be exercised in applying Engelberg's index of synchrony. Suppose that a population has been synchronized so that half the population divides with perfect synchrony at one point in time while sometime later the other half of the population also divides with perfect synchrony. The normalized rate of cell division consists of two delta functions. Hence, $e = 1$ even though only half the population divides simultaneously. Thus e is an appropriate measure of synchronization only when the r plot has a single peak during a given cell population doubling period.

(b) Zeuthen's phasing index [10]

This index was proposed by Zeuthen to measure the degree of synchrony in a burst of mitotic activity. It is based on the time interval required for one half the population to undergo mitosis. This interval will be short if mitotic activity is high. To define this index, note that to any time t_1 there corresponds a time t_2 such that

$$n(t_2) = 1.5\, n(t_1). \tag{15}$$

Consider a single burst of mitotic activity. Let t_{50} be the minimum of the time interval lengths t_2-t_1 where the pair (t_1,t_2) satisfies (15). Denoting the mean cycle time by t_c, Zeuthen's index

(16) $$z = (\tfrac{1}{2}\, t_c - t_{50})/(\tfrac{1}{2}\, t_c).$$

Considering the growth curve of a synchronous population to be the limit of continuous growth curves $t_{50} = 0$; hence, $z = 1$. On the other hand, a population in exponential growth will increase to 1.5 times its size in every time interval of length $(\log_2 1.5)t_c$. Hence,

$$z = 1-2 \log_2 1.5 = -.16992$$

To avoid negative values, Zeuthen's index can be normalized

(17) $$z^* = (t_c \log_2 1.5 - t_{50})/(t_c \log_2 1.5)$$

Note that $z^* = 0$ for exponential growth and $z^* = 1$ for synchronous growth.

(c) Scherbaum's synchronization index [9].

Consider an interval (t_1,t_2) during which synchronous multiplication takes place. Let $n_1 = n(t_1)$ and $n_2 = n(t_2)$. Scherbaum's synchronization is defined as

(18) $$s = [n_2/n_1 - 1][1-(t_2-t_1)/t_c].$$

For perfectly synchronous growth $s = 1$.

A drawback of this definition is that there is no criterion for choosing the interval of synchronous growth. The next example shows that the limit of s as asynchronous growth is approached depends on the choice of the Scherbaum interval. Consider an exponentially growing population with growth rate $\gamma(1+\delta)$ during Scherbaum's interval (t_1,t_2) and growth rate $\gamma(1+e)$ during the remainder of the doubling interval $(0,t_c)$. By using the fact that the total number of cells doubles during the doubling period it can be shown that

$$\varepsilon = p\delta(1-p)$$

where

$$p = (t_2 - t_1)/t_c.$$

Hence,

$$n_2/n_1 = \exp(\gamma(1+\delta)(t_2-t_1)) = 2^{(1+\delta)p}$$

and so

$$s = (2^{(1+\delta)p}-1)(1-p).$$

Letting $\delta \to 0$, keeping p fixed, the growth rate tends to the constant value γ since ε tends to zero. However, s tends to $(2^p-1)/(1-p)$, a function of p.

(d) Blumenthal and Zahler's index [1].

The Blumenthal and Zahler index

(19) $$b(t_1,t_2) = n_1^{-1} n_2 - 2^{(t_2-t_1)/t_c}$$

for a time interval (t_1,t_2). Note that b = 0 for exponential growth since $n_2 = n_1 2^{(t_2-t_1)t_c}$

A properly selected time interval is required if the index is to be meaningful. It sould be much less than a generation time, since $b(t_1,t_I+t_c) = 0$ for perfectly synchronized as well exponential populations and will be close to zero for all parasynchronous populations. However, for a short interval δt,

$$\begin{aligned} b(t,t+\delta t) &= \frac{n(t+\delta t)}{n(t)} - 2^{\delta t} \\ &= 1 + \frac{1}{n}\frac{dn}{dt}\ \delta t + 0(\delta t)^2 - 2^{\delta t} \\ &= (r(t) - \ell n\ 2)\delta t + 0(\delta t)^2, \end{aligned}$$

is proportional to δt. Hence, $b(t,t+\delta t)$ tends to zero as $\delta t \to 0$ for all populations populations except if t is one of the doubling times of a perfectly synchronized population. In the latter situation $b(t,t+\delta t) \to 1$.

5.5 The Decay of Synchronization [6]

A hypothetical population in which every cell has a division cycle of precisely T hours will remain perfectly synchronized indefinitely. By contrast consider a population for which the mean doubling time of a cell is T, the maximum doubling times if T+m and the minimum doubling time is T-m. Assume that the population has been perfectly synchronized prior to $t = 0$. From $t = 0$ to $t = T-m$ no cell division takes place; hence $r = 0$ during this period. Cell division will occur in the interval (T-m, T+m). The width of the r curve during this doubling period is 2m. During the next doubling period its width will be 4m. Note that the peaks of the r curve become progressively wider during subsequent divisions until the peaks of adjacent periods merge to form a continuous curve. Since the area under r must equal $\ell n2$ for each doubling period, r must progressively decrease in height as r becomes wider. Intuitively, it is clear that r approaches a horizontal line of height $(\ell n2)/T$. A horizontal line for an r curve signifies exponential growth. Hence, a synchronized population is eventually converted to an asynchronous (exponential) population because of the variation in doubling times of individual cells.

The analytic justification of the above arguments requires an equation for the rate of increase of the number of cells n(t) at time t. A cell which exists in the time interval between t and t+dt arose from a parent cell in some previous time interval between s and s+ds, where $-\infty < s < t$. The probability that a cell will divide in the infinitessimal interval of time dt that occurs t-s seconds after the cell originated is f(t-s)dt, where f is the density function of the cycle times. Hence, the total number of cells produced in the time interval (t,t+dt) from parents born in (s,s+ds) is

$$2\, f(t-s)dn(s)\, dt.$$

The total number of cells produced in (t,t+dt) is

$$dn(t) = 2\int_{-\infty}^{t} f(t-s)\, dn(s)\, dt$$

which can be rewritten as

$$\frac{dn(t)}{dt} = 2\int_{-\infty}^{t} f(t-s)\left(\frac{dn(s)}{ds}\right) ds. \tag{20}$$

Equation (20) appears in Engelberg [4].

The integral in (20) can be separated into two parts one for $s \geq 0$ and one for $s < 0$. Denoting the negative-time portion of the integral by $\frac{dn_-}{dt}$, equation (20) can be rewritten as

$$\frac{dn(t)}{dt} = 2 \int_0^t f(t-s) \frac{dn(s)}{ds} ds + \frac{dn_-(t)}{dt} \tag{21}$$

The contribution to the growth rate from cells whose parents were born before zero time is given by the initial state function, $\frac{dn_-}{dt}$. The integral in (21) represents the subsequent growth of the culture.

Equation (21) can be solved for $\frac{dn}{dt}$ by using Laplace transforms. The Laplace transform operator L is defined by

$$L[g(t)] = \int_0^\infty e^{-pt} g(t)dt$$

Taking Laplace transforms of both sides of (21) it follows from the convolution theorem that

$$L[dn/dt] = 2L[f] \; L[\frac{dn}{dt}] + L[\frac{dn_-}{dt}]$$

and so

$$L[\frac{dn}{dt}] = \frac{L[\frac{dn_-}{dt}]}{1-2\; L[f]} \tag{22}$$

For a culture that is perfectly synchronized at time zero the initial state function is

$$\frac{dn_-}{dt} = n(0-)\delta(t)$$

where $\delta(t)$ is the Dirac delta function and $n(0-)$ is the number of cells just before time $t = 0$. The transform of the initial state function is

$$L(\frac{dn_-}{dt}) = n(0-)$$

It is convenient to set $n(0-) = 1$. Equation (22) then becomes

$$L[\frac{dn}{dt}] = L[\frac{dn_+}{dt}] \tag{23}$$

(24) $$L[\frac{dn_+}{dt}] = \frac{1}{1-2L(f)}$$

From (22), the growth rate, dn/dt, of a culture having arbitrary initial synchronization is the convolution of $\frac{dn_+}{dt}$ and $\frac{dn_-}{dt}$.

Equations (23) and (24) will be used to study the decay of synchronization for two different cycle time distributions.

(a) The perpetually perfectly synchronized culture.

Assume that all cells have the same doubling time T. Then $f(t) = \delta[t-T)$ and $L(f) = e^{-pT}$.

The right side of equation (24) can then be expanded in a series to yield

(25) $$L[\frac{dn_+}{dt}] = \sum_{k=0}^{\infty} 2^k e^{-pk^T}$$

The inverse of equation (25) is

(26) $$\frac{dn_+}{dt} = \sum_{k=0}^{\infty} 2^k \delta(t-kT).$$

Thus the growth-rate function is the sum of a sequence of exponentially increasing impulses. Cells always double when t = kT. Perfect synchronization always exists, so the growth rate never returns to that of an exponential.

(b) The normal distribution.

The cycle time of each cell is assumed to be normally distributed with mean T and variance σ^2. It can be shown that $L[f] = e^{-pT+p^2\sigma^2/2}$. Hence (24) reduces to

(27) $$L[\frac{dn_+}{dt}] = \frac{1}{1-2\ e^{-pT}+p^2\sigma^2/2}$$

The inverse of (27) will be found by the method of partial fractions. To apply this method the roots, p_k, of the denominator of the right side of (27) are required. These roots satisfy the equation

(28) $$e^{-p_kT+p_k^2\ \sigma^2/2} = \frac{1}{2} = e^{-\ell n2-i2\pi k}.$$

Equating the exponents in (28) gives

(29) $$(\sigma^2/2)p_k^2 - Tp_k + \ell n2+i2\pi k = 0$$

Solving (29) with the quadratic formula gives

$$P_k = \frac{T}{\sigma^2}\left[1-\sqrt{1-2(\sigma/T)^2(\ell n2+i2\pi k)}\right] \tag{30}$$

The negative square root is used to insure that P_k remains finite as $\sigma \to 0$. Moreover, the values of P_k as $\delta \to 0$. Morover, the values of P_k as $\delta \to 0$ are the same as those for the impulse doubling-time distribution.

The denominator appearing on the right side of (27) can now be factored by using the roots P_k in (30). Equation (27) and hence (23) can be written as a sum of partial fractions:

$$L[\frac{dn}{dt}] = \sum_k \frac{c_k}{(p-p_k)} \tag{31}$$

The complex coefficients c_k are determined from

$$c_k = \lim_{p \to p_k} \frac{p-p_k}{1-2e^{-pT+p^2\sigma^2}} \tag{32}$$

By application of L'Hospital's rule

$$c_k = \frac{1}{T\sqrt{1-2(\sigma/T)^2(\ell n2+i2\pi k)}} \tag{33}$$

The inverse of (31) is

$$\frac{dn}{dt} = \sum_k c_k e^{p_k t} \tag{34}$$

since $(p-p_k)^{-1}$ is the transform of $e^{p_k t}$

It is convenient to write

$$c_k = |c_k| e^{i\theta_k} \tag{35}$$

where $|c_k|$ and θ_k are the magnitude and phase angle, respectively, of c_k and

$$p_k = \gamma_k + i\omega_k \tag{36}$$

where γ and ω are the real and imaginary parts of p. Substitution of (35) and (36) into (34) and using the fact that the complex terms occur in conjugate pairs, (34) can be reduced to the form

$$\frac{dn}{dt} = c_o e^{p_o t} + \sum_k |c_k| e^{\gamma_k t} \cos(\omega_k t + \theta_k) \tag{37}$$

As k increases, it follows from (30) that the real parts of p_k become more negative. Hence, (37) shows that for large values of t, $\frac{dn}{dt}$ is asymptotic to $c_o e^{p_o t}$. Therefore, the normalized growth rate r approaches a constant as $t \to \infty$. In other words the population eventually grows asynchronously. It is not known if this result holds for arbitrary distribution of cycle times which have a finite variance.

Approximate methods of plotting the normalized growth rate curve can be found in [6]. The partial fraction expansion method is used when growth-rates are calculated for large times, since the higher order terms of the series can be neglected. For small times the series expansion of $[1-2L(f)]^{-1}$ is valuable because higher order terms in it can be neglected.

5.6 Discussion

If the different definitions of synchrony are applied to the same data, different numbers will generally be obtained. Some numerical examples appear in Burnett-Hall and Waugh [3]. The difference in values is due to the fact that the definitions are related to different aspects of the growth curve data. Definitions are to some extent arbitrary and can only be judged by their usefulness in a context of a given application.

The indices of synchrony indicate to what extent the growth pattern of a population has been displaced from random (exponential) growth towards synchronized growth. However, consider a biologist who wishes to collect as many metaphase cells as possible. For this purpose the metaphase index (defined analogously to the mitotic index) is a better criterion for synchrony. The biologists wants to harvest the cells when the metaphase index in a maximum.

In this chapter methods for quantitating the degree of synchrony of a cell population with respect to cell division have been described. However, in many experiments other stages of the cell cycle are also of interest, for example, the transition from G_1 to S. In the next chapter the degree of synchrony will be generalized so that it is applicable to different stages of the cell cycle.

REFERENCES

1. Blumenthal, L.K. and Zahler, S.A. Index for measurement of synchronization of cell populations, Science, 135, 724, 1962.

2. Brown, B.W. and Thompson, J.R. A rationale for synchrony strategies in chemotherapy. In Epidemiology, 38-45, edited by Ludwig, D. and Cooke, K.L., SIAM Publications, Philadelphia, 1975.

3. Burnett-Hall, D.G. and Waugh, W.A.O'N. Indices of synchrony in cellular cultures, Biometrics, 23, 693-716, 1967.

4. Engelberg, J. A method of measuring the degree of synchronization of cell populations, Exp. Cell Research, 23, 218-227, 1961.

5. Harris, T.E. The Theory of Branching Processes, Springer-Verlag, Berlin, 1963.

6. Hirsch, H.R. and Engelberg, J. Decay of cell synchronization: solutions of the cell-growth equation, Bull. Math. Biophysics, 28, 391-409, 1966.

7. James, T.W. Cell Synchrony, a Prologue to Discovery. In Studies in Biosynthetic Regulation. Edited by Cameron, I.L. and Padilla, G.M., Academic Press, New York, 1966.

8. Sankoff, D. Duration of detectable synchrony in a binary branching process, Biometrika, 58, 77-81, 1971.

9. Scherbaum, O.H. A comparison of the degree of synchronous multiplication in various microbial systems, J. Protozool. Suppl. 6, 17, 1959.

10. Zeuthen, E. Artificial and induced periodicity in living cells, Advances in Biol. and Med. Phys. 6, 37-73, 1958.

11. Zeuthen, E. (ed.), Synchrony in Cell Division and Growth, Interscience, New York, 1964.

VI. FLOW MICROFLUOROMETRY

6.1 Introduction

The development of flow microfluorometry (FMF)[1] by Kametsky et al. [23] and Van Dilla et al. [38] allows Kinetic cell-cycle analyses to be used clinically. Cell cycle parameters can be determined in one or two days instead of weeks as formerly required by autoradiography.

The basis of FMF analysis is the DNA distribution of the cells under study. The technique for generating this DNA histogram is briefly described below. Some knowledge of the instrumentation is important in understanding and formulating the mathematical models required in FMF analysis.

Usually only one type of cell is studied. The population of cells is dispersed as well as possible into a suspension of intact, single cells. After fixation, the cells are stained with a stoichiometric DNA-specific fluorescent stain. The cell suspension flows, single file, past an intense beam of laser light at rates of 1,000-3,000 cells/second. The light emitted by the excited fluorescent stain is measured by a photomultiplier. Since the intensity of the emitted light is proportional to the amount of fluorescent stain in the cell, the resulting electric signal is proportional to the fluorescence and hence to the DNA content of the cell. The electric signal is then quantized on a multichannel analyzer. Accumulation of a large number of such measurements yields the DNA distribution for the cell population. The DNA histogram can be obtained in a matter of minutes. A typical result of an FMF run is shown in Fig. 6.1.

The shape of a DNA histogram and the factors influencing it will be examined for a steady-state population of cells with constant phase lengths..

---

1 Also known as flow cytometry (FCM) or cytofluorometry.

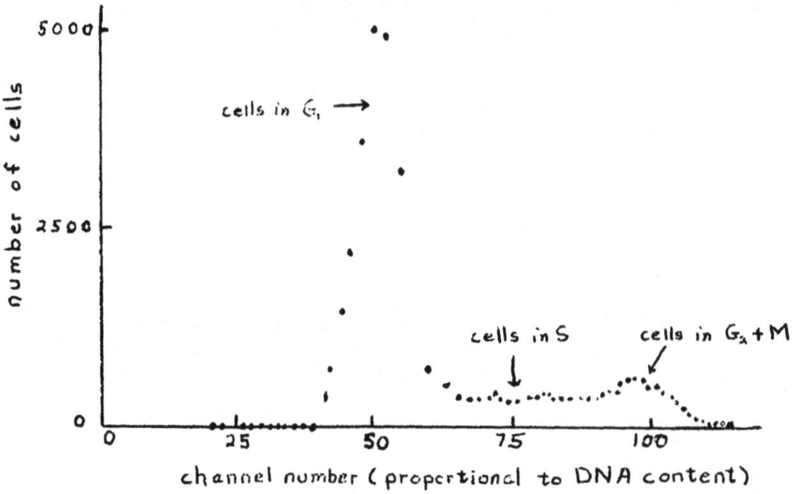

Figure 6.1 DNA histogram of a typical FMF run.

6.2 DNA Histogram: Steady-state and Constant Phase Length

Consider a steady-state population of 120,000 cells. Assume that the length of each phase is a constant D_i, $(i = 1,2,3,4)$. In particular, $D_1 = 6$ hr., $D_2 = 4$ hr., $D_3 = 1$ hr., $D_4 = 1$ hr. and the cycle time $C = 12$ hr. Recall that the age density function for each phase $n_i(a) = \frac{1}{D_i}$ $(0 < a < D_i)$ is a time independent, uniform density function. Similarly, the age density function for the whole cycle is $n(a) = \frac{1}{12}$ $(0 < a < 12)$. It follows that the expected number of cells in any interval of the ith phase is

$$\text{(1)} \qquad (\text{no. of cells in phase } i)\left(\frac{\text{length of interval}}{D_i}\right).$$

An analogous result holds for the whole cycle. Therefore, the number of cells in G_1, S, G_2 and M is 60,000, 40,000, 10,000 and 10,000, respectively.

The DNA content of a cell is shown in Fig. 6.2 for a constant rate of DNA synthesis

$$\text{(2)} \qquad r = \frac{d}{D_2} = \frac{d}{4}\,.$$

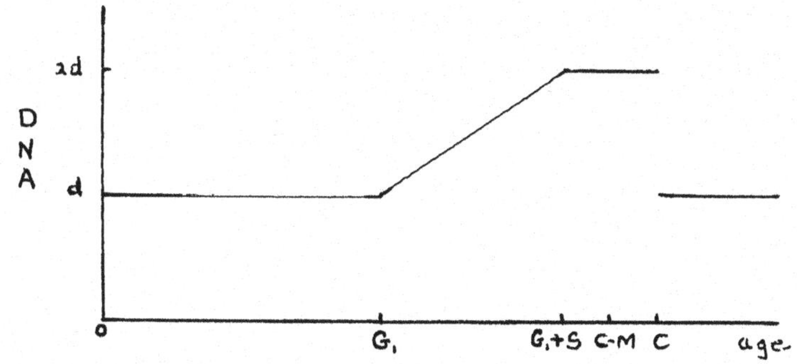

Figure 6.2 DNA content of a single cell.

Assume that cells whose DNA content lies in the interval (d-1,d] are recorded in channel 10, while cells whose DNA content lies in the interval $(\frac{9+i}{10}d, \frac{10+i}{10}d]$ are recorded in channel 10+i (i = 1,...,10).

Under the above assumptions the theoretical DNA histogram is completely determined. The only cells recorded in channel 10 are those in the G_1 phase. Hence, 60,000 cells will be recorded in channel 10. The age of a cell in the S phase whose DNA content lies in the interval $(\frac{9+i}{10}d, \frac{10+i}{10}d]$ is in the interval

$(\frac{(9+1)d}{10r}, \frac{(10+i)d}{10r}]$ for i = 1,2,...,9. The length of this age interval is $\frac{d}{10r}$.

Hence, from formula (1), the number of cells recorded in channel 10+i is

$[(\frac{d}{10r})/4] \times 40{,}000 = [\frac{1}{10}] \times 40{,}000 = 4{,}000$ since d/r = 4 from (2). When i = 10, 4,000 cells in S and all the cells in the G_2 and M phases (with DNA content 2d) will be recorded in channel 20. Hence, 24,000 cells will be recorded in channel 20. The resulting DNA histogram is shown in Fig. 6.3.

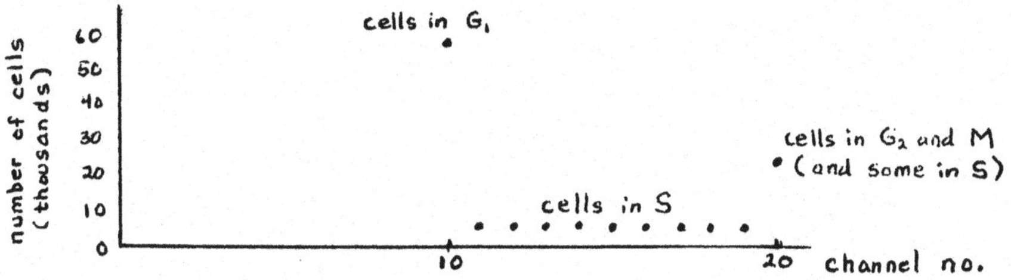

Figure 6.3 Theoretical DNA histogram (steady state, constant phase length and constant rate of DNA synthesis).

The same DNA histogram will be obtained if the length of each interval D_i is increased or decreased proportionately - that is, D_i is replaced by $\bar{D}_i = kD_i$ where $k > 0$. The expected number of cells in phase i remains the same since $\frac{D_i}{C} = \frac{\bar{D}_i}{\bar{C}}$. The ratio of the age interval, corresponding to the DNA interval $\frac{d}{10}$, to the length of the new S interval is still

$$\left(\frac{d}{10r}\right)/kD_2 = \frac{1}{10},$$

since $r = d/kD_2$ it follows from formula (1) that the same number of cells will be recorded in channel i as in the old DNA histogram. Suppose the increases are not proportionate and one of the new ratios $\frac{\bar{D}_i}{\bar{C}}$ is increased. Then the number of cells in the channels corresponding to that phase will also be increased.

Next the effect on the DNA histogram of varying the rate of DNA synthesis will be considered. Suppose the DNA content of a cell is now of the form illustrated in Fig. 6.4. The slope of

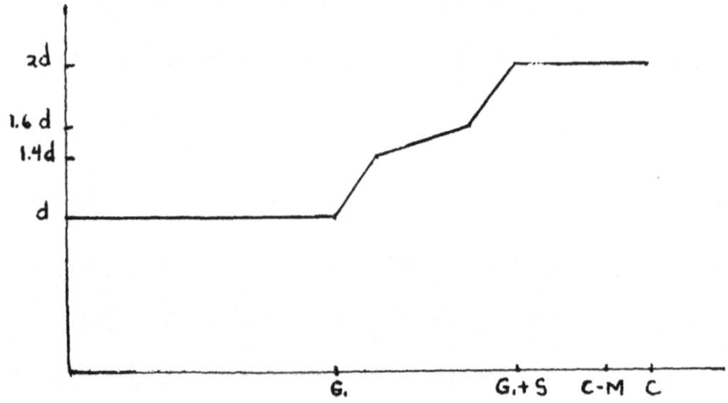

Figure 6.4 DNA content of a single cell.

the line segments or the rate of DNA synthesis in S is

$$r(a) = \begin{cases} \frac{4}{10} d & 0 < a \leq 1 \\ \frac{d}{10} & 1 < a \leq 3 \\ \frac{4}{10} d & 3 < a \leq 4 \end{cases}$$

where a is the age of a cell in S.

The length of the S interval corresponding to a given DNA interval can be found, as described previously for each of the 3 line segments. Applying formula (1) it can be shown that channels 11-14 and 17-20 will contain 2500 S-phase cells; channels 15 and 16 will each contain 10,000 cells. Hence, the DNA histogram will have the shape depicted in Fig. 6.5.

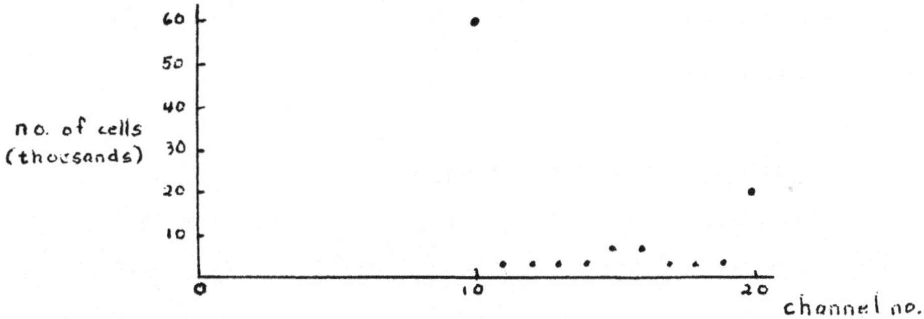

Figure 6.5 Theoretical DNA histogram (steady-state, constant phase lengths; variable DNA synthesis rate).

Note that as the DNA rate increases (decreases) the number of cells in the corresponding channel decreases (increases).

The theoretical histograms of Fig. 6.5 and Fig. 6.3 do not resemble an actual observed histogram. In the observed histogram of Fig. 6.1 the cells in G (and in other phases) are not always recorded in a single channel for a variety of reasons. There may be biological variability of DNA content, measurement errors arising from nonspecific DNA staining and instrumental errors in the measurement of the amount of stain present. Since the total variability is due to a large number of independent error sources it is reasonable to assume that it is normally distributed. Dean and Jet [6] verified this point experimentally, Accordingly, let

$$(3) \qquad P(j|i) = \frac{1}{\sqrt{2\pi}\,\delta_i} e^{-(j-i)^2/2\delta_i^2}$$

be the probability that a cell will actually be recorded in channel j given that its DNA content requires that it be recorded in chennel i. If the cells in each channel of the theoretical histogram are dispersed by using formula (3), the resulting histogram will resemble an experimental histogram. In fact, the observed number of cells in channel j, $n_{oj} = \sum_{i=10}^{20} n_{ti}\ p(j|i)$, where n_{ti} is the theoretical number of cells in channel i.

In summary, the factors that influence the DNA distribution include: biological, cytochemical and instrumental variability; the rate of DNA synthesis; the age density function of the cells which in turn depends on the distributions of the phase durations.

6.3 Generation of DNA Histogram for Random Phase Lengths

DNA histograms have been simulated using a digital computer by M. Eisen and N. Macri in order to model the action of drugs. The discrete model upon which the computer program is based is described below.

The biological model consists of the four standard phases G_1,S,G_2 and M, denoted by 1,2,3, and 4, respectively, and G_0 (a resting state with diploid DNA content). The time T_i spent by a cell in phase i(i = 1,2,3,4) is a random variable with an arbitrary distribution function F_i. It is assumed that the time spent by a cell in a phase is independent of the time spent in any other phase.

In order to obtain a discrete model, time t is measured in steps $\delta > 0$. That is, $t = k\delta$ where k = 0,1,2,3,... . A cell entering phase 1 is considered to have age 0 in that phase. After a time δ it either ages to δ with a certain probability or leaves that phase and enters phase 2 with another probability. These probabilities are determined from the distribution function F_1. For computational purposes it is convenient that the cell can only reach a maximum age $(M_1-1)\delta$. The cell must then leave phase 1 and enter phase 2. The integer M_1 is determined from the distribution function and depends on the desired degree of accuracy. In other words M_1 is the smallest integer such that

$$1 - F_1(M_1\delta) \leq 10^{-d}$$

where the constant d depends on the required number of decimal places. Analogous age compartment models are used for phases 3 and 4.

The model for the S phase is different. After leaving phase 1 a cell enters one of the delay lines with a probability determined from F_2. Delay line i consists of i+1 compartments ($i = 1,2,\ldots,M_2-1$). After a time δ it must pass to the next compartment in the delay line (if there is one). A cell in the last compartment must enter phase 3.

Cells then proceed through phases 3 and 4 in the manner described for phase 1. During M a cell may die or split producing 1 descendant and 1 dead cell or 2 descendants. A live cell then enters G_1 or passes into G_0. A cell in G_0 may eventually return to G_1. This sequence of events is repeated as illustrated in Fig. 6.6. The probabilities associated with the sequence of events occurring during and after passage through M are:

p_d: the probability that a cell dies in M

p_1: the probability that a cell has 1 descendant

p_2: the probability that a cell has 2 descendants

p_0: the probability that a live cell enters G_0. (Note that $1-p_0$ is the probability that a live enters G_1)

r: the probability that a cell in G_0 returns to G_1

It is assumed that p_0 is independent of p_d, p_1 and p_2. These probabilities are treated as constants. However, it will be seen that no extra complications arise in formulating or numerically solving the difference equations if these probabilities are functions of time or the number of cells.

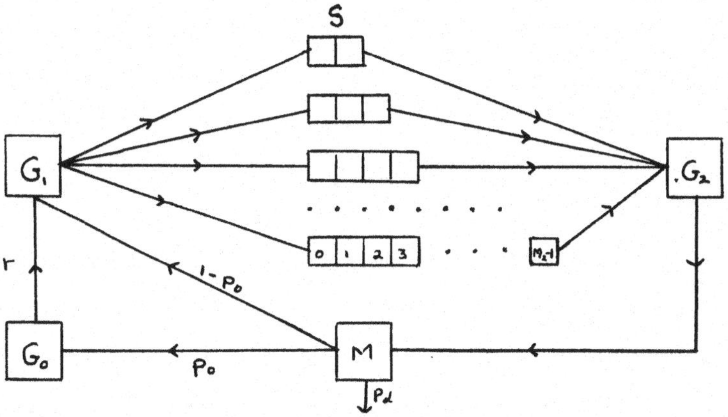

Figure 6.6 Discrete delay model of the cell cycle.

The following notation will be used

$n_j(k\delta,m\delta) \equiv$ number of cells in state j at time $k\delta$ having age $m\delta$

$(k = 0,1,2,\ldots\ ;\ m = 0,1,2,\ldots,M_j-1) \quad j = 1,3,4.$

$n_2(k\delta,m\delta|\ell) \equiv$ number of cells in state 2; having age $m\delta$ in delay line ℓ

$(1 \leq \ell \leq M_2-1,\ 0 \leq m \leq \ell;\ \ k = 0)$

$n_0(k\delta) \equiv$ number of cells in G_0 at time $k\delta$

$n_d(k\delta) \equiv$ number of cells which die at time $k\delta$

$\mu_j \equiv$ mean time spent by a cell in phase j

$\sigma_j \equiv$ standard deviation for the time spent by a cell in phase j

$F_j(t) = P_j(T \leq t) \equiv$ probability that the time spent by a cell in state j is $\leq t$.

$H_j(t) \equiv 1-F_j(t)$.

The number of cells in state j at time $k\delta$ $(k > 0)$ and age $m\delta$ $(m > 0)$ is equal to the number of cells of age $(m-1)\delta$ at time $(k-1)\delta$ multiplied by the probability that a cell survives to age greater than $m\delta$ given that it survived longer than age $(m-1)\delta$. Hence, for $k > 0$ and $m > 0$

(4) $$n_j(k\delta,m\delta) = n_j((k-1)\delta\ ,\ (m-1)\delta)\,\frac{H_j(m\delta)}{H_j((m-1)\delta)} \quad (j = 1,3,4)$$

In phase 2, the delay line model implies

(5) $$n_2(k\delta,m\delta|\ell) = n_2((k-1)\delta,\ (m-1)\delta|\ell).$$

Cells can enter state 1 at time $k\delta > 0$ at age 0 only if they were in state M or in state G_0 at time $(k-1)\delta$. Recall that if a cell is in G_0 the probability that it enters G_1 is r. A cell will leave M at age $i\delta$ $(i = 0,1,\ldots,M_4-1)$ provided that it does not reach age $(i+1)\delta$ given that its age is greater than $i\delta$. The probability of this occurring is $\dfrac{H_4(i\delta)-H_4((i+1)\delta)}{H_4(i\delta)}$. The cell may have 1 or 2 descendants. The probability that there is one descendant and it enters G_1 is $(1-p_0)p_1$. The probability that there are two descendants and one enters G_1 while the other enters G_0 is $2(1-p_0)p_0p_2$. The probability that there are 2 descendants and both enter G_1 is $(1-p_0)^2p_2$. Moreover, twice the number of such cells enter G_1. It follows from all of the foregoing facts that for $k > 0$

$$(6)\quad n_1(k\delta,0) = \sum_{i=0}^{M_4-1} n_4((k-1)\delta,\ i\delta)\ \frac{H_4(i\delta)-H_4((i+1)\delta)}{H_4(i\delta)}\ (1-p_0)(p_1+2p_2)+rn_0((k-1)\delta).$$

Analogous reasoning leads to the following equations:

$$(7)\quad n_4(k\delta,0) = \sum_{i=0}^{M_3-1} n_3((k-1)\delta,i\delta)\ \frac{H_3(i\delta)-H_3((i+1)\delta)}{H_3(i\delta)}$$

$$(8)\quad n_0(k\delta) = \sum_{i=0}^{M_4-1} n_4((k-1)\delta,i\delta)\ \frac{H_4(i\delta)-H_4((i+1)\delta)}{H_4(i\delta)}\ p_0(p_1+2p_2)$$

$$(9)\quad n_d(k\delta) = \sum_{i=0}^{M_4-1} n_4((k-1)\delta,i\delta)\ \frac{H_4(i\delta)-H_4((i+1)\delta)}{H_4(i\delta)}\ (p_d+p_1).$$

Since cells in the last compartment of a delay line in S must pass into G_2,

$$(10)\qquad n_3(k\delta,0) = \sum_{\ell=1}^{M_2-1} n_2((k-1)\delta,\ \ell\delta|\ell).$$

Finally, a cell leaving G_1 is placed in delay line ℓ, if and only if, its sojourn time in S is less than $(\ell+1)\delta$ but at least $\ell\delta$. The probability of this last event is

$$F_2((\ell+1)\delta) - F_2(\ell\delta).$$

Hence,

$$(11)\quad n_2(k\delta,0|\ell) = [\sum_{i=0}^{M_1-1} n_1((k-1)\delta,\ i\delta)\ \frac{H_1(i\delta)-H_1((i+1)\delta)}{H_1(i\delta)}][F_2((\ell+1)\delta = F_2(\ell\delta)]$$

$$\ell = 1,\ldots,M_2-1\ .$$

Once the initial conditions

$n_j(0,m\delta)\qquad j = 1,3,4;\quad m = 1,\ldots,M_j-1$

$n_2(0,m\delta|\ell\ )\quad \ell = 1,\ldots,M_2-1;\quad m = 0,\ldots,\ell\ .$

$n_0(0)$

$n_d(0)$

are specified, equations (4)-(11) can be solved. In particular,

$n_2(k\delta,\ m\delta|\ell)\quad \ell = 1,\ldots,M_2-1;\ m = 0,\ldots,\ell$

are determined for any time $t = k\delta$. Assume that DNA is synthesized at a constant rate for a fixed sojourn time in S. However, the rate can vary with the sojourn time. It follows that compartment i in delay line ℓ contains

$$(12) \qquad D(1+\frac{i}{\ell}) \qquad i = 0,\ldots,\ell$$

units of DNA. Since the number of cells in each compartment is known a DNA histogram can be constructed. The DNA content can be translated into channel numbers as described above. The resulting theoretical histogram can then be spread by using (3), or some other appropriate density, to produce an experimental type histogram.

The above model has been generalized to include blocked cells. Instead of (12), other distributions of DNA in each compartment can be used.

Computer simulations of sequential DNA distributions have also been carried out by Gray [15] based on the Takahashi-Kendall model. Gray used his technique to analyze sequential sets of experimental DNA distributions from perturbed cell populations. The parameters in the model were adjusted, by trial and error, until the simulated DNA histogram coincided with the particular experimental DNA histogram under consideration. This procedure determines the durations and dispersions of the phases of the cell cycle as well as the point in the cell cycle at which the perturbing agent exerts its effect. However, the answer may not be unique since different values of the parameters may give the same answer.

Kim [24,25] has also simulated DNA distributions using the formalism of Hahn but allowing cell death. Transition probabilities are also defined for transfer from a cycling to a noncycling state of the same maturity and vice versa.

In this section the simulation of DNA histograms has been described. The remainder of the chapter is concerned with the reverse process. Given an experimental histogram find the proportion of all cells in each phase, the mean sojourn time in each phase and its variance.

At present the "uniqueness" of the solution to the foregoing problem has not been investigated. Recall that the same histogram can be obtained for proportional phase lengths. Due to errors in measurement a different experimental histogram will be obtained the next time the experiment is performed. If the two histograms are "close" to each other will the two sets of parameters obtained by analyzing the histogram be "close" to each other?

6.4 Definition of an Asynchronous Population

Asynchrony was defined in Chapter 5. This term is also used to describe cell populations whose DNA histograms resemble Fig. 6.1. Such histograms have a large peak due to G_1-phase cells, a smaller peak at twice the DNA content due to (G_2+M)-cells and a continuum between, due to S-phase cells. Moreover, such DNA histograms are unchanging or very slowly changing with time. This time invariance is crucial. (Some methods of analysis of DNA histogram require that the histograms vary in time). These methods cannot be applied to time invariant histograms. It will be seen that these two concepts of asynchrony can be reconciled under certain assumptions.

A common assumption in FMF analysis is that the time spent by a cell in any phase is independent of the time spent in any other phase. (There is some controversy about the independence of the phases [36] and even about the kinetic and functional discreteness of the phases [31].) Trucco's model (Chapter III, Section 6) shows that an asynchronous population has an age density function which is time independent. Assume that the DNA content of a cell is a fixed function, which may depend on the sojourn time of the cell in the S phase. It follows that the theoretical DNA histogram will not vary with time. Hence, the observed DNA histogram will also be time invariant. Moreover, simulations of DNA histograms of asynchronous populations show that they resemble the DNA histogram of Fig. 6.1. Thus, the definition of an asynchronous population is consistent with common usuage of this term in FMF analysis.

6.5 Graphical Analysis of Asynchronous Populations

Two graphical methods are used to estimate the fraction of cells in the G_1, S and G_2+M phases. In the method of Baisch, et al. [1] the area under the DNA histogram is divided into 3 regions corresponding to the 3 phases as shown in Fig. 6.7(a).

The fraction of cells in each phase is then obtained by measuring the area of each of the regions with a planimeter.

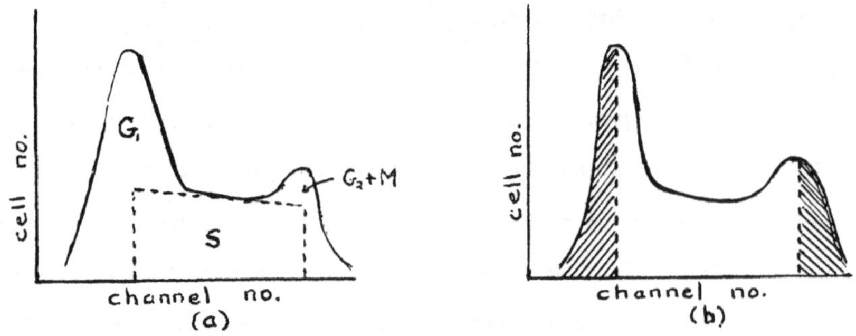

Figure 6.7 Graphical analysis of DNA histograms.

The graphical method of Barlogie et al. [2] it is assumed that the left half of the G_1 peak and right half of the G_2+M peak represent only G_1 and G_2+M cells, respectively. (See Fig. 6.7(b)). These halves are then folded around their respective peaks. Thus, two symmetrical peaks are formed whose areas yield the fraction of cells in each of these two phases. The S fraction is then obtained by subtraction.

As can be seen from examples, both of these graphical methods become unreliable as the S-phase fraction increases or the G_2+M phase fraction decreases.

6.6 Analytic Analysis of Asynchronous Populations

Several mathematical techniques for analyzing FMF data appear in the literature- Dean & Jett [6,7], Baisch, Gohde & Linden [1] and Fried [13,14]. In all three of these models the G_1 and G_2+M populations are assumed to be normally distributed. The models differ mainly in their treatment of the S population and in their technique for evaluating the statistical parameters (mean and standard deviation) for G_1 and G_2+M. Dean and Jett [6] use a second degree polynomial to estimate the portion of the curve corresponding to the S state. Baisch et al. employ a normal distribution function for S in the asynchronous case and a normal density function in the synchronous case. Fried subdivides S into compartments of increasing DNA content, each one normally distributed and all having the same coefficient of variation as the G_1 compartment. All three employ a least squares method to fit the theoretical curve to the observed data. Dean & Jett [6] and Baisch et al. [1]

treat the statistical parameters as well as the number of cells in each compartment as unknowns to be solved for in the least squares fit. Fried estimates the mean and standard deviation and uses least squares only to solve for the number of cells in each compartment. This device simplifies the least squares computation, but since the original estimates may not be sufficiently accurate, repeated ones may be necessary in order to obtain an acceptable fit. Also, the optimal choice of intercompartmental spacing for S may vary, especially in the case of chemically treated populations. Hence, the procedure requires considerable interaction between the computer and programmer.

Of the three, Fried's model appears to be the most general. A population totally synchronized in the S state is likely to be normally distributed and may not be adequately approximated by a second degree polynomial. Also, a normal distribution or density function would not be the best approximation to an S population having two or more significant peaks. Subdividing S into normally distributed compartments provides flexibility in approximating the S curve. This is especially so for populations having various degrees of synchrony.

Baisch's approximation and Fried's technique are special cases of the model of Eisen and Schiller [11] presented below.

A good approximation to the DNA fluorescence density function f can be obtained by dividing the number of cells in a channel by the total number of cells. That is, let Y be a random variable which given the channel in which a cell will be recorded; then

$$f(i) = P(Y = i) = \frac{n_i}{n} \; (i = 0,1,2,\ldots,c),$$

where c = the maximum number of channels used,

n_i = the number of cells in channel i,

$n = \sum_{i=0}^{c} n_t$ = the total number or cells.

The function f will be determined by decomposing it into three component density functions corresponding to cells in phases G_1, S and G_2+M.

Let X be a random variable corresponding to the channel in which a cell in phase G_1 will be recorded. Denote by f_{G_1} the probability density function of X.

In order to determine the density functions for cells in the other phases the following two assumptions are made.

Assumption 1

Cells just entering S will be recorded in channel X.

This seems reasonable since the DNA content of a cell in G_1 does not change.

Next, consider a cell whose DNA content was D when the cell was in G_1 and whose present DNA content is λD ($1 \leq \lambda \leq 2$). Then λ is called the relative DNA content.

Assumption 2

Cells with relative DNA content λ are recorded in channel λX.

For example, cells in G_2+M will be randomly distributed in channels according to the random variable 2X. By standard rules of transformation of random variables the density function of 2X is

(13) $$f_{G_2+M}(x) = \frac{1}{2} f_{G_1}(x/2).$$

From assumptions 1 and 2 it follows that a cell in S will be recorded in channel $Z = \lambda X$ for some λ ($1 < \lambda < 2$). For a fixed λ, the density function of Z is

$$f_s(x|\lambda) = \frac{1}{\lambda} f_{G_1}(x/\lambda).$$

Hence, if $g(\lambda)$ is the density function for the random variable λ, then the density function for Z is

(14) $$f_s(x) = \int_1^2 (1/\lambda) f_{G_1}(x/\lambda) g(\lambda) d\lambda.$$

The density function f_s depends on the G_1 distribution, the relative rate of DNA synthesis, and the relative age density of the S population. During the S state a cell doubles its DNA content. Hence, we assume that the DNA content D(a) of a given cell in S is an increasing function of the cell's age A in S which satisfies $D(t_s) = 2D(0)$, where t_s is the cell's total sojourn time in S. The values D(0) and t_s vary from cell to cell. However, we assume that the cells are relatively homogeneous in their DNA synthesis. To be more precise, let $\alpha = A/t_s$ and $\lambda = D(\alpha)D(0)$ denote the relative age ($0 \leq \alpha \leq 1$) and relative DNA content ($1 \leq \lambda \leq 2$), respectively, for a given cell in S. Then λ is a function of α

satisfying $\lambda(0) = 1$, $\lambda(1) = 2$. We add to assumptions 1 and 2 the following:

Assumption 3

$\lambda(\alpha)$ is the same increasing and continuously differentiable function for all the cells in S.

As a particular case, if the cells are synthesizing DNA at a constant rate r, then $r = D(0)/t_s$ and $\lambda(\alpha) = 1+\alpha$.

In general, the density function for α depends on the density function for λ. By assumption 3,

$$g(\lambda)d\lambda = g[\lambda(\alpha)]d\lambda(\alpha).$$

Hence, from equation (14)

$$f_s(x) = \int_0^1 \frac{1}{\lambda(\alpha)} f_{G_1}\left(\frac{x}{\lambda(\alpha)}\right) g[\lambda(\alpha)]\lambda'(\alpha)d\alpha \ , \tag{15}$$

which illustrates the dependence of the S distribution on the G_1 distribution and the relative rate of DNA synthesis.

If the relative age probability density function $\rho(\alpha)$ for the random variable α were known, then

$$\rho(\alpha)d\alpha = g[\lambda(\alpha)]d\lambda(\alpha)$$

and

$$f_s(x) = \int_0^1 \frac{1}{\lambda(\alpha)} f_{G_1}\left(\frac{x}{\lambda(\alpha)}\right) \rho(\alpha)d\alpha \ . \tag{16}$$

The probability density $\rho(\alpha)$ can be expressed in terms of the (absolute) age density function for the S population. This function is known when the cells are growing exponentially and when the population is in a steady state of renewal with cell production just enough to overcome cell loss. We now consider these two cases.

Let $n(t,a)$ denote the age density function for the S population:

(i) $n(t,a)da$ = the expected number of S cells at time t with S ages between a and a+da.

(ii) $\int_0^\infty n(t,a)da = N(t)$, the total number of S cells at time t.

If the population is growing exponentially, then

$$N(t) = N(0)e^{\gamma t}$$

and using (111) and (119) of Chapter 3,

$$n(t,a) = n(0,0)e^{\gamma(t-a)}\phi(a),$$

where $\phi(a)$ is the probability that the sojourn time t_S of a cell in S is greater than a. In this case the ratio $n(t,a)/N(t)$ is independent of time t. Hence, letting $K = n(0,0)/N(0)$,

$$Ke^{-\gamma a}\phi(a)da$$

is the probability that a cell in S will have age between a and a+da. If $\delta(\tau)$ is the probability density function for the sojourn time in S, then

$$\eta(t,\alpha) = Ke^{-\gamma a}\delta(\tau) \quad (\tau \geq a)$$

is the joint age, sojourn time probability density function. Also, if $\rho(\alpha|\tau)$ and $\eta(\alpha|\tau)$ denote the conditional probability density functions for relative age and absolute age, respectively, with respect to sojourn time, then

$$\rho(\alpha|\tau)d\alpha = \eta(\alpha\tau|\tau)d(\alpha\tau)$$

or

$$\rho(\alpha|\tau) = \tau\eta(\alpha\tau|\tau).$$

Hence

$$\rho(\alpha) = \int_0^\infty \tau\eta(\tau,\alpha\tau)\, d\tau = K\int_0^\infty \tau e^{-\gamma\alpha\tau}\delta(\tau)d\tau = -K\bar{\delta}'(\gamma\alpha),$$

where $\bar{\delta}(\gamma)$ denotes the Laplace transform of $\delta(\tau)$. Finally, since

$$\phi(a) = 1-\int_0^a\delta(\tau)d\tau \quad \text{and} \quad \int_0^\infty Ke^{-\gamma\alpha}\phi(a)da = 1,$$

it follows that

$$K = \frac{\gamma}{1-\bar{\delta}(\gamma)}.$$

$$\rho(\alpha) = \frac{\gamma\bar{\delta}'(\gamma\alpha)}{\bar{\delta}(\gamma)\ -1}$$

and equation (16) becomes

$$(17) \qquad f_S(x) = \frac{\gamma}{\bar{\delta}(\gamma)-1}\int_0^1 \frac{1}{\lambda(a)} f_{G_1}\left(\frac{x}{\lambda(a)}\right) \bar{\delta}'(\gamma\alpha)d\alpha\ .$$

When the growth rate $\gamma = 0$ (equivalently, $m = 1$, cf. p.118 Chapter 3), the population is in a steady state of balanced growth. The relative age probability density function becomes $\rho(\alpha) = 1$. Hence, the relative age α is uniformly distributed over [0,1]. Equation (15) for the S density function now becomes

$$(18) \qquad f_S(x) = \int_0^1 \frac{1}{\lambda(\alpha)} \; f_{G_1}\left[\frac{x}{\lambda(\alpha)}\right] d\alpha.$$

If the rate of DNA synthesis is constant, then $\lambda(\alpha) = 1 + \alpha$. Hence, for a steady state and constant rate of DNA synthesis, f_S is completely determined by f_{G_1}. These two conditions also imply that $g(\lambda) = \lambda$ for $1 \leq \lambda \leq 2$ - i.e., g is uniformly distributed.

The density functions f_{G_1}, f_S and f_{G_2+M} determines the DNA fluorescence density function f. The usual assumption is that f_{G_1} is a normal density function. The remaining functions can then be calculated from (13) and (18) under the assumption of a constant rate of DNA synthesis. These continuous density functions are discretized by assigning the value $f_i(x).1$ to the probability that a cell in phase i will be recorded in channel x. Let p_{G_1}, p_S, p_{G_2+M} be the probability that a cell from the population under study is in phase G_1, S, G_2+M, respectively. Then by the total probability theorem

$$(19) \qquad f(x) = p_{G_1} f_{G_1}(x) + p_S f_S(x) + p_{G_2+M}\, f_{G_2+M}(x).$$

Because of experimental errors, (19) is interpreted in the sense of least squares. This will give equations for μ, σ, p_{G_1}, p_S and p_{G_2+M}. These are non-linear transcendental equations and must be solved numerically.

The major difficulty in applying the techniques described in this section is that there is no general numerical method for solving nonlinear optimization problems. Hence, a program which will work for one data set may not work for another set. A new optimization subroutine may be required.

All optimization procedures are highly dependent on the initial values used for the unknown parameters. Initial estimates of these parameters can be obtained graphically or by using (19).

If μ and σ were known the equations for the least squares determination of the other unknowns would be linear. The parameters μ and σ can be determined if a pure G_1 population were available.

From (17) it is evident that f_S can be a very complicated function. It is doubtful that the described methods for analyzing FMF histograms can be trusted for arbitrary data sets. The results should be corroborated by other means (such as autoradiography or the procedure described in Section 5.9).

6.7 Estimation of Mean Phase Lengths for Asynchronous Populations

Let δ_j be the probability density function for the sojourn time in phase j of the cell cycle ($j = G_1, S, G_2, M$) and $\bar{\delta}_j$ its Laplace transform. It follows from equation (119) in Chapter 3 that the fraction of cells

$$p_j = N_j(t)/N(t)$$

in state j are [11]:

$$(20) \qquad \begin{aligned} p_{G_1} &= [1-\delta_{G_1}(\gamma)]/\bar{\delta}(\gamma) \\ p_S &= \bar{\delta}_{G_1}(\gamma)[1-\bar{\delta}_S(\gamma)]/\bar{\delta}(\gamma) \\ p_{G_2+M} &= \bar{\delta}_{G_1}(\gamma)\bar{\delta}_S[1-\bar{\delta}_{G_2}(\gamma)\bar{\delta}_M(\gamma)]/\bar{\delta}(\gamma) \end{aligned}$$

where $\bar{\delta}$ is the Laplace transform of the cell cycle density function. Since the phase durations are independent

$$\bar{\delta}(\gamma) = \bar{\delta}_{G_1}(\gamma)\bar{\delta}_S(\gamma)\bar{\delta}_{G_2}(\gamma)\bar{\delta}_M(\gamma).$$

Equations (20) hold in the limit for $\gamma = 0$. If $\gamma = 0$, then $\bar{\delta}_j(\gamma) = 1$ for each j. Equation (116) of Chapter 3 implies that $m = 1$. Therefore, the probability, $m/2$ that a cell, upon completion of mitosis, undergoes binary fussion is one-half. The proliferating population is then in a steady state of renewal with all production just enough to offset cell loss. By taking limits as γ approaches zero (20) reduces to

$$(21) \qquad \begin{aligned} p_{G_1} &= T_{G_1}/T, \\ p_S &= T_S/T, \\ p_{G_2+M} &= (T_{G_2}+T_M)/T, \end{aligned}$$

where T_j is the mean time spent in phase j ($j = G_1,S,G_2,M$) and $T = T_{G_1}+T_S+T_{G_2}+T_M$ is the mean cell cycle time.

The proportion of cells in each phase, p_i, can be determined from FMF analysis. If T and T_M are known the remaining mean phase times can be calculated from equations (21). The mean cycle time and the mean mitosis time can be found by time lapse photography. Other methods will be discussed in the next section.

It is important to realize that equations (21) are valid only in the steady state; they may not even be good approximations otherwise. For example, if δ and δ_j are gamma densities of the form

$$f(t) = \frac{1}{\Gamma(\gamma)} C^{\gamma} t^{\gamma-1} e^{-Ct} \quad (t \geq 0)$$

where

$$\Gamma(\gamma) = \int_0^{\infty} t^{\gamma-1} e^{-t} dt,$$

(20) reduces to

$$p_{G_1} = \frac{b}{b-1} \left(\frac{b^{\alpha}-1}{b^{\alpha}}\right) \quad (\alpha = T_{G_1}/T),$$

$$p_S = \frac{b}{b-1} \left(\frac{b^{\beta}-1}{b^{\alpha+\beta}}\right) \quad (\beta = T_S/T),$$

$$p_{G_2+M} = \frac{b}{b-1} \left(\frac{1}{b^{\alpha+\beta}} - \frac{1}{b}\right) \quad (b = M).$$

Equations (22) were also derived by Mode [27] using the theory of multitype age-dependent branching processes.

If the p_i are determined from FMF analysis, then equations (22) can be solved for α,β and b. Then if T and T_M can be determined by other means, $T_{G_1} = \alpha T$, $T_S = \beta T$ and T_{G_2} can easily be calculated.

6.8 Methods of Estimating Mean Cycle and Mitotic Time Asynchronous Populations

The mean cycle and mitosis time can be obtained by autoradiography. However, such procedures are time consuming and if used in conjunction with FMF analysis would result in a loss of speed. Therefore, two faster methods for finding these parameters are discussed.

(a) Cinemicrography

The micro ciné technique consists of time lapse photographs of cells growing in vitro taken through a microscope. Further details can be found in [28,34,35].

Since mitosis is a visible event the mean duration of mitosis can be estimated from the photographs. The mean cycle time can be found by following cells from mitosis to mitosis provided the time for exponential growth is much greater than the cycle time. Estimators of the parameters of the gamma density function can be obtained - for example, by the method of moments. This procedure consists of equating the sample moments to the corresponding equations for the population moments and then solving the resulting equations for the unknown parameters. Let $t_1,\ldots,t_n$ $(n > 2)$ be independently observed cycle times of n cells. The sample means and variances of these numbers are

$$\bar{t} = \frac{1}{n}\sum_{i=1}^{n} t_i \qquad S^2 = \frac{1}{n-1}\sum_{i=1}^{n}(t_i-\bar{t})^2 .$$

For the gamma density $ET = \gamma/C$ and $\mathrm{Var}\ T = \gamma/C^2$. To obtain estimates of γ and C by the method of moments set $\hat{\gamma}/\hat{C} = \bar{t}$ and $\hat{\gamma}/\hat{C}^2 = S^2$. This yields $\hat{\gamma} = \bar{t}^2/S^2$ and $\hat{C} = \bar{t}/S^2$. It follows that estimators for ET and Var T

$$ET = \hat{\gamma}/\hat{C} = \bar{t} \qquad \sigma_T^2 = S^2.$$

On the other hand, if the time for exponential growth is short as compared to the cycle time a different method is preferable. If cells are followed from mitosis to mitosis part of a cycle may occur during a period of nonexponential growth. Few conclusions can be drawn from such samples [29].

A method of overcoming the above difficulty is provided by using the stable age distribution which is a consequence of exponential growth. Cells are sampled randomly and their time to division is recorded provided the cell does not die. These recorded times will be shorter than the times obtained by following cells from mitosis to mitosis and can be used to estimate the mean cycle time.

Consider a cell aged u at time $t = 0$. The probability of a cell having such an age[2] is determined from the stable age density function

---

[2] Formally the age lies between u and u + du.

$$Ke^{-\gamma u}\phi(u)$$

where

$$\phi(u) = 1- \int_0^u \delta(t)dt$$

and

(23) $$K = \frac{\gamma}{1- \bar{\delta}(\gamma)} .$$

Given that this cell does not disintegrate the probability that it will not divide before t is

$$\frac{\phi(t+u)}{\phi(u)} .$$

Integrating this last expression with respect to the stable age density yields

(24) $N(t) = K \int_0^\infty \phi(t+u)e^{-\gamma u}du = Ke^{\gamma t} \int_t^\infty \phi(x)e^{-\gamma x}dx,$

the probability that a nondisintegrating cell, randomly sampled from the stable age distribution, will not divide during time t. Differentiating (24) gives

(25) $$N' = \gamma N - K\phi.$$

Let $\bar{t}$ be the expected time to division, given there is no disintegration, of a cell with a stably distributed age. Integrating (25) from 0 to infinity yields

(26) $$-1 = \gamma \bar{t} - KT$$

where T is the mean cycle time. Solving (26) for T and using (23) yields

$$T = [1- \bar{\delta}(\gamma)][\bar{t} + \frac{1}{\gamma}]$$

which can be written as

(27) $$T = \frac{m-1}{m} [\bar{t} + \frac{1}{\gamma}]$$

by using (115) of Chapter 3. Using the doubling time, t_d, (27) can be rewritten as

(28) $$T = \frac{m-1}{m} [\bar{t} + t_d/\ell n\ 2].$$

If there is no cell death (27) and (28) reduce to

(29) $$T = \frac{1}{2} [\bar{t} + \frac{1}{\gamma}] = \frac{1}{2} [\bar{t} + t_d/\ell n\ 2]$$

The mean cycle time is estimated in vitro as follows. Estimate γ according to (19) of Chapter 2 and m according to (22). Select cells randomly and record the times until they divide (provided they do so). Estimate $\bar{t}$ by the sample mean of these times. Finally insert these values into (27).

Multiplication of (25) by t before integration leads to a formula for σ^2 = Var T:

$$\sigma^2 = \frac{m-1}{m}\,[S^2-\bar{t}^2+2\gamma\bar{t}]-T^2. \tag{30}$$

Here S^2 is the sample variance of times to division of cells that divide in a stable population.

A similar method was applied to estimate expectation and variance of the cycle time of some human cells in [20]. The result was that tumorous cells seemed to have longer cycle times than their normal counterparts!

(b) RCS_i analysis [16]

In the RCS_i (radioactivity per cell in a narrow window, S_i, in the S phase) method cells are pulse labelled with ^{3}H-TdR and also stained specifically and stoichiometrically for DNA. The window in the S phase, S_i, is defined by a narrow range of DNA content which in turn corresponds to a narrow range of fluorescence. Cells in the S_i window are separated from the rest of the population by means of a cell sorter described in detail by Van Dilla et al. [39]. Cell sorting is accomplished by forming a suspension of cells. The dispersed cells are carried single file by a liquid jet across an intense beam of laser light. Each cell crosses the laser beam and emits a pulse of fluorescent light proportional to its stain content. The pulse of fluorescence is detected photoelectrically. It is compared with a DNA fluorescent distribution of such cells. This DNA histogram is obtained earlier and is stored in the memory of a multichannel pulse height analyzer. If the measured value for a cell falls in the preset range, S_i, the instrument delays long enough for the cell to arrive at the point where the liquid jet is breaking up into droplets. Droplets containing cells in the desired fluorescence range are electrically charged and subsequently deflected by a DC electric field into a container. About 10^4 cells can be collected in a few minutes. The average radioactivity per sorted cell for each sample is then determined by liquid scintillation counting. By sampling the population periodically and plotting the time variation of the RCS_i sequence and RCS_i curve is obtained as shown in Fig. 6.8 (b).

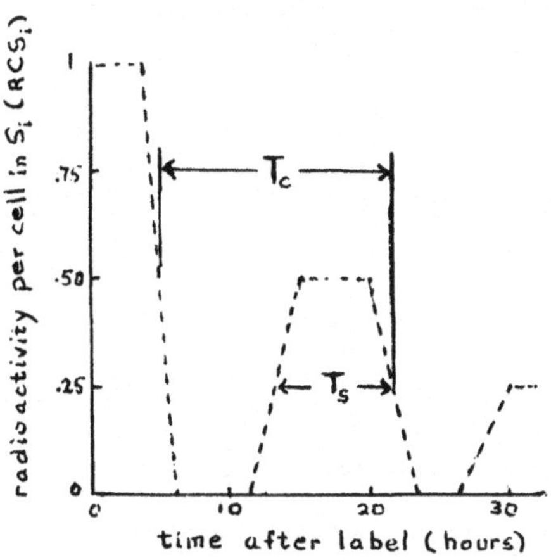

(a) constant phase sojourn time

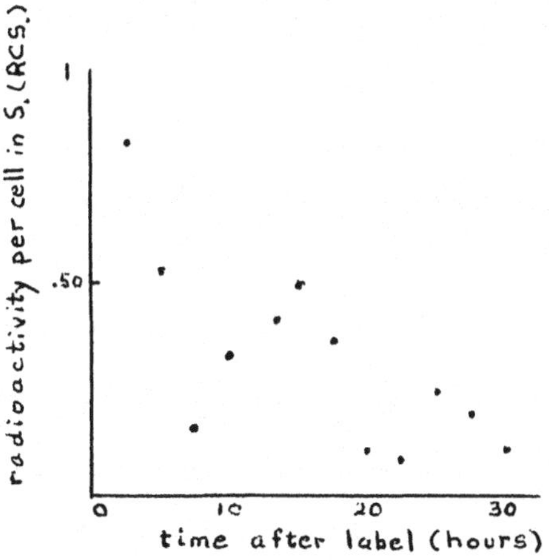

(b) actual experimental data

Figure 6.8 RCS_i curves.

Some insight into the reason for the shape of an RCS_i curve can be obtained by considering a population of cells with constant phase sojourn times. Immediately after pulse labeling ($t = 0$) it is assumed that all of the cells in the window S_i are labelled. For convenience, $RCS_i(0)$ is assigned the value unity. As time increases the labelled cells mature and leave the S_i part of the S phase. The RCS_i curve drops to zero as the labeled cells traverse the G_2, M, G_1 and part of the S-phase since unlabelled cells pass into the S_i window. When the labelled cohort next passes through S_i the maximum height of the RCS_i curve is one half because of the dilution of label with cell division in the preceding M phase. Thus each cell cycle produces one wave but with a halving of the amplitude as shown in Fig. 6.8 (a).

The length of the cell cycle is the time from the .50 point on the descending limit of the first wave to the .25 point on the descending portion of the succeeding wave. The length of the S phase is the time interval between .25 points on the second wave. When the sojourn time in each phase is random the RCS_i curve becomes rounded and progressively dampled as shown in Fig. 6.8 (b). The above rules for estimating the mean cycle time and the mean length of the S phase are only approximations. It becomes necessary to extract cell parameters from the curve according to a mathematical model.

The mathematical model that will be used is similar to the Kendall-Takahashi model described in section 2 of Chapter 3. The equations describing the number of cells in each compartment is given by (3) and (4) with $\beta = 2$ and $\mu_i = 0$. It is also assumed that $\lambda_1 = \lambda_{G_1}, \lambda_S, \lambda_{G_2}, \lambda_M$ are constants during the corresponding phases.

The evolution of radioactivity around the cell cycle can be described by the same equations as those for the number of cells in each compartment except in compartment 1. This exception arises since the radioactivity does not double as does the cell number at mitosis. Hence, the equation describing the flux of radioactivity through the first state is

$$\frac{dR_1(t)}{dt} = \lambda_k R_k(t) - \lambda_1 R_1(t)$$

where $R_1(t)$ is the radioactivity associated with state at time t and the λ's are defined as above. Otherwise the equations describing the radioactivity change are identical with those describing the change in cell number.

The asymptotic solution for the number of cells in each compartment is (see page 78)

$$n_i(t) = n_i(0)e^{Et}$$

where E is the largest real root of

$$(1+\frac{x}{\lambda_{G_1}})^{g_1}(1+\frac{x}{\lambda_S})^{s}(1+\frac{x}{\lambda_{G_2}})^{g_2}(1+\frac{x}{\lambda_M})^{m} = 2.$$

Since the radioactivity will be divided by a corresponding number of cells, the number of cells in each compartment need only be determined up to a constant of proportionality. A convenient choice is

$$n_i(0) = \prod_{j=0}^{i=1} \lambda_j \prod_{j=i+1}^{k} (\lambda_j+E) \qquad (\lambda_0 = 1;\ i = 1,\dots,k-1)$$

$$n_k(0) = \prod_{j=0}^{k} \lambda_j .$$

The radioactivity is initially distributed only across the S phase so that

$$R_i(0) = n_i(0) \qquad g_i < i \le g_1+s$$

$$R_i(0) = 0 \qquad \text{otherwise.}$$

The initial conditions, $R_i(0)$, were chosen so that the theoretical radioactivity per cell in S_i at time $t = 0$ is

$$rcs_i(0) = 1.$$

The radioactivity at any time t can then be calculated from the system of differential equations describing the R_i. Then the theoretical radioactivity per cell in a window in S_i is defined to be

$$rcs_i(t) = \sum R_i(t)/\sum N_i(t)$$

with both sums all compartments corresponding to the DNA content of the S_i window. To find the corresponding compartment it is assumed that the rate of DNA synthesis is constant . This assumption can easily be generalized. It follows that for an S phase composed of s states, the DNA content of compartment i is

$$NDA(i) = (G_1 \text{ amount})\ (1 + i/s) \qquad 1 \le i \le s\,.$$

Hence, the compartments corresponding to a given S_i window can be found.

In [16] it is assumed that the phase duration and their coefficient of variation are known. This determines the parameters of the Takahashi model for the of cells in each compartment. The theoretical rcs_i curve could then be calculated and compared with the experimental RCS_i curve with good agreement.

Further research is required to automate this procedure just as autoradiography has been automated. It should then be possible to determine the mean phase duration and their variances. As suggested in Fig. 5.8 (a) some sort of simple model might yield a good estimate of T_c. This would be a valuable adjunct to FMF analysis since the mean duration of each phase could then be found quite rapidly (RCS_i analysis only takes a few days).

6.9 Analysis of Synchronous Populations. Single Histogram

An approximation to the right side of equation (19) can be obtained by replacing the density $g(\lambda)$ by a step function as follows. Let

$$\lambda_i = 1 + i/n \qquad i = 0,1,\ldots,n$$

and define

$$G(\lambda) = \int_1^\lambda g(x)\, dx.$$

Then, for n sufficiently large,

$$f_S(x) = \sum_{i=1}^{n} \int_{\lambda_{i-1}}^{\lambda_i} (1/\lambda) f_{G_1}(x/\lambda) dG(\lambda) \simeq \sum_{i=1}^{n} p_i/\lambda_i f_{G_1}(x/\lambda_i),$$

where

$$p_i = G(\lambda_i) - G(\lambda_{i-1}) \text{ and } \sum_{i=1}^{n} p_i = 1.$$

Equation (19) can be replaced by

$$(31) \qquad f(x) \simeq p_{G_1} f_{G_1}(x) + \sum_{i=1}^{n} q_i (1/\lambda_i) f_{G_1}(x/\lambda_i) + (1/2) p_{G_2+M} f_{G_1}(x/2),$$

where $q_i = P_S p_i$ $(i = 1,2,\ldots,n)$. Approximation (31) is essentially equivalent to the one introduced by Fried [13] when the f_i are normal.

The unknowns are now P_{G_1}, P_{G_2+M}, q_i $(i = 1,2,\ldots,n)$, and the parameters μ and σ (note that $\sum q_i = P_S \sum p_i = P_S$). Theoretically, these unknowns could be determined by the method of least squares. Practically, this is not feasible for all data sets since the problem is non-linear with respect to μ and σ. However, for any fixed values of μ and σ, the remaining unknown can be found by solving a system of linear equations. Bounds for μ and σ can be obtained either by inspection of the FMF histogram or from (31). For each value of μ and σ, χ^2 can be calculated by using the right side of equation (3) for the theoretical values. By varying μ and σ, an optimal choice can be obtained according to a χ^2 criterion. This can be done automatically by using optimization programs.

The major difficulty in applying all the previously described analytic approximations is that there is no general numerical technique for solving nonlinear optimization problems. By using Bayes' theorem in conjunction with the method of maximum likelihood this problem is avoided. The parameters are estimated by iteration. This new approach of Eisen et al. [12] could be applied to any of the previous approximations. However, an approximation similar to (31) will be used to illustrate the method. Note that the coefficient of variation is not assumed constant. By allowing the variance of the normals used to approximate the S phase to vary, greater accuracy may be obtained.

Let f be the DNA fluorescence density function. Let

$$f_i(x|\mu_i,\sigma_i) = \frac{1}{\sqrt{2\pi}\sigma_i} e^{-1/2\left(\frac{x-\mu_i}{\sigma_i}\right)^2} \qquad i = 1,2,\ldots,m$$

where

f_1 = fluorescence density function of G_1 cells with $\mu_1 = \mu$

f_i = fluorescence density function of S cells with $\mu_i = (1+\frac{i-1}{m-1})\mu$

$i = 2,\ldots,m-2$

f_m = fluorescence density function of G_2 + M cells with $\mu_m = 2\mu$.

No assumptions are made about the σ_i - they are arbitrary. The probability that a cell belonging to population i will be recorded in channel x will be approximated by $f_i(x)[(x+\frac{1}{2}) - (x-\frac{1}{2})] = f_i(x)$. Assume that

(32) $$f = p(1)f_1 + p(2)f_2 + \ldots + p(m-1)f_{m-1} + p(m)f(m)$$

where p(i) is the probability of a cell belonging to population i.

The experimental histogram gives a random sample $x_1,x_2,\ldots,x_n$ of the channel numbers in which cells 1,2,...,n, respectively, are registered. The problem is to estimate the p(i), μ and $\sigma(i)$.

Assume that initial estimates for

$$p(j) = P(\text{cell belongs to j-th population}) \qquad j = 1,2,\ldots,m$$

μ_i and σ_i are known. Methods for estimating these quantities will be given at the end of this section. Then $p(i|j)$ = probability that a cell belongs to population i given that it was recorded in channel j can be found. By Bayes' theorem (Eisen and Eisen [10])

(33) $$p(i|j) = \frac{f_i(j|\mu_i,\sigma_i)p(i)}{\sum_{i=1}^{m} f_i(j|\mu_i,\sigma_i)p(i)} .$$

The likelihood function for the given sample $x_1,\ldots,x_n$ is

$$\ell n\left(\prod_{i=1}^{n} f(x_i)\right)$$

where f is defined by (1) [40]. Setting the derivatives of the likelihood function with respect to μ, σ_i^2 $(i = 1,2,\ldots,m)$ and $p(i)$ $(i = 1,2,\ldots,m-1)$ equal to zero and solving the resulting equations yields the following maximum likelihood estimators

$$(34) \qquad \mu = \frac{\sum_{i=1}^{n} p(1|x_i)x_i}{\sum_{i=1}^{n} p(1|x_i)} = \frac{\sum p(1|j)j(n_j/n)}{\sum p(1|j)(n_j/n)}$$

$$(35) \qquad \sigma_i^2 = \frac{\sum_{j=1}^{n} p(i|x_j)(x_j-\mu_i)^2}{\sum_{j=1}^{n} p(i|x_j)} = \frac{\sum p(i|j)(j-\mu_i)^2(n_j/n)}{\sum p(i|j)(n_j/n)}$$

$$(36) \qquad p(i) = \frac{1}{n}\left(\sum_{j=1}^{n} p(i|x_j)\right) = \sum p(i|j)(n_j/n)$$

where n_j is the number of cells recorded in channel j. The summations in the right sides of equations (34) - (36) range from the number of the first channel having a nonzero number of cells to that of the last channel having a nonzero number of cells.

To apply this result assume that initial estimates for p(i), μ_i and σ_i are known. Then $p(i|j)$ can be calculated from (33). These values are then substituted in the right sides of equations (34) - (36) to obtain better estimates for the μ, σ_i^2, and p(i). If another set of data were available these calculated values could be used as the initial estimates and the whole procedure repeated to obtain still better estimates of the parameters. However, there is no need to use a new set of data. By the law of large numbers the ratios n_j/n will be approximately the same. Therefore the above procedure can be repeated with the same set of data until the estimates converge to the desired degree of accuracy.

To actually apply the procedure the S component is first approximated by two subpopulations ($m = 4$). The calculated parameters are then used to obtain a theoretical histogram by using (33). The theoretical and experimental histograms are compared by means of some goodness of fit test (such as, chi square). If the desired degree of accuracy is obtained stop; otherwise repeat the procedure for $m = 5,6\ldots$ and so on.

The initial estimates can be obtained by examining the DNA histogram. However, assuming complete ignorance the procedure can be started with the following values obtained by using the method of moments:

$$p(i) = \frac{1}{m} \qquad i = 1,2,\ldots,m$$

$$\mu_i = \mu(1+ \frac{i-1}{m-1}) \qquad i = 1,2,\ldots,m$$

where

$$\sum j(n_j/n) = \frac{\mu}{m} [1+(1+ \frac{1}{m-1}) + \ldots + (1 + \frac{m-1}{m-1})]$$

and so

$$\mu = \frac{2}{3} \sum j(n_j/n).$$

The standard deviations are

$$\sigma_i = \sigma_1 \ (1 + \frac{i-1}{m-1}) \qquad i = 1,2,\ldots,m$$

where

$$\sum j^2(n_j/n) = \frac{1}{m} [\sigma_1^2 + \mu_1^2 + \ldots + \sigma_m^2 + \mu_m^2]$$

and so

$$\sigma_i^2 = \frac{6(m-1)}{14m-13} \sum j^2(n_j/n) - \mu_1^2.$$

Dr. Macri has written a program based on the above method which was applied in [12] to analyze an experimental DNA histogram. The given example also illustrates illustrates that the decomposition of DNA histograms is not unique!

6.10 Analysis of Synchronous Populations from Multiple Histograms

In this section some methods of analyzing time sequences of DNA histograms will be described. These techniques are based on the fact that in synchronous populations, the age distribution of cells changes with time. Since the DNA content of a cell is determined by its maturity, the DNA histogram will also vary with time. In particular, if the number of cells whose DNA content lies in a specific range is plotted as a function of time, the graph will have a wave-like appearance. It will be seen that such waves contain kinetic information about the cells.

A. Graphical Method

To illustrate the above ideas consider the following simplified model of the cell cycle. The cell cycle is divided into 10 compartments. The G_1, S, G_2 and M phases contain 3,4, 2 and 1 compartments, respectively. A cell begins its life in G_1 in compartment 1. After 1 time unit it passes to the next compartment and so on. Finally, the cell reaches compartment 10 or the M phase. The cell then divides into 2 daughter cells in a unit of time and both of these daughter cells enter compartment 1. Suppose that at time zero, compartments 4,5 and 6 (the first 3 compartments of the S phase) contain 2,000, 3000 and 1000 cells, respectively. Fig. 6.9 is the graph of the number of cells in the first and last compartments of the S phase plotted as a function of time.

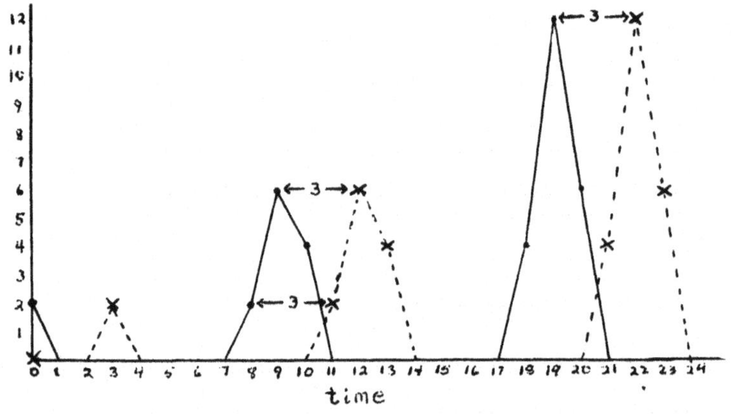

x... no. of cells in the last S phase compartment.

•— no. of cells in the first S phase compartment.

Figure 6.9 npi curves.

Each curve in Fig. 6.9 can be interpreted as the number of cells of the population with DNA content i[3] plotted as a function of time (or npi curves) because of the correspondence between maturity and DNA content. The use of such curves for finding the mean length of the phases of the cell cycle was proposed by Scherr and Zietz [33]. Actually, they considered the fraction of cells of the population with DNA content I or FPI curves. However, the above method was used for finding the mean cycle time in an earlier paper by Bronk et al [4].

The cell cycle time, T_c, is the time between two corresponding points (e.g., peaks) of two successive waves (see Fig. 6.9) of a single npi curve.

Note that two different curves corresponding to different DNA contents have the same shape. However, corresponding points on the curves are separated by a time Δt which measures how long it takes the cell to travel between the two compartments. If the compartments are chosen near the beginning and end of the S phase, Δt approximates the mean traverse time of the cells through S. In the above example, $\Delta t = 3$ (see Fig. 6.9). Since 1 unit of time is required for a cell to traverse the first compartment of the S phase, the length of the S phase is four.

The oscillatory behavior of our simple model is also exhibited by NPI curves obtained from an experimentally synchonized population of cells. Such experimental NPI curves are determined from a series of DNA histograms recorded at times t_j $(j = 1,2,\ldots)$. The histogram corresponding to $t = t_j$ yields the number of cells, $N(t_j)$, with DNA content lying in an interval I. The plot of N versus t is the NPI curve. The actual appearance of or typical NPI curve is shown in Fig. 6.10. It has been observed that the later peaks decay exponentially. There is a loss of synchrony with an eventual return to exponential growth (cf. Chapter 5, Section 5). The same behavior is observed if the number of cells whose DNA content lies in a certain interval is plotted as a function of time.

---

[3] Actually, because of the discrete number of compartments, the DNA content lies in a certain range.

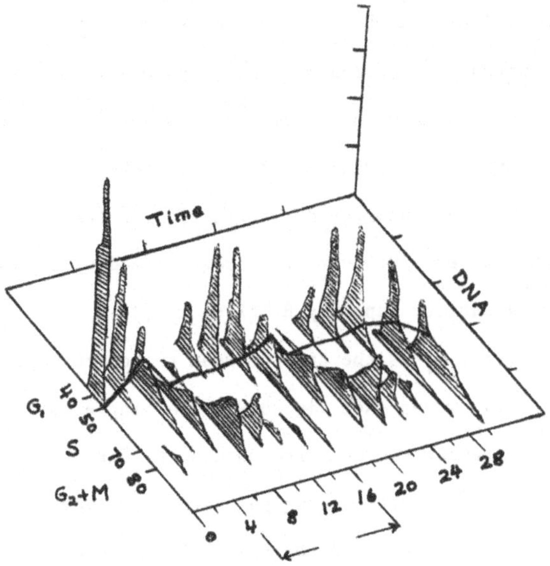

Figure 6.10 Three dimensional DNA histograms and NPI curve.

To explain the above behavior a more refined model of the cell cycle is required. The simple Takahashi model of the cell cycle will be used (See Chapter 3, Section 2). Let $n_i(t)$ be the number of cells in compartment i $(i = 1,2,\ldots,k)$ at time t. Then $n_i(t)$ satisfy (5) and (6) of Chapter 3. Assume that initially there is one cell in a compartment, say s, of the S phase. This compartment corresponds to a fixed range of DNA content. Hence, the initial conditions are

$$n_i(0) = 0 \qquad (i \neq s); \quad n_s(0) = 1. \tag{37}$$

The Laplace transform of $n_i(t)$ is

$$\bar{n}_i = \int_0^\infty e^{-pt}\, n_i(t)\, dt.$$

Taking Laplace transforms of the defining equations (cf. (5) and (6) of Chapter 3) yields

$$\begin{aligned} &(\lambda+p)\bar{n}_1 - 2\lambda\bar{n}_k = 0 \\ &(\lambda+p)\bar{n}_i - \lambda\bar{n}_{i-1} = 0 \qquad i = 2,\ldots,s-1,\ s+1,\ldots,k \\ &(\lambda+p)\bar{n}_s - \lambda\bar{n}_{s-1} = 1 \end{aligned} \tag{38}$$

from which

$$\bar{n}_s = \frac{(\lambda+p)^{k-1}}{(\lambda+p)^k - 2\lambda^k} .$$

The roots of the characteristic equation

$$(\lambda+p)^k - 2\lambda^k = 0$$

are

$$p_n = 2^{1/k}\lambda e^{\frac{2\pi i n}{k}} - \lambda \qquad n = 0,1,2,\ldots,k-1 .$$

Hence, taking inverse transforms of $\bar{n}_s$ gives the solution

$$(39) \qquad n_s(t) = \sum_{m=0}^{k-1} \frac{(\lambda+p_m)^{k-1}}{k(\lambda+p_m)^{k-1}} e^{p_m t} = \frac{1}{k} e^{-\lambda t} \sum_{m=0}^{k-1} \exp(2^{1/k}\lambda e^{2\pi i m/k})t$$

It is evident from (39) that the number of cells with a given range of DNA content corresponding to compartment S will oscillate.

Further insight into the behavior of $n_s(t)$ can be obtained by expanding the terms in (31) which involve k in a power series in $\frac{1}{k}$. The number of compartments is assumed to be so large that term of order $\frac{1}{k^3}$ or higher are negligible. It is also assumed that t is sufficiently large so that the sum in (39) is approximated by the terms for which m = 0,1 and k-1. These terms have the largest real part as can be seen from the Argand diagram of $e^{2\pi i m/k}$, noting that $e^{2\pi i(k-1)/k} = e^{-2\pi i/k}$ has the same real part as $e^{2\pi i/k}$. Hence, when combined with the term $e^{-\lambda t}$ they will produce the largest negative powers of the exponential and so dominate the sum in (39). Thus, (39) is approximated by

$$(40) \qquad n_s(t) = \frac{1}{k} e^{-\lambda t}[\exp(2^{1/k}\lambda t) + \exp(2^{1/k}\lambda e^{\frac{2\pi i}{k}} t) + \exp(2^{1/k}\lambda e^{-2\pi i/k}t)].$$

Expanding $2^{1/k}$ in powers of 1/k yields $2^{1/k} = e^{\ln 2/k} \approx 1 + \frac{\ln 2}{k} + \frac{1}{2}\cdot\frac{(\ln 2)^2}{k^2}$.

Multiply this last series by the series expansions for $e^{2\pi i/k}$ and $e^{-2\pi i/k}$ and calculat all terms of degree two or less in $\frac{1}{k}$. The resulting series are then substituted in (40). After some simplification (40) reduces to

$$(41) \qquad n_s(t) \approx \frac{\sigma^2}{\mu^2} e^{\gamma t}[1+e^{-\beta t/\mu}(e^{2\pi i t/T_o}+e^{-2\pi i t/T_o})]$$

where

(42) $\gamma = \ln 2/T_D \qquad T_D = \mu(1+\frac{1}{2}(\ln 2)\frac{\sigma^2}{\mu^2})^{-1}$

(43) $T_o = \mu(1+\frac{\sigma^2}{\mu^2}\ln 2)^{-1}$

(44) $\beta = 2\pi^2\sigma^2/\mu^2.$

Here $\mu = k/\lambda$ and $\sigma^2 = k/\lambda^2$ are the mean and variance of the cycle time which has a gamma distribution (see page 185).

Equation (41) shows that the number of cells in compartment S has the usual asymptotic exponential growth. However, there is an additional term which multiplies the growth term by an exponentially damped oscillating function. From (43), the period of this oscillating functions is slightly less than the generation time. This general behavior was obtained by Hirsch and Engelberg [19]. The above results were also derived by Bronk et al [4] using a renewal equation to describe cell growth.

The number of cells in compartment s+r at time t can also be found by solving (38) for

$$\bar{n}_{s+r} = \frac{\lambda^r(\lambda+p)^{k-1-r}}{(\lambda+p)^k-2\lambda^k}$$

and finding the inverse Laplace transform. This process yields

(45) $n_{s+r}(t) = \frac{\sigma^2}{\mu^2} e^{\gamma(t-T_D r/k)}[1+e^{-\beta t/\mu}(e^{2\pi i(t)T_o-r/k)}+e^{-2\pi i(t/T_o-r/k)}]$

Note that the number of cells in compartment s+r have the same period of oscillation as the number of cells in compartment s. However, there is a time lag of $(r/k)T_o$.

Similar results hold for any initial distribution in which there may be more than one cell in each compartment. The number of cells in a given compartment s oscillate with period T_o and the number cells in compartment s+r exhibit a time lag of $(r/k)T_o$. These results are a consequence of the fact that the general solution is a linear combination of particular solutions in which there is a single cell in one compartment while the other compartments are empty.

The hypothesis formulated from the simple model introduced at the beginning of the section are valid for the Takahashi-Kendall model for sufficiently large t.

If $n_s(t)$ is plotted as a function of t (cf. Fig. 6.9) the distance between peaks is T_o defined in (43). Therefore the distance between two successive peaks can be used to estimate the mean cycle time μ.

Next suppose that the subscript s denotes the first compartment of the S phase which is decomposed into a total of c compartments. If $n_{s+r}(t)$ where $r = c - 1$ is plotted as a function of t on the same graph as $n_s(t)$ (cf. Fig. 6.9) then the distance between two successive peaks of n_s and n_{s+r+1} is $(c/k)T_o$. This result follows from (41) and (45). However, the sojourn time in the S phase is also a gamma distribution with parameters λ and c. Let the mean sojourn time in S be denoted by μ_s; then $\mu_s = c/\lambda$. It follows that the mean time required to travel from the first compartment of S to the first compartment of G_2 is

$$(46) \qquad \frac{c}{k} T_o = \frac{c/\lambda}{k/\lambda} T_o = \frac{\mu_s}{1+\sigma^2/\mu^2 \; \ln 2} .$$

Therefore the distance between two successive peaks of the number of cells in the first compartment of S and the number of cells in the first compartment of G_2 can be used to estimate the means sojourn time in S.

The Takahashi-Kendall model will now be related to experimentally determined DNA fluorescent histograms. Each compartment in the S phase is assigned a mean DNA content which is a nondecreasing function of the compartment number. Compartments of the G_1 phase have DNA content d. The first compartment of the S phase has DNA content d while the last compartment of the S phase has DNA content 2d. The compartments of the G_2+M phases also have DNA content 2d.

For the moment, dispersion because of variability in DNA staining and machine detection are ignored. The effects of these factors will be considered later. Hence, cells are recorded in a given channel if their DNA content lies in a certain range. Consequently, an interval of the channel axis corresponds to a DNA interval which in turn corresponds to a certain number of compartments. Therefore, according to our model the number of cells recorded in an interval of the channel axis in the S phase is just the sum of the number of cells in the corresponding compartments of the S phase.

Suppose that the number of cells in a given interval of the channel axis is recorded as a function of time. The time variation of this function is just the time variation of the number of cells in the corresponding compartments of the Takahaski-Kendall model. However, the sum of functions with period T_0 still has period T_0. Thus, after a sufficient elapse of time, the distance between two successive peaks of NPI curves obtained from a series of DNA fluorescent histograms can be used to estimate μ.

The last statement remains true even when there is variability in DNA staining and recording of cells. These variations can be accounted for by postulating the existence of a time independent probability density. This density, $p(x|y)$, gives the probability of a cell being recorded in channel x given that the cell would be recorded in channel y if there were no variability in staining or recording. This density induces a similar density for compartments. By applying the total probability theorem, it is easily seen, from the form of (45), that the only effect will be a change of phase in the expression for the number of cells in each compartment. The period, T_0, is unchanged. Thus, the distance between successive peaks from experimental NPI curves can still be used to estimate the mean cycle time.

The graphical method of finding the mean sojourn time in the S phase is not directly applicable to NPI curves obtained from DNA fluorescent histograms. The channel corresponding to the beginning and end of the S phase must be determined. The channel analyzer must be adjusted so that an interval I of the channel axis at the beginning of the S-phase corresponds to the first compartment of the S-phase in the mathematical model. This correspondence can be established by relating channels to DNA content and so to compartments (since DNA content is an increasing function of the compartment number). Then the number of cells in the channel interval equals the number of cells in the first compartment of the Takahashi-Kendall model.

Unfortunately the rate of DNA synthesis may not be the same at the beginning and end of the S phase. Thus the mean DNA content of two compartment at the beginning and end of the S phase may be different. However, in mammalian cells the rate of DNA synthesis usually vary at most by a factor of two. Hence, between $\frac{1}{2}$ and 2 compartments may correpsond to the interval I when it is at the end of the S phase. The period of the number of cells in I, when plotted as a function of time, is the same as the period of the number of cells in the last compartment of the S phase. However, there may be a slight difference in phase. This difference decreases as the number of compartments in the S phase increases. Hence, after a sufficient time lapse, the distance between two successive peaks of the number of cells in an interval of the channel axis at the beginning and end of the S phase can be used to estimate the mean sojourn time in S.

The rate of DNA synthesis is not accurately known for most mammalian cells and may change after treatment with drugs. Thus, the actual number of compartments associated with an interval I of the channel axis lies between half and double the assigned number of compartments. Nevertheless, the period of oscillation of the number of cells in I remains the same irregardless of the number of compartments. There is only a slight change in the phase. Therefore, the graphical method described in the preceding paragraph still gives an approximation for the mean sojourn time in S. Research is required to estimate the errors in using these graphical methods.

The mean sojourn time in the G_1 phase and the combined G_2+M phase cannot be accurately determined by analogous graphical methods. A major problem is that early and late stages of the G_1 and G_2+M phase cannot be accurately distinguished from the DNA fluorescence histogram. Further it is not obvious how to relate compartments of the mathematical model to intervals of the channel axis of the G_1 and G_2+M phases. Therefore other methods will be described below to find the mean sojourn time in these phases.

B. Flow of Cells Across Boundaries in the Cell Cycle.

The flow rate of cells across certain physiological boundaries in the cell cycle is useful in studying the action of drugs. For example, suppose a drug blocks cells at the G_1-S interface. It would be desirable to know the fraction of cells released from the block at any time. Another important application is that the density functions of the length of the phases of the cell cycle can be expressed in terms of certain flow rates.

The number of cells flowing from G_1 to S, S to G_2+M and G_2+M to G_1 will be denoted by f_{12}, f_{23} and f_{31}, respectively. These flows will be calculated from the generalized Takahashi-Kendall equations (3) and (4) of Chapter 3. It is assumed that $\mu_i = 0$ for all i. Let N_1, N_2, N_3 be the total number of cells in G_1, S and G_2+M, respectively and

$$N = N_1 + N_2 + N_3$$

be the total number of cells. Suppose i and j are the numbers associated with the last compartments in the G_1 and S phases, respectively. Then, by adding the equations corresponding to all the compartments in each of the 3 phases, the following equations are obtained:

(47) $$\dot{N}_1 = \beta\lambda_k n_k - \lambda_i n_i = \beta f_{31} - f_{12}$$

(48) $$\dot{N}_2 = \lambda_i n_i - \lambda_j n_i = f_{12} - f_{23}$$

(49) $$\dot{N}_3 = \lambda_j n_j - \lambda_k n_k = f_{23} - f_{31}$$

Note that (47) - (49) can be derived by using a 3 compartment model representing the 3 phases as described in Zietz [41].

The number of cells flowing between the phases can be found by solving (47) - (49). However, analysis of the DNA fluorescent histogram usually yields the probability or fraction of cells in each state

$$p_i = N_i/N.$$

Substituting the values of $\dot{N}_i$ from (47) - (49) into

$$\dot{p}_i = \frac{\dot{N}_i}{N} - p_i \frac{\dot{N}}{N} \qquad (i = 1,2,3)$$

yields

(50) $$-\left(\frac{f_{12}}{N}\right) + \beta\left(\frac{f_{31}}{N}\right) = \dot{p}_1 + p_1 \frac{\dot{N}}{N}$$

(51) $$\left(\frac{f_{12}}{N}\right) - \left(\frac{f_{23}}{N}\right) = \dot{p}_2 + p_2 \frac{\dot{N}}{N}$$

(52) $$-\left(\frac{f_{31}}{N}\right) + \left(\frac{f_{23}}{N}\right) = \dot{p}_3 + p_3 \frac{\dot{N}}{N} \quad .$$

Solving equations (50) - (52) for $\frac{f_{12}}{N}$ yields

(53) $$\frac{f_{12}}{N} = -\dot{p}_1 + [\frac{\beta}{\beta-1} - p_1](\frac{\dot{N}}{N})$$

For the steady state (53) reduces to

(54) $$(\frac{f_{12}}{N})_{SS} = (\frac{\beta}{\beta-1} - p_1)(\frac{\dot{N}}{N})_{SS} .$$

Equation (54) is a generalization of a result of Dean and Anderson [5] who assumed that all cells move synchronously through the cycle, $(\frac{\dot{N}}{N})_{SS} = \gamma$ the growth rate constant and $\beta = 2$. In this situation all cells flow from G_1 to S at time T_1, the sojourn time in G_1. Thus, $N = N(T_1)$ is the total number of cells when $t = T_1$.

Let f_{G_1}, f_S, f_{G_2+M} be the density functions for the sojourn time in phases G_1, S and G_2+M, respectively. Cells flow out of a compartment at time t if and only if the cells entered the compartment at time t-x and spent time x in the compartment. Translating this conservation of flow into mathematical terms via the total probability theorem gives the following equations:

(55) $$f_{12}(t) = \int_0^t f_{31}(t-x)f_{G_1}(x)dx$$

(56) $$f_{23}(t) = \int_0^t f_{12}(t-x)f_S(x)dx$$

(57) $$f_{31}(t) = \beta\int_0^t f_{23}(t-x)f_{G_2+M}(x)dx.$$

The flows can be determined from (50) - (52). Their value depends only on quantities which can be obtained from FMF analysis of a series of histograms and $\dot{N}/N$. Therefore, the density functions for the mean sojourn time in each phase can be determined from (55) - (57). However, a great deal of numerical analysis is involved. A simpler method of estimating the moments of these densities appears in part C.

C. Calculation of the Mean Transit-time Through a Compartment Via the Fraction of Cells in the Compartment [3].

The mean sojourn time through the phases of the cell cycle can be calculated from the fraction of cells in these phases. It is assumed that cells do not leave the cell cycle but after division enter G_1 at twice the original number.

The cell cycle is divided into k compartments. Let

$$h_{ij}(t) = \begin{cases} 1 & \text{if cell j is in compartment i at time t and has not divided} \\ 0 & \text{otherwise} \end{cases}$$

The total time spent by cell j in compartment i is the random variables

$$T_{ij} = \int_0^\infty h_{ij}(t)dt^4.$$

Let N(t) denote the total number of cells in the population at time t. An estimator for the mean sojourn time in compartment i is

$$(58) \qquad T_i = \frac{1}{N(0)} \sum_{j=1}^{N(0)} T_{ij} = \frac{1}{N(0)} \sum_{j=1}^{N(0)} \int_0^\infty h_{ij}(t)dt = \int_0^\infty h_i(t)dt$$

where

$$(59) \qquad h_i(t) = \frac{1}{N(0)} \sum_{j=1}^{N(0)} h_{ij}(t) = \frac{\bar{n}_i(t)}{N(0)} \quad .$$

Here $\bar{n}_i(t)$ is the number of cells in compartment i at time t which have not yet completed the first cell cycle. The quantity h_i is not known, since all cells are registered by the flow cytometer and not only cells in their first cycle. It is convenient to introduce another function used to express the h_i. Let

$$(60) \qquad g_i(t) = \frac{n_i(t)}{N(0)}$$

where n_i is the number of cells in compartment i. The fraction of cells in compartment i at time t is

$$(61) \qquad p_i(t) = \frac{n_i(t)}{N(t)} = \frac{N(0)}{N(t)} g_i(t).$$

This is a quantity which can be calculated from an analyses of DNA fluorescent histograms. The following considerations show how the h_i can be derived from the p_i. However, some simplifying assumptions are necessary.

After some cells have divided the distribution of cells in the compartment is assumed to be of the following form. For some integer r, compartments 1,2,...,r only contain the divided cells while the remaining compartments contain only cells

---

4 This is an integral of a random variable.

which have not divided. Further, it is assumed that any cell which has divided does not enter compartment r+1 until the number of cells in that compartment decreases to a minimum. Let t_{r+1} be the time at which this minimum occurs and let $t_o = 0$.

When the cycle time is constant the above assumptions are fulfilled. Otherwise the results deduced below are only approximations which improve the closer the variance of the cycle time is to zero.

For any arbitrary time t, either $t > t_k$ or there exists a integer r $(0 \leq r \leq k_{r-1})$ such that

$$t_r \leq t \leq t_{r+1} \, .$$

It follows from the definition of t_r and r that

$$N(t) = N(0) + \frac{1}{2} \sum_{m=1}^{r} n_m(t) \, . \tag{62}$$

From (60) and (62) it follows that

$$g_i(t) = p_i(t) + \frac{1}{2} \sum_{m=1}^{r} p_i(t) g_m(t) \qquad (i = r+1,\ldots,k). \tag{63}$$

If the g_m $(1 \leq m \leq r)$ were known, then g_i $(i = r+1,\ldots,k)$ could be calculated from (63). However, the following equation can be deduced from (63).

$$g_i(t)(2-p_i(t) - \sum_{\substack{m=1 \\ m \neq i}}^{k} p_i(t) g_m(t) = 2p_i(t) \tag{64}$$

with $i = 1,\ldots,r$. This set of r linear equations can be solved for the r unknowns g_i. The remaining g_i $(r < i \leq k)$ can be calculated from (63).

The unknowns

$$h_i(t) = \begin{cases} 0 & i \leq r \\ g_i(t) & r \leq i \leq k \end{cases} \tag{65}$$

since cells in compartments corresponding to $i \leq r$ have already completed their first cycle while those in compartments corresponding to $r < i \leq k$ are still in their first cycle. Note that $h_i(t) = 0$ for $t > t_k$. Inserting the h_i into (56) yields the mean transit time T_i for cells in compartment i. This procedure can be modified to yield the variance of the transit times.

The above method will be illustrated by deriving formulas for the mean duration of the cell cycle phases. The cell cycle is divided into compartments 1,2,3 representing the G_1, S and G_2+M phases, respectively. Three situations must be considered depending on the time t.

(a) no cells have divided ($0 \leq t \leq t_1$)

It follows from the definitions that

$$h_i = g_i = p_i \qquad (i = 1,2,3)$$

(b) only compartment 1 contains divided cells ($t_1 \leq t \leq t_2$)

Equations (63) become

$$g_2 = p_2 + \frac{1}{2} p_2 g_1$$

$$g_3 = p_3 + \frac{1}{2} p_3 g_1 .$$

These equations determine g_2 and g_3 once the unknown function g_1 is obtained from (64) - namely,

$$g_1(2-p_1) = 2p_1 .$$

Hence, from (65),

$$h_1 = 0$$

$$h_2 = g_2 = p_2 + \frac{p_1 p_2}{2-p_1}$$

$$h_3 = g_3 = p_3 + \frac{p_1 p_3}{2-p_1}$$

(c) only compartments 1 and 2 contained divided cells ($t_2 \leq t \leq t_3$). Here (63) reduces to

$$g_3 = p_3 + \frac{1}{2} p_3 g_1 + \frac{1}{2} p_3 g_2 .$$

The function g_1 and g_2 are obtained from (64) which gives

$$g_1(2-p_1)-p_1 g_2 = 2p_1$$

$$g_2(2-p_2)-p_2 g_1 = 2p_2 .$$

Hence,

$$h_3 = g_3 = p_3 + \frac{p_1 p_3}{2-p_1-p_2} + \frac{p_2 p_3}{2-p_1-p_2}$$

while $h_1 = h_2 = 0$. From (58)

$$T_1 = \int_0^\infty h_1(t)dt = \int_0^{t_1} p_1(t)dt$$

$$T_2 = \int_0^\infty h_2(t)dt = \int_0^{t_2} p_2(t)dt + \int_{t_1}^{t_2} \frac{p_1(t)p_2(t)}{2-p_1(t)} dt$$

$$T_3 = \int_0^{t_3} p_3(t)dt + \int_{t_1}^{t_2} [\frac{p_1(t)p_2(t)}{2-p_1(t)} + \frac{p_1(t)p_3(t)}{2-p_1(t)-p_2(t)}]dt + \int_{t_2}^{t_3} \frac{p_2(t)p_3(t)}{2-p_1(t)p_2(t)} dt.$$

As another example consider two compartments, the first representing interphase and the second mitosis. The mean time for mitosis, $T_M = T_2$.

6.11 Rate of DNA Synthesis

Finding the rate of DNA synthesis by means of autoradiography is a very time consuming procedure. Flow-cytometry offers a simple and rapid method of approximating the rate of DNA synthesis.

(a) Asynchronous Populations

Let m(D,t) be the density of the number of cells with DNA content D at time t. In the steady state, let m(D) be the density of the number of cells having DNA content D. The DNA content is assumed to double over the S period and so $1 \le D \le 2$. The values m(1) and m(2) only include cells in S. In the S phase the function m is assumed to satisfy the Rubinow maturity equation (21) of Chapter 3 with $\alpha = D$. Using the steady state solution (cf. (70), Chapter 3) the maturity equation reduces to

$$\gamma m(D) = - \frac{d}{dD} (V(D)m(D)) \tag{66}$$

where γ is the exponential growth rate constant per unit time and

$$v(D) = \frac{dD}{dt}$$

is the rate of change of DNA content. Integrating (64) from 1 to D and solving for v(D) leads to

$$v(D) = \frac{v(1)m(1)-\gamma M(D)}{m(D)} \tag{67}$$

where

$$M(D) = \int_1^D m(x)dx$$

is the integral distribution function, the number of cells in S with DNA content at most D.

The flux of cells past D = 1, $v(1)n(1)$, must equal f_{12} the flux of cells into S from G_1. Substituting the value of f_{12} from (54) with $\beta = 2$ and $N(T_1) = N_1$ into (67) gives

$$v(D) = \frac{\gamma[N_1(2-p_1)-M(D)]}{m(D)} . \tag{68}$$

This approximation for the rate of DNA synthesis appears in Dean and Anderson [5].

Another approximation to the rate of DNA synthesis can be found by determining the relative DNA content from (18) when f_S and f_{G_1} are known. The functions f_S and f_{G_1} can be found by the Bayesian type analysis described in Section 6.9.

(b) Synchronous Populations

The interval on the channel axis corresponding to the S phase is divided into n equal subintervals, which define the subcompartments. It is assumed that $\frac{1}{n}$ of the DNA is replicated in each subcompartment. The method described in Section 6.10, part C, is applied to find the mean transit time T_i in compartment i (i = 1,2,...,n). The mean transit time, T_i, is inversely proportional to the rate of DNA synthesis R_i in compartment i. Hence,

$$R_i = \frac{1}{nT_i} . \tag{69}$$

6.12 Determination of Percentage of Cells in G_0 [9]

An FMF histogram for $G_0 \cup G$ cells is given. As a control, an FMF histogram of a population of G_1 cells is used. It is assumed that:

(a) the population of G_1 cells is nearly pure

(b) the density of a cell in the G_1 population being in channel x is

$$n(x) = \frac{1}{\sqrt{2\pi}\sigma} e^{-\frac{(x-\mu)^2}{\lambda\sigma^2}}$$

(μ and σ are unknowns and will be determined later by least squares.) The problem is to determine the percentage of cells in the G_0 and G_1 states from the first histogram using the control.

The following notation will be used:

M = total number of channels used

v_i = number of G_o cells in channel i. $v = \sum_{i=1}^{M} v_i$.

w_i = number of G_o G_1 cells in channel i. $w = \sum_{i=1}^{M} w_i$.

$g_i = \frac{v_i}{v}$ = the probability of a G_o cell being in channel i.

$t_i = \frac{w_i}{w}$ = the probability of a $G_o \cup G_1$ cell being in channel i.

p = probability of a cell being in G_o (Note that 1-p is the probability of a cell being in G_1).

$n_i = n(i)$, the approximate probability of a G_1 cell lying in channel i.

If all the above assumptions were correct and the date were exact then, by the total probability theory,

(70) $$pg_i + (1-p)n_i = t_i \qquad i = 1,2,\ldots,M.$$

Since there are variations in the data due to the experimental procedure and the fact that the assumptions are not exactly satisfied (1) will not hold exactly. Hence we try to satisfy these equations by the method of least squares. Thus we minimize

(71) $$Q = \sum_{i=1}^{m} A_i^2$$

where

$$A_i = pg_i + (1-p)n_i - t_1.$$

Theoretically, this is accomplished by solving the equations $\frac{\partial Q}{\partial p} = 0$, $\frac{\partial Q}{\partial \mu} = 0$ and $\frac{\partial Q}{\partial \sigma} = 0$. This procedure yields

(72) $$\sum_{i=1}^{M} A_i(g_i - n_i) = 0$$

(73) $$\sum_{i=1}^{M} A_i(i-p)\left(\frac{i-\mu}{\sigma^2}\right)n_i = 0$$

(74) $$\sum_{i=1}^{M} A_i(1-p)\left(-\frac{n_i}{\sigma} + \frac{(i-\mu)^2}{\sigma^3} n_i\right) = 0.$$

Equations (72) - (74) are nonlinear equations and are too complicated to solve analytically. They can be solved numerically.

Another method of decomposition is via the Bayesian approach, described in Section 6.9.

6.13 Generalization of the Degree of Synchrony

Various definitions of the degree of synchrony with respect to cell division appear in Chapter 5. Engelbergs degree of synchrony (see (13) in Chapter 5) can be generalized so that it is applicable to different stages of the cell cycle.

Assume that the cell cycle is divided into k compartments as described in §6.10 part C. The degree of synchrony with respect to compartment i can be calculated by the following artifice. The growth curve (n vs. t) of the population is constructed by counting every cell in compartment i as two cells while counting every other cell as one cell. This growth curve can then be converted into an r plot as previously and Engelberg's degree of proportional synchrony determined.

Instead of using the doubling time in the above index, the mean time that a cell needs after division to enter compartment i can be used. This progression-time τ_{ij} is defined as the time that cell j needs after division to enter compartment i. Using the notation introduced in §6.10, part C

$$\tau_{ij} = \sum_{k=1}^{i-1} T_{kj}.$$

The mean progression time can be calculated once the density function G_i of the progression time is found.

The density function G_i can be determined from h_i defined by (59). At any time t, cells with progression-time $\tau_i = t$ are entering compartment i while those with $\tau_i = t-T_i$ are having compartment i (this is an approximation). Hence, the probability of a cell which has not yet completed its first cycle being in compartment i at time t is approximately

$$h_i(t) = \int_{t-T_i}^{t} G_i(\tau)d\tau.$$

Upon differentiation the last equation becomes

$$G_i(t) = G_i(t-T_i) + \frac{dh_i}{dt}.$$

If $G_i(t)$ is known at some time (say, t = 0) then $G_i(t)$ can be calculated at arbitrary times from the above formula. Note that $G_i(t) = 0$ if $t < 0$.

A new time-interval index of synchrony, which can be used with confidence α, is defined as follows. For a fixed α, choose T_α so that

$$P(\tau_i > T_\alpha) = \int_{T_\alpha}^{\infty} G_i(t)dt = 1-\alpha .$$

Then the index of synchrony for the time interval $[t_0, t_0+T_\alpha]$ is

$$i_\alpha = \max_{t_0 \leq t \leq t_0+T_\alpha} p_i(t)$$

where p_i is defined by (61).

Another definition of synchrony is used in Beck [3]. His index is

$$S = \frac{1}{2\,\ell n\,2} \int_0^{t_r} \left|G_i(t) - \frac{\ell n\,2}{T_\alpha}\right| dt \quad (T_\alpha = \text{mean doubling time})$$

where $t_r \leq t$ is the time for the cells in compartment r to reach a minimum (see §6.10 C).

6.14 Discussion

Increasing the reliability and accuracy of FMF analyses requires improvements in cell preparation and staining and in mathematical techniques.

(a) Cell Preparation and Staining

One of the major difficulties in flow-system measurements is the requirement that each cell sample be reduced to a suspension of properly stained single cells (or nuclei) before analysis. The dispersal process is usually straightforward for cultured cells or cells grown singly. However, the dispersal of solid tissue into representative samples of simple cells is still a difficult problem. Selecting an unbiased sample of cells is not easy. If the cell dispersal procedure is too harsh some cells may be degraded and left with a fraction of their original DNA content; if too mild some cells may be clumped. The DNA distribution is skewed to lower fluorescence values if the dispersal is too harsh. Cell clumps increase the area of certain portions of the distribution. Work is currently underway in several laboratories on improving dispersal techniques for solid normal tissue and tumors. The perfection of these techniques will greatly expand the role of flow analyses in kinetic studies.

The application of cell-dispersal techniques to tissues usually results in a heterogeneous population of cells. At present normal cells can be separated from some tumor cells on the basis of DNA content when the tumor cells are heteroploid. The separation of cells with identical or similar DNA contents is more difficult. New separation methods based on cytochemical, morphological and immunological properties of cells will have to be developed. Some examples of such methods appears in Table 6.1.

TABLE 6.1 SOME CELL SEPARATION METHODS.

Cell Types	Separation Method	Investigators
leukocytes	cell volume	Van Dilla et al. [37]
leukocytes	two-angle light scatter	Salzman et al. [30]
leukocytes	3 fluorescent dyes	Shapiro et al. [32]
T-lymphocytes B-lymphocytes	immunofluorescent markers	Julius et al. [22]
T-lymphocytes B-lymphocytes	fluorescent enzyme substrates	Dolbeare and Phares [8]
transformed and nontransformed mouse $3T_3$	dye	Hawkes and Bartholomew [18]

Stains highly specific for DNA should be developed. Faint cytoplasmic staining in cells with a relatively large cytoplasmic mass can result in a displacement of the G_1 peak to higher fluorescent values. Thus, inaccurate results will be obtained from FMF analysis.

(b) Analytical Methods

Nearly all of the techniques used in FMF analysis are based on a physical assumption which intuitively is an approximation to reality. Research is required to see if these approximations can be justified mathematically. Estimates of the errors involved should also be investigated. The interpretation of DNA histograms is difficult when cell decycling and cell death occurs. More refined mathematical models should be developed.

In the interpretation of flow-cytometric data the phases are assumed to be independent and discrete. However, there is some controversy about the independence of the phases [36] and the kinetic and functional discreteness of the phases [31].

Fiting flow-cytometric data to a model requires the solution of nonlinear equations or some sort of optimization procedure. Unfortunately, there is no general technique for solving such problems. Hence, some effort should be concentrated on developing techniques for estimating parameters which are model independent. Quantitative estimates of goodness of fit should also be developed.

It seems unlikely that completely automatic and rigorous methods of decomposing DNA histograms that require no knowledge of the G_1-peak parameters, will be discovered. Therefore standards should be developed and analyzed simultaneously with the cell sample so that an estimate of the G_1-peak parameters can be found. Once these parameters are known the nonlinear estimation problem is reduced to a linear problem which can easily be solved.

The parameter estimates obtained by flow analysis should be compared with similar estimates obtained by more conventional techniques. Baisch et al. [1] have compared the lengths of the phases of the cell cycle, determined by flow cytometry, with estimates obtained by FLM analysis, with good agreement. The data presented in [17] for CHO cells analyzed by various kinetic techniques serves to further corroborate flow techniques. It remains to be seen whether the agreement between flow and classical analyses is maintained when tissues and solid tumors are investigated.

Simultaneously staining cells with several different stains has the added advantage of improving resolution and multidimensional sorting based on several measured cell parameters.[5] For example, by staining for proteins it may be possible to distinguish the early, middle and late parts of the G_1 phase or discriminate between several different cell types. Graphical techniques for visualization of multiparameter data have been developed in many computer centers.

[5] See the abstracts of Dean and Pinkel on Dual Laser Flow Cytometry and Evaluation of Eight Fluorochrome Combinations for Simultaneous DNA-Protein Flow Analysis by Stohr et al. in Flow Systems Newsletter published by Los Alamos Scientific Laboratory, Los Alamos, New Mexico, Nov. 1978.

A two parameter analysis of fluorescence and light scatter is contained in the paper by Jensen et al. [21]. Salzman et al. [30] have developed methods for the analysis of multiangle light scattering data.

Conditions for the uniquess of the decomposition of DNA histograms and the extracted kinetic parameters should be investigated. Recall that the same DNA histogram can be obtained for different phase lengths. If only an approximate solution is sought, for example, fitting data using a χ^2 criterion, then two decompositions may exist which differ in other parameters.

Cell dispersal and staining are discussed in a forthcoming book on flow cytometry and sorting [26]. This book also contains applications of flow-cytokinetic methods to cultured cell and clinic. There is also a chapter on quantitative cell analysis by Gray et al. which is essentially the material in [17].

REFERENCES

1. Baisch, H., Göhde, W. and Linden, W. Analysis of PCP-data to determine the fraction of cells in the various phases of cell cycle, Rad. Environm. Biophys. 12, 31-39, 1975.

2. Barlogie, B., Drewinko, B., Johnston, D.A. et al. Pulse cytophotometric analysis of synchronized cells in vitro, Cancer Res., 36, 1176, 1976.

3. Beck, H.P. A new analytical method for determining duration of phases, rate of DNA-synthesis and degree of synchronization from flow-cytometric data or synchronized cell populations, Cell and Tissue Kin., 1978.

4. Bronk, B.V., Dienes, G.J. and Paskin, A. The stochastic theory of cell proliferation, Biophys. J., 8, 1353-1397, 1968.

5. Dean, P.N. and Anderson, E.C. The rate of DNA synthesis during S phase by mammalian cells in vitro. In Pulse Cytophotometry (ed. Haanen, C.A.M., Hillen, H.F.P. and Wessels,J.M.C.) Eurp. Press Medikon, Ghent, Belgium, 1975.

6. Dean, P.N. and Jett, J.J. Mathematical analysis of DNA distributions derived from flow microfluorometry, J. Cell Biol., 60, 523-27, 1974.

7. Dean, P.N. and Jett, J.J. Flow microfluorometric data analysi . In Annual Report of the Biological and Medical Research Group (H-4) of the LASL Health Division, Los Alamos Scientific Laboratory Report LA-5227-PR, 1972.

8. Dolbeare, F.A. and Phares, W. Hydrolases in cells and nuclei measured by flow cytometry using fluorogenic substrates, J. Histochem. Cytochem. (submitted).

9. Eisen, M. Determination of percentage of cells in G_0 using an FMF, Division of Biophysics, Temple University, Internal Report, 6/76.

10. Eisen, M. and Eisen, C. Probability and Its Applications, Quantum Publishers, Inc., New York, N.Y., 1975.

11. Eisen, M. and Schiller, J. Microfluorometry analysis, J. Theor. Biol., 66, 799-809, 1977.

12. Eisen, M., Macri, N. and Mehta, J. Microfluorometry Analysis II. A Bayesian approach, to appear, J. Theor. Biol., 1979.

13. Fried, J., Method for the quantitative evaluation of data from flow microfluorometry, Comp. Biomed. Res. 9, 263-276, 1976.

14. Fried, J., Yataganas, X., Kitahara, T., Perez, A., Ferguson, R., Sullivan, S. and Clarkson, B., Quantitative analysis of flow microfluoremetric data from synchronous and drug-treated cell populations, Comp. Biom. Res. 9, 277-290, 1976.

15. Gray, J.W., Cell-cycle analysis of perturbed cell populations: computer simulation of sequential DNA distributions, Cell Tissue Kinet., 9, 499-516, 1976.

16. Gray, J.W., Carver, J.H., George, Y.S. and Mendelsohn, M.L., Rapid cell cycle analysis by measurement of the radioactivity per cell in a narrow window in S phase (RCS_i), Cell Tissue Kinet, 10, 97-109, 1977.

17. Gray, J.W., Dean, P.N. and Mendelsohn, M.L., Quantitative cell cycle analysis: flow cytometry and sorting. Lawrence Livermore Laboratory, preprint UCRL-79482, 1977.

18. Hawkes, S.P. and Bartholomew, J.C., Quantitative determination of transformed cells in a mixed population by simultaneous fluorescence analysis of cell surface and DNA in individual cells, Proc. Natl. Acad. Sci. (USA),

19. Hirsch, H.R. and Engelberg, J.J., Determination of the cell doubling-time distribution from culture growth-rate data, J. Theoret. Biol. 9, 297-302, 1965.

20. Jagers, P. and Norrby, K. Estimation of the mean and variance of cycle times in cinemicrographically recorded cell populations during balanced exponential growth, Cell Tissue Kinet., 7, 201-211, 1974.

21. Jensen, R.H., King, E.B. and Mayall, B.H. Cytological detection of cervical carcinoma with new cytochemical markers and flow microanalysis, Third International Symposium on Detection and Prevention of Cancer, UCRL-77469.

22. Julius, M.H., Masuda, T. and Herzenberg, L.A. Demonstration that antigen-binding cells are precursors of antibody-producing cells after purification with a fluorescence activated cell sorter, Proc. Natl. Acad. Sci. (USA), 69, 1934, 1972.

23. Kamentsky, L.A., Melamed, H.R. and Derman, H. Spectrophotometer: new instrument for ultrarapid cell analysis, Science, 150, 630, 1965.

24. Kim, M., Bahrami, K. and Woo, K.B. A discrete time model for cell age, size and DNA distributions of proliferating cells and its application to movement of labelled cohort, IEEE Trans. Biomed. Eng., BME-21, 387, 1974.

25. Kim, M. and Woo, K.B. Kinetic analysis of cell size and DNA content distributions during tumor cell proliferation: Ehrlich ascites tumor study. Cell Tissue Kinet., 8, 197, 1975.

26. Mendelsohn, M.L. (ed.). Flow Cytogenetics and Sorting, John Wiley and Sons, New York (to be published).

27. Mode, C.J. Multitype age-dependent branching processes and cell cycle analysis, Math. Biosciences, 10, 177-190, 1971.

28. Norrby, K. Population kinetics of normal, transforming and neoplastic cell lines Acta Path. Microbiol. Scand., 78, Supp., No. 214, 1970.

29. Norrby, K., Johannison, G. and Mellgren, J. Proliferation in an established cell line. An analysis of birth, death, and growth rates Exptl. Cell Res., 48, 582-594, 1967.

30. Salzmann, G.C., Crowell, J.M., Hansen, K.M. et al. Gynecologic specimen analysis by multiangle light scattering in a flow system, J. Histochem. Cytochem. 24, 308, 1976.

31. Shackney, S.E. A cytokinetic model for heterogeneous mammalian cell populations. II. Tritiated thymidine studies: The percent labeled mitosis (PLM) curve, J. Theoret. Biol. 44, 49, 1974.

32. Shapiro, H.M., Schildkraut, E.R., Curbelo, R., et al. Combined blood cell counting and classification with fluorochrome stains and flow instrumentation, J. Histochem. Cytochem. 24, 396, 1976.

33. Scherr, L. and Zietz, S. FP_i analysis of time sequences of DNA distribution yields length of G1, S and G2 phases of cell cycle, Rad. Res., 67, 585, 1976.

34. Showacre, J.L. In Methods of Cell Physiology, Chapter 7, Academic Press, New York, 147-157, 1968.

35. Sisken, J.E. Analysis of variations in intermitotic time. Cinemicrography in Cell Biology, Academic Press. New York, 1963.

36. Sisken, J.E. and Morasca, L. Intrapopulation kinetics of the mitotic cycle, J. Cell Biol., 25, 179, 1965.

37. Van Dilla, M.A., Fulwyler, M.J. and Boone, I.V. Volume distribution and separation of normal human leukocytes, Proc. Soc. Exp. Med., 125, 367, 1967.

38. Van Dilla, M.A., Trujillo, T.T., Mullaney, P.F., et al. Cell microfluorometry: a method for rapid fluorescence measurement, Science, 163, 1213, 1969.

39. Van Dilla, M.A., Steinmetz, L.L., David, D.T., Calvert, R.N. and Gray, J.W. High speed cell analysis and sorting with flow systems: biological applications and new approaches, IEEE Trans. Nucl. Sci., NS-21, 714, 1974.

40. Wani, J.K. Probability and statistical inference. Appleton-Century-Crofts, New York, 1971.

41. Zietz, S., Mathematical Modeling of Cellular Kinetics and Optimal Control Theory in the Service of Cancer Chemotherapy, PhD. Thesis, Dept. of Math., Univ. of California, Berkeley, California, 1977.

VII. CONTROL THEORY

7.1 Introduction

Control theory arose in two fields, biology and engineering. At the time when biologists were first beginning to formulate some of the concepts of control theory in a descriptive and qualitative manner, engineers already had made considerable progress in developing and applying quantitative methods. However, these two lines of development did not mingle until the 1940's. It now appears that control theory will be one of the major conceptual frameworks for biological research in the near future.

Many modern books on control theory appear to be advanced treatises on mathematics. A student may lose sight of the fact that the motivating factors in the development of control theory were practical problems. To emphasize the role played by practical problems and to indicate the intertwining of biology and engineering a brief history of control theory will be presented.

Some highlights of the early history of control theory are shown in

TABLE 7.1. EARLY HISTORY OF CONTROL THEORY

Approximate Date	Person	Device
300-0 B.C.	Ktesibios	Water clock based on float regulator
	Philon	Oil lamp with fuel oil level maintained by float regulator.
	Heron of Alexandria	Pneumatica, a book, described several forms of water level regulation using floats.
1600	Cornelis Drebbel	temperature regulator
1681	Dennis Papin	pressure regulator for steam boilers
1765	I. Polzunov	water float regulator for a boiler
1769	James Watt	fly ball govenor for controlling the speed of a steam engine.

As seen from Table 7.1, the early history of control theory was characterized by simple practical devices developed in an intuitive manner. Efforts to increase the accuracy of the control systems led to slower attenuations of the transient oscillations and even unstable systems. It then became necessary to develop a theory of automatic control.

In 1868 Maxwell [45] formulated a differential equation model of governors. A few years later Vyshnegradskii [66] proposed a mathematical theory of regulators. The major emphasis in engineering for the next 140 years was on practical problems of design of automatic regulating devices.

Possibly the first to recognize the prevalence of regulatory processes in biology was Claude Bernard. The following quotation from his book [9] published in 1878 reveals his use of modern concepts.

> "It is the fixity of the "milieu interieur" which is the condition of free and independent life,... all the vital mechanisms however varied they may be, have only one object, that of preserving constant the conditions in the internal environment."

Seven years later Fredericq [23] also emphasized that living organisms have control systems which tend to minimize deleterious disturbances imposed by the environment.

> "The living being is an agency of such sort that each disturbing influence induces by itself the calling forth of compensatory activity to neutralize or repair the disturbance. The higher in the scale of living beings, the more numerous, the more perfect, and the more complicated do these regulatory agencies become. They tend to free the organisn completely from the unfavorable influences and changes occurring in the environment."

These ideas were further extended by Cannon [18], who coined the word homeostasis, which he defined as the "coordinated physiological reactions which maintain most of the steady states of the body."

Meanwhile engineers, motivated by practical problems, slowly developed the theory of automatic regulating devices. In 1922, Minorsky's basic paper [49] on the theory of automatic steering of ships appeared. Another noteworthy paper was that of Hazen [31] on the theory of shaft-positioning types of servomechanisms. In the 1920's the development of radio and the subsequence rise of the electronics industry were the major factors for the further theoretical development of control theory in the U.S.A. The problems of stability of amplifiers and the design of the feedback amplifier led to the basic work of Nyquist [50]. In 1933, Lee and Wiener published their investigations of systems of differential equations descriptive of many physical systems. These results were applied by Bode [11] and his coworkers at the Bell Laboratories to the design of telephone systems and electronic feedback amplifiers. The frequency domain was used primarily to describe the operation of the feedback amplifiers.

Prior to World War II control theory developed in a different manner in the U.S.S.R. Many eminent Russian mathematicians were interested in the theoretical problems generated by practical problems in control. These mathematicians dominated the field of control theory. Hence, the Russian theory tended to utilize a time-domain formulation based on differential equations.

During World War II the theory and practice of automatic control received a tremendous impetus from the necessity of designing and constructing automatic pilots, radar tracking systems and automatic gun sighting and fire control systems. In general, before 1940, the design of control systems was an art involving trial and error. During the 1940's mathematical methods increased in number and control engineering became an established branch of engineering.

The paper by Rosenblueth et al. [55], appearing in 1943, was one of the first attempts to refine the work of Cannon. The book by Wiener [67] expressed the view that human beings were complicated control systems. He coined the term cybernetics.

Following and during World War II frequency-domain techniques based on the Laplace transform and complex frequency domain continued to dominate the theory of control. The advent of Sputnik and the space age imparted a new impetus to control engineering. Missiles and space probes required the design of highly complex and accurate control systems. The necessity to minimize the weight of satellites and control their trajectories very accurately gave birth to the important field of optimal control. Due to these requirements, the time-domain methods of Liapunov, Minorsky and others became increasingly popular. One of the major reasons for this surge in popularity was the use of high speed computers for control problems in the 1950's. Time-domain theorems could now be applied by utilizing numerical techniques which were not feasible by hand. At present both time-domain and frequency-domain approaches are utilized in the analysis and design of control systems.

The interdisciplinary exchanges between systems theory and biology have become increasingly apparent. The texts by Guyton et al. [29], Grodins [27], Yamamoto and Brobeck [69], Kalmus [37], and Bayliss [4] constitute significant contributions from biologists and physiologists to a growing borderland between biological and technological regulation. An engineering oriented approach to this territory are the texts by Milhorn [48] and Milsum [47]. Other relevant books are [13, 14, 17, 26, 33, 40, 54, 60, 63]. A perusal of the literature reveals that the physiologist is using engineering methods for the analysis of physiological control problems. On the other hand, the engineer and mathematician are investigating adaptive control problems which are similar to the control problems encountered in biology [5].

A more detailed history of some early regulators can be found in Mayr [44]. Aldolph [1] describes the origins of physiological regulation.

7.2 External Description of Systems (input output relations)

A. Systems and Block Diagrams

Webster's Dictionary defines a system as "a group of diverse units so combined by nature or art as to form an integral whole, and to function, operate, or move in unison, often in obedience to some form of command." Systems theory is the study of the behavior of such a group of units when subjected to certain conditions or inputs. Mathematics is the natural language of systems theory since it is concerned with abstract properties rather than the physical form of the contituent parts. Hence, the positioning of a missle launcher or the distribution of iodine in man can be studied by similar systems techniques.

A control system is a system whose purpose is regulating or controlling the flow of hormones, energy, information or other quantities in some desired way. Control theory is concerned with all questions related to the design and analysis of control systems.

To utilize mathematics in the design and analysis of systems, it is necessary to formulate a mathematical model of a system. An informal representation of a system will be presented in order to motivate the rigorous definition appearing at the end of this section.

The following notation will be used to describe systems:

t(time): a real number or a discrete variable

u(input variables or signals): controls chosen by the operator or mechanism controlling the system or perturbations in the system.

y(output variables or signals): describe either

(1) the behavior of the system and the outputs correspond to the observations that can be made on the system or

(2) what is produced by the system.

Note that both the input and output variables can be vectors.

The concept of a (dynamical) system is a (dynamic) transformation of inputs into outputs. The term dynamic (as opposed to static or memoryless) is used because past input values $(u(t), t_0 \leq s < t)$ as well as the present value $u(t)$ determine the output $y(t)$. In other words

$$y(t) = f(u_{[t_0,t]},t) \tag{1}$$

where f is a function that describes the output of the system at time t for an input function u whose past and present values are denoted by $u_{[t_0,t]} = \{u(s) | t_0 \leq s \leq t\}$. This direct description in terms of the input/output relation is called the external description of the system.

A system can be represented by a block diagram as shown in Fig. 7.1. For simplicity, only a single input and a single output have been shown. However, many systems have multiple inputs and outputs - that is, u and y can be vectors. In cybernetic terminology a block diagram is called a black box. This nomenclature is intended to emphasize the transformation of u into y rather than the corresponding mechanism. The omission of operational details may be deliberate or because of ignorance.

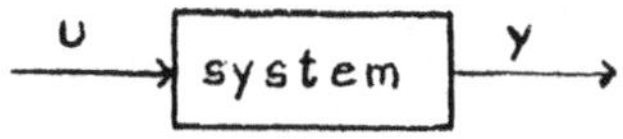

Figure 7.1 Block diagram of a system.

Some examples of systems appear in Table 7.2.

TABLE 7.2 EXAMPLE OF SYSTEMS

System	Input	Output
respiratory	CO_2	respiratory rate
telephone	microphone vibration	earphone vibration
boat	position of rudder	heading
heart	flow through coronary artery	cardiac output
automobile	1. accelerator 2. brake	speed

Frequently, complex systems are decomposed into subsystems by representing each subsystem by a block diagram. The blocks are joined by directed line segments which represent the flow of signals between the blocks. The output signals of some subsystems may be merged to produce an input for another subsystem. The device performing this service is called a junction. A junction with two inputs

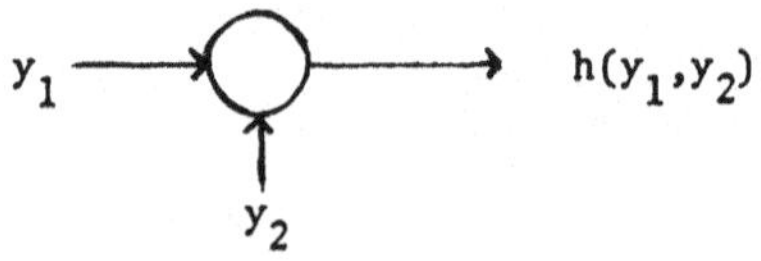

Figure 7.2 Two-input junction.

is diagrammed in Fig. 7.2. For example, if $h(y_1,y_2) = y_1-y_2$, the junction is called a differential.

Consider the system driver plus automobile. The objective of the driver is to keep the automobile in the center of a chosen lane on the road. The input to this system is the desired path of the car and the output is the actual path of the car. This system can be decomposed into subsystems as illustrated in Fig. 7.3. An error in the difference between the output and the input is detected by the

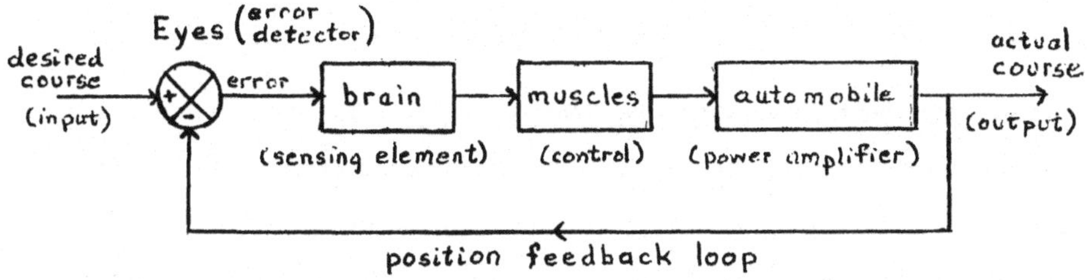

Figure 7.3 Automobile steering control system.

driver's eyes. This in turn activates the brain. The brain transmits signals to the driver's muscle which control the steering wheel. The automobile's steering mechanism provides power amplification for turning the front wheels. Of course, much more detail would be required to construct a mathematical model of this process.

A mathematical definition of a dynamical system from an external point of view can be formulated and a formal theory of systems can be developed. However, in this monograph an informal presentation will be given, since the purpose of this monograph is to give some applications of control theory.

B. Open and Closed Loops. Positive and Negative Feedback.

Suppose the water level, y, in a tank is to be held above zero when the outlet flow through the value v_o is varied. Fig. 7.4, shows how the tank level is controlled manually by an operator at a remote location. The operator cannot see the tank. He adjusts the inflow rate by valve v_i at irregular intervals. Precise determinations of the inflow rate, the outflow rate or the height of the water are not used in operating v_i.

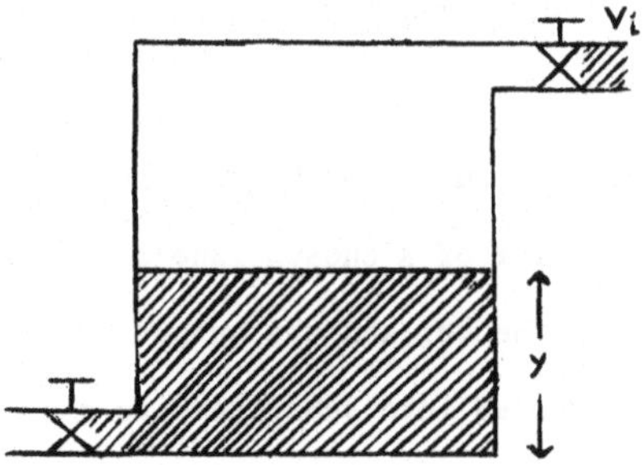

Figure 7.4 Tank level control system.

The example above is an <u>open-loop control system</u>. The control u (open or close v_i to alter the input flow of water) is determined at any time based on the goal for the system (y > 0) and all available prior knowledge about the system. The input is not influenced by the output, y, of the system at time t. A block diagram of an open-loop control system is shown below.

An open-loop control system.

If the behavior of an open-loop system is not completely understood or if unexpected disturbances act upon it, then there may be considerable, unpredictable variations in the output. Open-loop systems are not used when accuracy is required.

Fig. 7.5 illustrates a tank level control version of the system shown in Fig. 7.4. Even though the output flow rate through value v_o is varied, it can maintain the tank level y within specified limits. If the tank level is not at the specified height, an error voltage, e, is developed. This voltage is amplified and applied to a motordrive which adjusts value v_i thereby restoring the desired tank level.

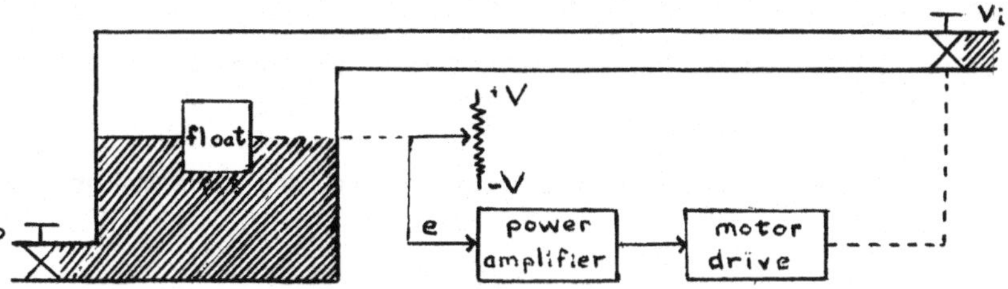

Figure 7.5 Automatic tank level control system.

The automatic tank level control system is an example of a closed-loop or feedback control system. Fig. 7.6 is the block diagram for a closed-loop system.

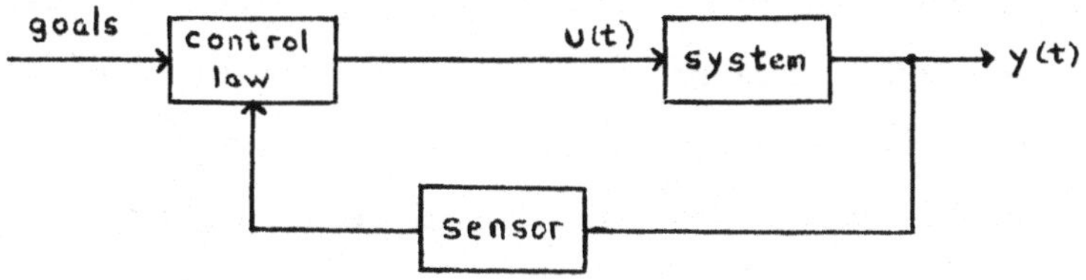

Figure 7.6 A closed-loop control system.

In the closed loop system, the control u = e is modified in some way by information about the behavior of the system output. A feedback system can often be designed to cope with lack of knowledge of the systems behavior, inaccurate components and unexpected disturbances. With all their many advantages, feedback systems have a very serious disadvantage. Closed-loop systems may inadvertently develop unstable oscillations.

In a feedback system the output y can influence the input u. More generally, let v_1 and v_2 be any two variables which can affect each other. The following notation will be used.

$v_1 \rightarrow v_2$ means that a primary increase (↑) or decrease (↓) in v_1 causes a secondary ↑ or ↓, respectively, in v_2. In other words v_1 and v_2 change in the same direction.

$v_1 \dashrightarrow v_2$ means that a primary ↑ or ↓ in v_1 causes a secondary ↓ or ↑, respectively, in v_2. In other words v_1 and v_2 change in the opposite direction.

Using this notation two types of feedback can be defined:

(i) <u>Negative Feedback</u> exists between two variables v_1 and v_2 if and only if

$v_1 \underset{\dashleftarrow}{\overset{\longrightarrow}{}} v_2$ or $v_1 \underset{\longleftarrow}{\overset{\dashrightarrow}{}} v_2$.

(ii) <u>Positive Feedback</u> exists between two variables v_1 and v_2 if and only if

$v_1 \underset{\longleftarrow}{\overset{\longrightarrow}{}} v_2$ or $v_1 \underset{\dashleftarrow}{\overset{\dashrightarrow}{}} v_2$.

A properly designed closed-loop system with negative feedback minimizes the influence of unexpected disturbances so that the system's output is maintained close to a desired level. Consider, for example, the feedback control system symbolized by $u \underset{\dashleftarrow}{\overset{\longrightarrow}{}} y$. Suppose a disturbance causes the input to increase. This results in y increasing. However, the output signal on traversing the loop causes u to decrease and so the system's output decreases. Hence, the total effect is that the system tends to return the output to its previous value. An analogous result holds if the disturbance causes the input to decrease. Self-regulation in the body is a consequence of negative feedback control systems.

A positive feedback system (for example, $v_1 \underset{\longleftarrow}{\overset{\longrightarrow}{}} v_2$) causes its output to depart unindirectionally from the desired level. In the body a <u>vicious circle</u> is a positive feedback loop whose operation tends to increase the pathological changes responsible for a particular syndrome. Not all positive feedback loops produce vicious circles. The condition for a vicious circle to develop will be discussed in subsection C.

An interesting example of the above concepts is provided by hemorrhagic shock. Circulatory shock is a state of the circulation in which the cardiac output is too low to supply the normal nutritional needs of the body's tissues. Guyton and Crowell [28] noted the following results when the arterial pressure of dogs was reduced rapidly by acute hemorrhage. The animals whose pressures were only slightly reduced recovered rapidly, whereas those whose pressures were greatly reduced died rapidly. However, at intermediate pressure levels it was impossible to determine for many hours whether the animal would live or die.

Guyton and Crowell [28] explained these results in terms of feedback. They identified six separate types of feedback that further depress the cardiac output in shock. In addition, they stated four types of negative feedback mechanisms that help to oppose the progression of shock. Three of these processes are shown in the symbol-and-arrow diagram in Fig. 7.7.

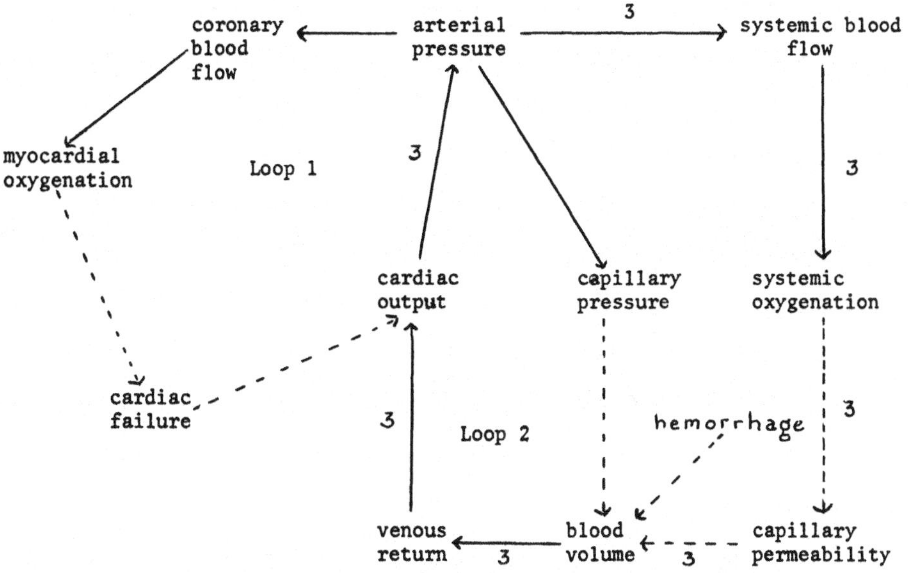

Figure 7.7 Some factors in hemorrhagic shock.

Loops 1 and 3 are positive feedback loops. Decreased cardiac output leads to decreased arterial pressure in both these loops. In Loop 1,decreased coronary blood flow results in decreased myocardial oxygenation thereby producing heart failure. Heart failure still further decreases the cardiac output. In Loop 3 the decrease in blood flow leads to an increase in capillary permeability due to the decrease in oxygenation of the peripheral tissues. Consequently, blood volume, venous result and cardiac output all decrease.

Loop 2 is a negative feedback loop. A decrease in capillary pressure in shock causes fluid to be absorbed into the circulatory system. Hence, the blood volume and the cardiac output both increase. This raises the capillary pressure back to normal.

The experimental results can now be explained. Above a critical pressure level, the negative feedback eventually took charge and caused recovery. Below this critical level the positive feedback mechanisms of deterioration took charge and caused death.

C. Linear Control Systems and Transfer Functions.

Historically, the first examples of the external description of systems were linear systems. Linear systems provided the background for the modern state space approach presented in the next section. Many nonlinear systems can be approximated by linear systems for certain values of their variables in their operating range [22]. The resulting linear systems can be studied by powerful linear systems techniques described below. Diverse applications of linear systems theory to physiology and biology appear in [13, 14, 17, 27, 29, 33, 47, 48].

The system (1) is linear if the function f satisfies the following property:

If u_1 and u_2 are admissible functions and a_1 and a_2 are real numbers, then

$$f(a_1u_1(t)+a_2u_2(t),\ t) = a_1f(u_1,t)+a_2f(u_2,t) \tag{2}$$

By induction, the linearity property continues to hold for a finite number, say n, of input functions:

$$f(\sum_{i=1}^{n} a_iu_i(t),t) = \sum_{i=1}^{n} a_if(u_i(t),t). \tag{3}$$

It is assumed that the linearity property described in (3) remains valid for an infinite number of inputs - that is,

$$f(\int a(s)u(s)ds,t) = \int a(s)f(u(s),t)ds \tag{4}$$

Equation (4) follows from (3) under appropriate conditions on f, for example, continuity.

Suppose the output of a system is known for every function in a given set of inputs. It follows from (4) that the output of the system is completely determined for any infinite linear combination of the given functions. In particular, any continuous scalar input can be written as the infinite linear combination

$$u(t) = \int_0^\infty u(s)\delta(t-s)ds \tag{5}$$

where δ is the Dirac delta function[1] defined by

$$\int_{t_0}^{t_1} \delta(t-s)ds = \begin{cases} 1 & t_0 < t-s < t_1 \\ 0 & \text{otherwise} \end{cases} \tag{6}$$

Hence, the output of a linear system will be completely determined once its response to $\delta(t-s)$, the unit impulse at time s, is known. The mathematical details

[1] Actually δ is a distribution.

follow.

The <u>impulse</u> <u>response</u> of the system (1), when f is a scalar, is defined to be the output at time t in response to a unit impulse input applied at time s. Mathematically, the impulse response is

(7) $$g(t,s) = \begin{cases} f(\delta(t-s),t) & s < t \\ 0 & \text{otherwise} \end{cases}$$

Note that a zero level response is required before the impulse is applied. If energy were already stored in the system from previous inputs, then the response to any new input would not be determined solely by the inherent nature of the system itself but in part by its particular past history. Applying f to both sides of (5) it follows from (4) that

(8) $$y(t) = f(u(t),t) = \int_0^\infty u(s)\, f(\delta(t-s),t)ds.$$

Using (7), equation (8) can be rewritten as

(9) $$y(t) = \int_0^t u(s)g(t,s)ds.$$

If the linear system is time-invariant or stationary it can be shown that

$$g(t,s) = g(t-s,0) \equiv h(t-s)$$

Relation (9) becomes

(10) $$y(t) = \int_0^t u(s)h(t-s)ds,$$

the <u>convolution</u> of the two functions u and h. The convolution of u and h is often denoted by

$$y = u*h$$

The <u>Laplace</u> <u>transform</u> of a function f, denoted $\mathcal{L}f$ or F is defined by

(11) $$F(s) = \mathcal{L}f(s) = \int_0^\infty f(t)e^{-st}dt.$$

The <u>transfer</u> <u>function</u> of a linear, stationary system (1) is defined by

(12) $$H(s) = \mathcal{L}\, h(t)$$

where h, the impulse response, is the function appearing in (10). The set of t values is called the <u>time domain</u> while the set of s values is called the <u>frequency domain</u>. It will be seen that the transfer function can be used to translate certain time problems into frequency domain problems which can then be conveniently solved.

Note that for a linear system with m inputs and n outputs, the impulse response h is a n × m matrix. The (i,j)th element is the ith output when the jth input is $\delta(t-s)$ and the remaining inputs are zero. The <u>transfer matrix</u> is defined by (12), where $\mathcal{L}h$ symbolizes the matrix formed from h by Laplace transforming each of its elements.

A practical working definition of the transfer function can be obtained from (10) using the result[2]

$$\mathcal{L}(u * h) = (\mathcal{L}u)(\mathcal{L}h)$$

Hence, Laplace transforming both sides of relation (10) and solving for $H = \mathcal{L}h$ yields

$$H(s) = \frac{Y(s)}{U(s)} \tag{13}$$

The following definition is a verbal translation of (13).

<u>Definition</u> 1

The <u>transfer function</u> of a linear stationary system is

$$\frac{\text{Laplace transform of the output}}{\text{Laplace transform of the input}}$$

with zero initial conditions for the input and output.

The transfer function of a system can be determined experimentally by using (13). If a unit impulse, $u(t) = \delta(t)$, is applied at time $t = 0$, then $U(s) = 1$, since $\mathcal{L}\delta = 1$.

It follows from (13) that the Laplace transform of the output is H(s), the transfer function. Hence, the response of the system to a unit impulse need only be measured. Its Laplace transform is the transfer function.

A transform function is an input-output description of the behavior of linear stationary systems. The block diagram of such a system using a transfer function

---

[2] Provided $\mathcal{L}u$ and $\mathcal{L}h$ converge absolutely.

Figure 7.8 Block diagram in terms of transfer function.

of Fig. 7.8 is shown in Fig. 7.8. The interpretation of Fig. 7.8 is

output = (input)(system transfer function)

or

$$Y(s) = U(s)H(s). \tag{14}$$

The transfer function description does not include any information concerning the internal structure of the system and its behavior.

In classical control theory, linear time-invariant systems are frequently described by linear differential equations with constant coefficients. The next two examples illustrate the use of Def. 1 to find the transfer function of such systems.

A simple spring-mass-damper system is shown in Fig. 7.9. Its equation of motion cam be found

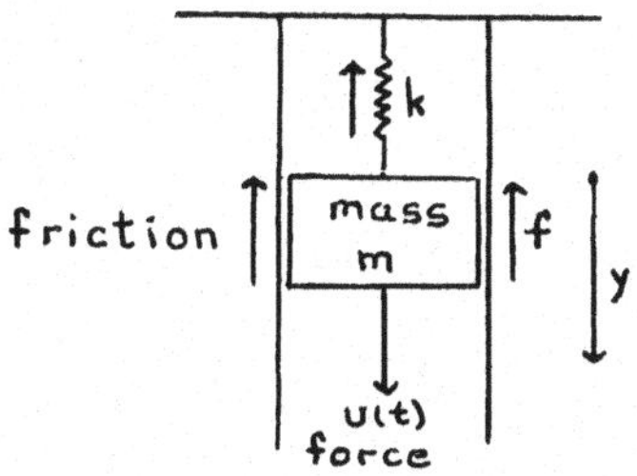

Figure 7.9 Spring-mass-damper system.

by using Newton's law, the describing equation for an ideal dashpot (friction) and an ideal spring. Denoting the displacement of the mass from its equilibruim position by y, the equation of motion is

$$m \frac{d^2y}{dt^2} + f \frac{dy}{dt} + ky = u(t), \tag{15}$$

where k is the spring constant and f is the friction constant.

The transfer function of the spring-mass-damper system is obtained by taking the Laplace transform of both sides of (15) using the linearity of the Laplace operator.

(16) $$\mathcal{L}(c_1f_1+c_2f_2+\ldots+c_nf_n) = c_1\mathcal{L}f_1+c_2\mathcal{L}f_2+\ldots+c_n\mathcal{L}f_n$$

and the fact that

(17) $$\mathcal{L}(f^{(n)}(t)) = s^nF(s)-s^{n-1}f(0^+)-s^{n-2}f^{(1)}(0^+)- \ldots - f^{(n-1)}(0^+).$$

The transformed equation (15), written with zero initial conditions is

$$ms^2Y(s)+fsY(s)+kY(s) = U(s).$$

Thus the transfer function

(18) $$H(s) = \frac{Y(s)}{U(s)} = \frac{1}{ms^2+fs+k}$$

The transfer function for the RC network shown in Fig. 7.10 is obtained from

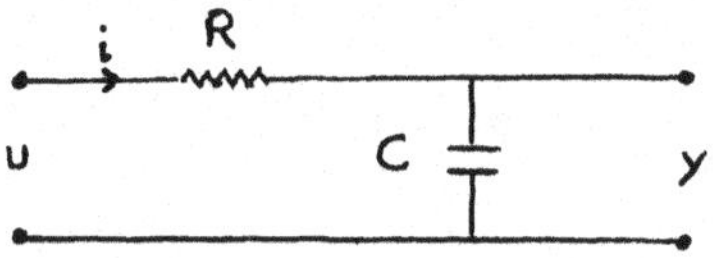

Figure 7.10 An RC network.

Kirchoff's voltage equations which yield

(19) $$u = iR + \frac{1}{C}\int i\, dt$$

(20) $$y = \frac{1}{C}\int i\, dt$$

Laplace transform equations (19) and (20) using linearity and the fact that

(21) $$\mathcal{L}\left(\int_a^t f(x)dx\right) = \frac{1}{s}F(s) + \frac{1}{2}\int_a^0 f(x)dx$$

Eliminating I(s) from the transformed equations yields

$$U(s) = sRCY(s)+Y(s)$$

Hence, the transfer function

(22) $$H_1(s) = \frac{Y(s)}{U(s)} = \frac{1}{\tau s+1}$$

where τ = RC, the time constant of the network.

The prototype of an open-loop system is shown in Fig. 7.8. For comparison, a closed-loop system with negative feedback appears in Fig. 7.11.

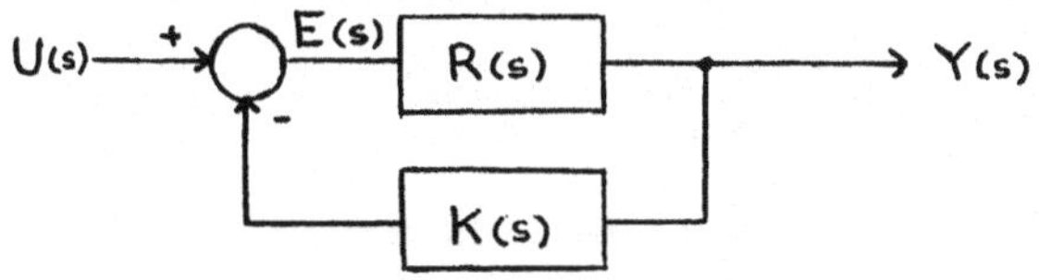

Figure 7.11 Closed-loop negative feedback system.

The significant difference between open-and closed-loop systems is the generation and utilization of the error signal

(23) $$E(s) = U(s) - K(s)\ Y(s).$$

The closed-loop system, when operating correctly, reduces the error to a minimum value.

The output of the open-loop system is given by (14). The output of the closed-loop system is defined as the limiting value of the output after a large number of feedbacks have occurred. To find this limiting value let Y_0 denote the output signal before any feedbacks have occurred and let Y_n be the output signal after n feedbacks have occurred.

Then

$$Y_0 = RU,$$

$$Y_n = RU - Y_{n-1}K \qquad (n = 1,2,3,\ldots)\ ;$$

(24) $$Y_n = RU(1-HK+(HK)^2 + \ldots + (-1)^n(HK)^n).$$

The output of the closed-loop system is by definition $Y = \lim\limits_{n \to \infty} Y_n$. If $|HK| < 1$ this limit exists and is given by

(25) $$Y(s) = [\frac{R(s)}{1+R(s)K(s)}]\ U(s) \qquad (|RK| < 1).$$

Hence, the transfer function of a closed-loop negative feedback system is

(26) $$\frac{R(s)}{1+R(s)K(s)} \qquad (|RK| < 1).$$

A closed-loop positive feedback system has the same block diagram representation as appears in Fig. 7.11 except that the minus sign at the two point junction is replaced by a plus sign. Hence, the corresponding formulae for a positive feedback system can be obtained from (23) - (26) by changing signs. For example, the transfer function for a closed-loop positive feedback system is

$$\frac{R(s)}{1-R(s)K(s)} \qquad (|RK| < 1). \tag{27}$$

The gain of a system at any time t is defined by

$$\text{system gain} = \frac{y(t)}{u(t)} = \frac{\text{output}}{\text{input}} .$$

The gain is a measure of the magnification of the input by the system. Frequently the gain of the system is desired after the system has been operating for some time and its transient response has disappeared. By convention, the steady-state response to the system to a constant input function is desired. The input used to define the gain is $U(t) = 1$ for $t > 0$. For a negative feedback system the transform of the output is, from (25),

$$Y(s) = \left[\frac{R(s)}{1+R(s)K(s)}\right] \frac{1}{s} \tag{28}$$

since $\mathcal{L}(1) = \frac{1}{s}$. The steady-state output can be obtained from (28) by utilizing the final-value theorem:

$$\lim_{t \to \infty} f(t) = \lim_{s \to 0^+} sF(s), \tag{29}$$

if f has a piecewise continuous derivative. It follows from (28) and (29) that

$$\lim_{t \to \infty} y(t) = \frac{R(0^+)}{1+R(0^+)K(0^+)} .$$

This result motivates the following definition.

Definition 2

The steady-state

$$G_c = \frac{R_0}{1+R_0K_0}$$

where

$$R_0 = R(0^+); \quad K_0 = K(0^+).$$

The steady-state open-loop gain or homeostatic index is

$$G_0 = R_0K_0.$$

The name open-loop gain is derived from the fact that H_oK_o would be the steady-state gain of the system depicted in Fig. 7.11 if the loop were interrupted before the minus sign and the output taken at the break point. The transfer function of the resulting system would be R(s)K(s).

Consider the example on hemorrhagic shock appearing at the end of subsection B. Let G_0 be the open loop steady state gain of a positive feedback loop whose output is the cardiac output. For simplicity assume that R = 1. Then the total change in cardiac couput caused by an initial stimulus S is (c.f. (24))

$$S+SG_0+SG_0^2+\ldots+SG_0^n+\ldots$$

It follows that the response has a finite value and does not lead to a vicious cycle if $G_0 < 1$.

Since negative and positive feedback act in exactly opposite directions, the net result of a system containing two loops with a common block whose transfer function is K is

$$S+S(G_p-G_n)+\ldots+S(G_p-G_n)^j+\ldots$$

where G_p is the open-loop gain of the positive feedback system and G_n is the open-loop gain of the negative feedback system. Hence, G_p-G_n determines whether a vicious cycle will occur.

D. Measures of Performance and Accuracy.

Classical control theory is concerned mainly with the analysis of linear time-invariant systems. The typical model of such systems is the negative feedback control system depicted in Fig. 7.11. The Laplace transform of the output of this system appears in (25). Theoretically, the denominator, 1+R(s)K(s), can be factored and partial fraction expansion can be used to obtain a sum of easily invertible terms. This would determine the output, y(t), in analytical form and the system's performance could be assessed.

Practically, the denominator in equation (25) is often a high degree polynomial. Moreover, an infinite number of possible inputs u(t) could be considered. Consequently, a complete analytic solution for the output is impractical.

To overcome this problem standard desirable features were proposed for evaluating system performance. Furthermore, only certain aperiodic and periodic signals were to be used as test inputs. The methods developed for analyzing system performance based on the above proposals did not require analytic solutions, but stressed graphical techniques[3]. These easily applicable methods are still used today[4] because of the insight they provide into the analysis and design of systems. They are discussed below.

1. Stability.

The location of the roots of the so-called closed-loop characteristic equation

$$1+R(s)K(s) = 0 \tag{30}$$

determines the stability of linear time invariant systems described by (25). Suppose the roots of the characteristic equation are denoted by

$$\lambda_i = \beta_i + j\omega_i$$

where $j = \sqrt{-1}$. Then, the system is

(i) unstable if $\beta_i > 0$ for any simple root or $\beta_i \geq 0$ for any repeated root;

(ii) stable in the sense of Lyapunov if $\beta_i \leq 0$ for all simple roots and $\beta_i < 0$ for all repeated roots;

(iii) asymptotically stable if $\beta_i < 0$ for all roots.

The above criteria for stability follow directly from the analytic form of the output obtained from (25) by partial fraction expansion and use of the inverse Laplace transform. Methods of determining stability, without actually finding the characteristic roots, are listed below.

(a) Routh's criterion

Hurwitz [32] and Routh [56] independently determined the necessary and sufficient conditions for stability from the coefficients of the characteristics equation. A useful form of their approach is described in Dorf [22].

[3] These techniques were developed before the wide-spread availability of digital computer.

[4] Many of these methods have been computerized.

(b) The Nyquist criterion

The physical ideas underlying frequency response analysis are extremely simple and are well known to anyone who has tried to rock a car out of a rut. One has discovered that little physical effort, applied at a certain frequency and with a certain phase relation to the car's motion, can produce large oscillations. These oscillations are at the system's natural frequency. Although oscillations can be built up at other frequencies it is readily found that much greater effort is required to do so. Graphical and analytical techniques based on frequency response have been devised for studying the dynamic characteristics of all types of linear systems and some nonlinear systems.

The frequency response of a stable linear time-invariant system is the steady-state (nontransient) response of the system to an input which is a sine (or cosine) function. The basic result is that the stead-state output is also a sine (or cosine) function of the same frequency but generally of a different amplitude and phase than the input. To verify this result for both sine and cosine inputs let

(31) $$u(t) = \sigma(t)e^{j\omega t}$$

where $\sigma(t) = 0$ if $t < 0$ and 1 otherwise. The Laplace transform of the corresponding output is, from (14),

$$Y(s) = H(s)\,\frac{1}{s-j\omega} = \frac{H(j\omega)}{s-j\omega} + Y_1(s),$$

where only the term $H(j\omega)/(s-j\omega)$ in the partial fraction expansion of $H(s)/(s-j\omega)$ is shown explicitly. The remaining terms are denoted by $Y_1(s)$. Since

$$H(j\omega) = |H(j\omega)|\,\exp(j\,\mathrm{Arg}\,H(j\omega))$$

it follows that

$$y(t) = \sigma(t)|H(j\omega)|e^{j\omega t}e^{j\,\mathrm{Arg}\,H(j\omega)} + y_1(t).$$

The inverse transform $y_1(t)$ of $Y_1(s)$ is composed of decreasing exponentials because of the stability assumption. Hence, the transient tends to 0 and as $t \to \infty$ the steady state solution is

(32) $$y(t) = \sigma(t)|H(j\omega)|\,e^{j\omega t}e^{j\,\mathrm{Arg}(j\omega)}.$$

which proves the basic result.

The reader might wonder why more complicated waveforms are not used as test inputs. Fourier's theorem decomposes any irregular but repetitive waveform into a sum of sine and cosine waves of various amplitudes, comprising in general the fundamental and all its harmonics. Since the response of the system to sine and cosine waves is known the response of the system to more complicated waveforms can be determined using Fourier's theorem and linearity.

Note that absolute value of the transfer function describing the sinusoidal steady-state behavior of a linear time-invariant system can be obtained by replacing s by $j\omega$ in the system transfer function H(s). This assertion can be proved easily by using (30) and (31). A frequency domain stability criterion based on the function $H(j\omega)$ was developed by Nyquist [50] in 1932.

To investigate the stability of a control system consider its transfer function

$$H(s) = \frac{N(s)}{D(s)} .$$

Assume that the denominator of the transfer function

$$D(s) = 1+F(s) \tag{33}$$

For example,

$$F(s) = R(s)K(s) \tag{34}$$

is the open-loop transfer function for the negative feedback system in Fig. 7.11. The system will be stable if and only if all zeros of D lie in the left-hand s-plane, where $s = \sigma+j\omega$. Instead of finding the zeros of D directly, which is usually difficult, the zeros are found indirectly as described in Dorf [22].

(c) <u>The Bode Plot.</u>

In the Bode diagram [22] the amplitude of frequency response transfer function $H(j\omega)$ is expressed in decibels[5] ($20 \log_{10}|G(j\omega)|$) and is plotted versus the frequency on semilogarithmic paper (or versus the logarithm of the frequency on ordinary paper). The phase of $G(j\omega)$ is plotted separately in degrees versus the frequency on semilogarithmic paper. This graphical method is used for determining

[5] Often called <u>db gain</u>.

the stability of minimum phase systems (systems with no open-loop poles or characteristic zeros in the right half plane   It is easier to sketch than the Nyquist diagram. Moreover, the effect of individual poles and zeros can be studied.

To see the reason for the advantage of a logarithmic plot consider the transfer function

$$(35)\qquad H(j\ ) = \frac{b\prod_{i=1}^{q}(1+j\omega\tau_i)}{(j\omega)^m\prod_{k=1}^{n}(1+j\omega\tau_k)\prod_{\ell=1}^{r}(1+(2\zeta_\ell/\omega_\ell)j\omega+(j\omega/\omega_\ell)^2)}$$

having q zeros, m poles at the origin, n poles on the real axis and r pairs of complex conjugate poles (as a function of s). Plotting the Nyquist diagram of such a function would be a formidable task. However,

$$(36)\qquad 20\log|H(j\omega)| = 20\log b+20\sum_{i=1}^{q}\log|1+j\omega\tau_i| - 20\log|(j\omega)^m|$$
$$- 20\sum_{k=1}^{n}\log(1+j\omega\tau_k)-20\sum_{\ell=1}^{r}\log\left|1+\left(\frac{2\zeta_\ell}{\omega_\ell}\right)j\omega + \left(\frac{j\omega}{\omega_\ell}\right)^2\right| .$$

The Bode diagram can be obtained by adding the plot of each individual factor[6]. The phase angle is

$$\phi(\omega) = \sum_{i=1}^{q}\tan^{-1}\omega\tau_i-m(90^\circ)-\sum_{k=1}^{n}\tan^{-1}(\omega\tau_k)-\sum_{\ell=1}^{r}\tan^{-1}\left(\frac{2\zeta_\ell\omega_\ell\omega}{\omega_\ell^2-\omega^2}\right).$$

Hence, the phase angle plot is simply the summation of the phase angles due to each individual factor of the transfer function. The procedure can be further simplified by using asymptotic approximations to the factors and only obtaining the actual curves at specific important frequencies. Details can be found in Bode [11], Dorn [22] and Milhorn [48].

One of the most important uses of the Bode plot is that of determining an unknown transfer function from experimental data of gain and phase shift obtained when a sinusoidal input is applied to the system.

---

6 Actually there are only four different types of factors. Time delays introduce another factor of the form $e^{-j\omega T}$.

2. Steady-state accuracy.

The Laplace transform of an open-loop system with transfer function R(s) is

(37) $$E_0(s) = U(s)-Y(s) = U(s)(1-R(s)).$$

Using (23) and (25) it can be shown that the Laplace transform of the error for the closed-loop system shown in Fig. 7.11, when K(s) = 1, is

(38) $$E_c(s) = \frac{U(s)}{1+R(s)} .$$

Steady state accuracy requires that these error signals approach a sufficiently small value for large values of time. The final-value theorem (28) allows this requirement to be analyzed without actually finding inverse transforms.

The position, velocity and acceleration (or step, ramp and parabolic) error constants are the asymptotic values of e(t) when step (u(t) = 1), ramp (u(t) = t) and parabolic ($u(t) = t^2/2$), respectively, are used as test inputs.

For example, the position error constant for the open-loop system is, from (37),

$$e_0(\infty) = \lim_{s \to 0^+} [u_0/s(1-R(s))] = 1-R(0^+).$$

From (3) the closed-loop position error constant is

$$e_c(\infty) = \lim_{s \to 0^+} s[\frac{u_0/s}{1+R(s)}] = \frac{1}{1+R(0^+)} .$$

$R(0^+)$ is often called the dc-gain and is normally greater than one. Therefore, the open-loop system can have a steady-state error of significant magnitude. By contrast, the closed-loop system will have a small steady-state error.

3. Transient response.

The inverse Laplace transform of a complicated transfer function may be difficult to determine. Hence, instead of a complete analytic solution for the transient response, it is often necessary to settle for a few of its characteristics when a unit step input, $U(s) = \frac{1}{s}$, is applied: These characteristics are defined in terms of the steady state response

$$e_{ss} = H(0^+),$$

where H is the transfer function of the system. These characteristics and their interpretation can be found in Brogan [12] and Dorf [22].

4. Frequency response

Frequency domain methods can also be used to study transient performance. Both Nyquist and Bode diagrams are frequency response methods which deal with open-loop transfer functions. They can be used to define relative stability which can be correlated with transient behavior. The relevant definitions appear in Dorf [22]. Some approximation techniques can be found in Saucedo and Schiring [58].

The application of classical control theory to cell biology is discussed in Appendix G.

7.3 Internal Description of Systems (State Space Description)

The state space approach is the central theme of modern control theory. By using vector and matrix theory multiple inputs and outputs can be treated. The state space approach is characterized by the direct use of differential equations which are frequently solved with the aid of digital or analogue computers. Many of the computer algorithms for solving differential equations are applicable to nonlinear, time-varying, systems. In the next section it will be seen that the internal description of systems is ideal for studying optimal control problems. This approach is also suitable for studying stochastic control systems. Since many biological systems are multi-input-output and nonlinear, the state space approach is extremely useful for biologists.

The state space description of systems is not a black box approach. Consequently, it requires detailed information about the system being analyzed. First, a model of the system must be formulated. Next, the model must be represented mathematically. In circuit theory many representations are possible-for instance, node equations, mesh equations, cut set equations or loop equations. In mechanics positions and velocities (Lagrangian mechanics) or positions and momenta (Hamiltonian mechanics) can be used as variables. A thermodynamic system can be modelled by choosing temperature and heat flow or temperature and entropy flow as the variables to describe heat conduction. In all of these systems it is possible to choose variables so that given the present state of the system the future state can be predicted. This last property of deterministic systems leads to the concept of state variables.

Definition 3

The state variables, $x_i(t)$ $(i = 1,2,\ldots,k)$, consist of a minimum set of functions which completely summarize the system's status in the following sense. If at any initial time t_0, the values of the state variables $x_i(t_0)$ are known, then the output[7] $y(t)$ and the values $x_i(t)$ can be uniquely determined for any time $t > t_0$ provided the input $u_{[t_0,t]}$ is known.

The term state space is used to denote the vector space in which the column vector

$$x = (x_1,x_2,\ldots,x_k)^T$$

lies for a fixed value of t.

Some simple examples of systems and their corresponding states appears in Table 7.3.

TABLE 7.3 EXAMPLES OF SYSTEMS AND THEIR STATES

System	States
two position light switch	on, off
mechanical system	position and velocity of elements
passive LCR network	currents through inductors voltage across capacitors and resistors.

According to Definition 3, $t_0, x(t_0)$, $u_{[t_0,t]}$ uniquely determine the state x at any time $t > t_0$. In other words, there is a function f, called the state transition function, such that

$$x(t) = f(t,t_0,x(t_0),u_{[t_0,t]}). \tag{39}$$

Furthermore, since y(t) is uniquely determined, a second function g exists with

---

[7] May be a vector.

(40) $$y(t) = g(t,x(t),u(t)).$$

The indirect description of the output in terms of the state and input by means of (39) and (40) is called the internal description of a system. For multiple inputs and outputs, (39) and (40) are interpreted as vector equalities.

Many systems can be described by a set of differential equations[8] (41) and a set of single-valued algebraic output equations (42):

(41) $$\dot{x} = h(t,x,u)$$

(42) $$y = g(t,x,u).$$

Under certain assumptions on h it can be shown that equations (41) and (42) are equivalent to an internal description of the system. Assume that h satisfies a Lipschitz condition with respect to x, is continuous with respect to u and is piecewise continuous with respect to t. It can be shown that there will be a unique solution x(t) for any t_0, $x(t_0)$ provided u is piecewise continuous [19]. The unique solution of (41) defines the function f of (39).

The following example illustrates some of the above concepts. Consider a freely fallying body subject to a constant gravitational input force a. Assume that the air resistance is negligible. The output y is the altitude of the bodies center of gravity. The equation of motion is

(43) $$\ddot{y} = -a.$$

Integrating twice gives

$$y(t) = y(t_0) + \dot{y}(t_0)(t-t_0)-a(t-t_0)^2/2.$$

The variable y is not the state since its initial value $y(t_0)$ does not determine y(t) uniquely. The set of variables y, $\dot{y}$, $\ddot{y}$ are not state variables since they are not the minimum set of variables required to uniquely determine y(t). The parameter $\ddot{y}$ is not required. It will be shown that $x(t) = (y(t),\dot{y}(t))^T$ is a state

---

8 A difference equation is used for discrete-time systems. Recall that higher order systems can be reduced to first order systems.

vector by replacing the input-output equation (43) by an equivalent system of equations of the form (41) and (42). Since, $x_2 = \dot{y}$, (43) can be written as

(44) $$\dot{x}_2 = -a.$$

Obviously,

(45) $$\dot{x}_1 = x_2,$$

(46) $$y = (1,0)x.$$

Clearly (44) and (45) are of the form (41) while (46) is of the form (42). The above approach of introducing new variables can be extended to higher-order input-output differential equations.

Alternatively, the reader can verify that the form of equations (39) and (40) for the above example is

$$x(t) = \begin{bmatrix} 1 & 1-t_0 \\ 0 & 1 \end{bmatrix} x(t_0) - \begin{bmatrix} (t-t_0)^2/2 \\ t-t_0 \end{bmatrix} a, \qquad y(t) = (1,0)x(t).$$

As another example consider the RC network described in Fig. 7.10. Differentiating (19) and (20) with respect to t gives

(47) $$\frac{di}{dt} = -\frac{i}{RC} + \frac{1}{R}\frac{du}{dt}$$

(48) $$\frac{dy}{dt} = i/C$$

Note that i and y cannot be used as state variables. The state equation must express $\dot{x}$ as a function of x,u and t (and not $\dot{u}$). Instead let

$$x_1 = i - \frac{1}{R}u \qquad x_2 = y.$$

Then (47) and (48) can be written in the form of (41) and (42) as follows:

$$\frac{dx_1}{dt} = -\frac{x_1}{RC} - \frac{u}{RC}$$

$$\frac{dx_2}{dt} = \frac{x_1}{C} + \frac{u}{RC}$$

$$y = (0,1)x.$$

Other choices of state variables are possible. The selection of state variables is not a unique process.

For linear continuous-time dynamical systems (41) and (42) can be written in the form:

$$x = A(t)x(t) + B(t)u(t) \tag{49}$$

$$y(t) = C(t)x(t) + D(t)u(t) \tag{50}$$

where A,B,C,D are matrices of dimension $k \times k$, $k \times m$, $n \times k$ and $n \times m$, respectively.

If a state space description of a system is given, then the input-output relation for the system can be found by eliminating x(t) from (40) using (39). Conversely, given the input-output description of a system can a state-space representation be found? Note that, in general, the state-space representation if it exists will not be unique since the state variables can be selected in an infinity of different ways. It has been shown that given a proper rational transfer matrix H(s) of a linear time invariant system, constant matrices A,B,C and D can be found such that (49) and (50) hold. Hence, the transfer function approach of classical control theory can be replaced by the state space approach of modern control theory. Nevertheless, transfer function methods are still popular for the design and analysis of linear systems. Methods of finding the state space representation of systems appear in Brogan [12].

The important concepts of controllability and observability were introduced by Kalman [35]. In a typical optimal control problem it is necessary to find a control input which optimally steers the system from a given initial state to a prescribed final state. Obviously it is necessary to discover whether or not the final state can be reached which is the problem of controllability. Frequently, an observer can only measure the output of a system. The problem of observability arises in determining the state of the system from these measurements. The precise definition of these two concepts appears below.

Definition 4

A system is controllable at t_0 if it is possible to find an admissible input function u(t) which will transfer the given initial state $x(t_0)$ to a desired final state x' at a finite time t_1. If this is true for all initial times t_0 and all initial states $x(t_0)$, the system is completely controllable.

The system described by (39) is controllable at t_0 if and only if

$$x' = f(t_1, t_0, x(t_0), u_{[t_0, t_1]}).$$

A simple example of an uncontrollable system is the system (49) and (50) when $B = 0$, because then the input has no influence on the state.

To illustrate the above concept consider the circuits shown in Fig. 7.12. The state of both systems is represented by the voltage across the capacitor. If a

Figure 7.12 Examples of (a) controllability and (b) uncontrollability.

voltage source of sufficient magnitude is connected to the terminals of the circuit depicted in (a), it is possible to drive the capacitor voltage to any desired value in a finite time. In the circuit shown in (b) the application of a voltage signal to the terminals has no effect on the capacitor voltage because of the balanced bridge. The system in (a) is controllable while the system in (b) is not.

Controllability for linear time invariant systems of the form (49) and (50) is a property of coupling between the input and state and thus involves only the matrices A and B as shown formally by the following criterion.

Controllability Criterion 1.

A constant coefficient union system of the form (49) and (50) is completely controllable if and only if

$$\text{rank } [B, AB, A^2B, \dots, A^{k-1}B] = k.$$

Observe that the given rank conditions are independent of the choice of initial and final states. This independence is more natural if the following equivalent definition of complete controllability is known. A linear system is completely controllable if any initial point can be transferred to the origin in a finite time.

For linear time dependent systems the controllability criteria are more complicated. The so-called transition matrix, $\Phi(t,t_0)$, is required. This matrix is the solution of the matrix differential equation

$$\frac{d\Phi(t,t_0)}{dt} = A(t)\Phi(t,t_0) \tag{51}$$

which satisfies the initial condition

$$\Phi(t_0,t_0) = I, \tag{52}$$

where I is the k × k identity matrix. The conditions for controllability are defined in terms of the matrix

$$G(t_1,t_0) = \int_{t_0}^{t_1} \Phi(t_1,s)B(s)B^T(s)\Phi^T(t_1,s)ds \tag{53}$$

Controllability Criterion 2

The system described by (49) and (50) is controllable on $[t_0,t_1]$ if any of the following equivalent conditions are satisfied:

(i) $G(t_1,t_0)$ is positive definite.

(ii) $|G(t_1,t_0)| \neq 0$.

(iii) zero is not an eigenvalue of $G(t_1,t_0)$.

To illustrate Controllability Criterion 1 consider the system

$$\dot{x} = \begin{bmatrix} -3 & -1 \\ -1 & -1 \end{bmatrix} x(t) + \begin{bmatrix} 1 \\ 1 \end{bmatrix} u(t), \qquad y(t) = [3,2]x(t). \tag{54}$$

It is completely controllable since

$$\text{rank } [B,AB] = \begin{bmatrix} 1 & -4 \\ 1 & -2 \end{bmatrix} = 2.$$

Analogous criteria hold for observability which is defined formally as follows.

Definition 5

A system is observable at t_0 if the initial state can be determined from $y_{[t_0,t_1]}$ where t_1 is some finite time and $t_1 \geq t_0$. If this is true for all initial times t_0 and initial states $x(t_0)$, the system is completely observable.

Consider, once again, the systems in Fig. 7.12. The output of both systems is measured by connecting an ideal voltmeter which draws no current to the terminals which are depicted by dots in the figure. The state of C can be deduced from the voltmeter reading in (a) but not in (b). The voltmeter reading for system (b) will always be zero. System (a) is observable; (b) is not.

Observability is a property of coupling between the state and the output. Hence, observability criteria for linear systems described by (49) and (50) involve the matrices A,C and

(55) $$H(t_1,t_0) = \int_{t_0}^{t_1} \Phi^T(s,t_0)C^T(s)C(s)\Phi(s,t_0)ds.$$

Observability Criterion 1

The constant coefficient linear system described by (49) and (50) is completely observable if and only if

$$\mathrm{rank}[C,CA,CA^2,\ldots,CA^{k-1}]^T = k.$$

Once again the plausibility of the fact that complete observability is independent of the final and initial states follows from the alternate definition of complete observability. In this definition for linear systems it is sufficient to consider the origin as the initial point.

Observability Criterion 2

The system (49) and (50) is completely observable at t_0 if there exists some finite t_1 for which any one of the following equivalent conditions holds:

(a) $H(t_1,t_0)$ is positive definite

(b) $|H(t_1,t_0)| \neq 0$

(c) zero is not an eigenvalue of $H(t_1,t_0)$.

The system (54) is completely observable by Criterion 1 since

$$\mathrm{rank}[C,CA]^T = \mathrm{rank}\begin{pmatrix} 3 & -11 \\ 2 & -7 \end{pmatrix} = 2.$$

Controllability and observability are discussed in greater depth in Kalman [36], Gilbert [25] and Zadeh and Desoer [71]. These concepts for nonlinear systems are active areas of research.

7.4 Optimal Control Theory

(a) Introduction

Modern technology often requires optimizing the behavior of systems. Some practical examples are maximizing the range of a rocket, minimizing the fuel used by a rocket on its journey, and maximizing the profit of a business. Methods of solving such problems are studied in optimal control theory.

Many problems in biology can also be formulated in terms of optimizing the behavior of systems. A recent survey of biological applications of optimal control theory appears in Banks [3]. Kernevez and Thomas [39] and Kernevez [38] have applied optimal control theory to biochemical problems involving enzyme kinetics. The endocrine system has been considered by Stear [61]. The optimal management of biological renewable resources has been studied by Swan [62] and Clark [16]. Wickwire [68] describes the application of optimal control theory to the spread of infectious diseases and pest control. Longini et al. [43] have formulated an optimization model for distributing a limited supply of influenza vaccine. Optimal control theory has also been applied to the study of the immune system (Appendix D) and to the study of radiation therapy (Appendix F). Applications to cancer chemotherapy will be discussed in Chapter 8.

If optimal control problems are stripped of the jargon peculiar to their originating discipline, then the following common features emerge. A system must reach some goal or accomplish some task. Usually the task can be completed in a number of different ways by altering the controls of the system. Each way involves a certain cost or has some numerical measure of efficiency associated with it.

To translate these features into a mathematical form, the system is described by a state transition function (39). The values of the state variables are used to determine the goal or task. The goal is preset by the modeller. Naturally, the system's controls are represented by the control functions appearing in the state transition function. As stated above, the goal can frequently be achieved by using any one of several different control functions. The set of all control functions that will be considered by the modeller is called the <u>admissible</u>

control set. The fact that the accomplishment of the task in a given time entails a certain cost, determines a cost function or performance criterion. This performance criterion can depend on the time interval required to accomplish the given task, the initial time, the final time and the evolution of the state and control functions during this time interval.

Using the preceding mathematical representation, a typical optimal control problem can be formulated abstractly as follows:

> Determine an admissible control which maximizes (or minimizes) the cost function for a system whose behavior is governed by its state transition function or by its state equations, which will be differential equations in the work below.

A more detailed mathematical description of optimal control problems and their solution will be presented below.

(b) The Pontryagin Minimum Principle

In the introduction an abstract formulation of an optimal control theory problem was developed. The mathematical statement of this dynamic optimization problem is the following. Determine an admissible control function u in order to minimize the performance criterion

$$J = \theta(x,t)\Big|_{t_0}^{t_f} + \int_{t_0}^{t_f} \phi(x(t), u(t), t)dt \tag{56}$$

for the system described by the state differential equation (41), $\dot{x} = h(t,x,u)$, where t_0 is the initial time, t_f is the final time; θ and ϕ possess continuous partial derivatives in x and u. Recall that the state x and the control u are vector-valued functions; for a fixed value of t they are column vectors with k and m components, respectively.

It is assumed that h has continuous partial derivatives in all of its variables. Frequently, such smoothness assumptions guarantee that for any piecewise continuous function u, there exists a unique solution to the differential equation (41) satisfying $x(t_0) = x_0$. Hence, the set of admissible control functions is defined to be the class of piecewise continuous functions over the interval of interest. It is assumed that for an admissible u and the given initial condition $x(t_0)$, the differential equation (41) defines a unique, admissible solution over the control interval of interest. An admissible solution is called an <u>admissible trajectory</u>.

The following examples illustrate the form of the performance criterion. Suppose the system under consideration is a space vehicle and it is desired that the vehicle reach its destination in a short a time as possible A suitable performance criterion is obtained by setting $\theta = 0$, $\phi = 1$, $t_0 = 0$ and $t_f = T$, the time required for the vehicle to reach its destination. In this situation, $J = T$ as required. Next, consider a system described by (41) and (42). Frequently, the objective is to determine an input u that causes the output y to follow a given reference signal r with a certain accuracy. Hence, a desirable quantity to minimize would be

$$J_1 = \int_{t_0}^{t_f} |y-r|^2 dt = \int_{t_0}^{t_f} |g(t,x,u)-r|^2 dt.$$

To include the effects of excessive control effort, J_1 could be revised to

$$J_2 = \int_{t_0}^{t_f} (|g(t,x,u)-r|^2 + C|u|^2) dt.$$

A compromise between small errors and large control amplitudes can be obtained by a proper selection of C.

To motivate Pontryagin's method consider the problem of extremizing a function $\phi : R^p \to R$, subject to the equality constraint $f(z) = 0$, where $f : R^p \to R^q$ and $p > q$. To solve this problem the Lagrangian is defined as

$$L(z,\lambda) = \phi(z) + \lambda^T f(z),$$

where $\lambda \in R^q$. In most advanced calculus books it is shown that, under certain conditions, if z^* is a constrained maximum or minimum, then there is a $\lambda^* \in R^q$ such that $\frac{\partial L}{\partial z} = 0$ and $\frac{\partial L}{\partial \lambda} = 0$ [9]. Hence, an extreme value of the function f subject to the equality constraint $\phi = 0$ can be found by extremizing the Lagrangian L with no constraints.

9

$$\frac{\partial L}{\partial z} = (\frac{\partial L}{\partial z_2}, \ldots, \frac{\partial L}{\partial z_p})^T \text{ and so } \frac{\partial L}{\partial z} = 0 \text{ is equivalent to } \frac{\partial L}{\partial z_i} = 0,\ i = 1,2,\ldots,p.$$

Pontryagin's method for solving dynamic optimization problems is analogous to Lagrange's method. The minimum value of the performance criterion J subject to the differential constraint (41) can be found by minimizing the unconstrained performance criterion

(57) $$J_\lambda = \theta(x,t)\Big|_{t_0}^{t_f} + \int_{t_0}^{t_f}[\phi(x,t) + \lambda^T(t)(h(x,u,t)-\dot{x}]dt$$

where $\lambda(t)$ is a column vector-valued function of t with k components.

Defining a scalar function, the Hamiltonian, by

(58) $$H(x(t),u(t),\lambda(t),t) = \phi(x(t),u(t),t) + \lambda^T(t)h(x(t),u(t),t)$$

the performance index J_λ becomes

(59) $$J_\lambda = \theta(x,t)\Big|_{t_0}^{t_f} + \int_{t_0}^{t_f} [H(x,u,\lambda,t)-\lambda^T\dot{x}]dt$$

Integrating the last term in the integrand of (59) by parts, yield

(60) $$J_\lambda = [\theta(x,t)-\lambda T_x]_{t_0}^{t_f} + \int_{t_0}^{t_f} [H(x,u,\lambda,t)+\dot{\lambda}^T x]dt.$$

Necessary conditions for an admissible control to be optimal will now be developed. Let u* be an admissible optimal control and x* the corresponding admissible trajectory. Consider an admissible, but not necessarily optimal control

(61) $$u(t) = u^*(t) + e\ \delta u(t) \qquad (t \in [t_0,t_f])$$

where $\delta u(t)$ is a variation in u(t) and e is a small number. Let

(62) $$x(t) = x^*(t) + e\ \delta x(t)$$

be the admissible trajectory corresponding to u. Substituting for x and u in (60) from (61) and (62), it is obvious that $J_\lambda = J_\lambda(e)$ is minimized when $e = 0$. Hence, a necessary condition for an admissible control to be optimal is

(63) $$\frac{\partial J_\lambda}{\partial e}\Big|_{e=0} = 0$$

independent of the value of $\delta u(t)$ and $\delta x(t)$. Differentiating (60) yields

(64) $$\frac{\partial J_\lambda}{\partial e}\Big|_{e=0} = [\delta x^T(\frac{\partial\theta}{\partial x} - \lambda)]_{t_0}^{t_f} + \int_{t_0}^{t_f} (\delta x^T[\frac{\partial H}{\partial x} + \dot{\lambda}] + \delta u^T[\frac{\partial H}{\partial u}])dt$$

Recall that the multipliers λ were introduced to remove the differential constraint. Hence, a necessary condition for a minimum is that (63) hold for arbitrary variations δx and δu. Thus, from (64), the following result is obtained.

Theorem 1 (Pontryagin's minimum principle)

For the control u and the related trajectory x, which satisfies

$$\dot{x} = h(x,u,t),$$

to minimize the performance criterion

$$J = \theta(x,t) \Big|_{t_0}^{t_f} + \int_{t_0}^{t_f} \phi(x,u,t)dt,$$

it is necessary that a vector $\lambda(t)$ exist such that:

(i) $\lambda(t)$ satisfies the adjoint equation

(65) $$\dot{\lambda} = -\frac{\partial H}{\partial x}$$

where the Hamiltonian

$$H = \phi + \lambda^T h;$$

the initial and final conditions (transversality conditions) are

(66) $$\delta x^T \left(\frac{\partial \theta}{\partial x} - \lambda\right) = 0 \qquad (t = t_0, t_f)$$

(ii) for any $t \in (t_0,t_f)$

(67) $$\frac{\partial H}{\partial u} = 0.$$

A heuristic proof of Theorem 1 will be outlined below.

Note

(a) Maximizing J is equivalent to minimizing -J.

(b) In order for (67) to hold δu must be completely arbitrary. If the admissible control is bounded, δu cannot be completely arbitrary and (67) may not hold. However, it will be seen that an optimal admissible u minimizes H. In fact, the advantage of the minimum principle is the replacement of the minimization of the functional J by an infinity of scalar function H minimization for $t \in (t_0,t_f)$

(c) The minimum principle only gives a necessary condition. Therefore, when a control verifying the conditions of Theorem 1 has been found, it should be verified that the control corresponds to the minimum of the performance criterion.

(d) The minimum principle is often called the maximum principle since the optimal control is found by maximizing a Hamiltonian H_1 where $H_1 = -H$.

(e) $$\dot{x} = h(x,u,t) = \frac{\partial H}{\partial \lambda} .$$

The transversality conditions in (66) can be specialized in several ways. For example, if the initial state of the system is specified but the terminal state is unspecified

$$x(t_0) = x_0, \quad \lambda(t_f) = \frac{\partial \theta(x(t_f),t_f)}{\partial x_{t_f}}$$

since $\delta x(t_0) = 0$ ($x(t_0)$ is fixed) and $\delta x(t_f)$ is completely arbitrary. In another class of problems both the initial and final states are specified and so $\delta x(t_0) = \delta x(t_f) = 0$. Thus λ is not determined from (66) at t_0 and t_f; $x(t_0)$ and $x(t_f)$ are the boundary conditions for the two-point boundary value problem.

Different forms of the Pontryagin minimum principle for different initial and terminal constraints and auxilliary conditions are summarized in Table 7.3. The derivation of the necessary conditions are similar to the method used for Theorem 1 and can be found in Sage and White [57]. For example, given the manifold $M(x,t) = 0$ - that is, the set of points $\{(x,t)|M(x,t) = 0\}$, where $M:R^{k+1} \to R^r$, contains the trajectory initially, that is, $M(x(t_0),t_0) = 0$ and the manifold $N(x,t) = 0$, where $N:R^{k+1} \to R^q$, contains the trajectory terminally, that is, $N(x(t_f),t_f) = 0$. The constraints are adjoined to the θ function by means of Lagrange multipliers ξ and ν to give the performance index

$$J = \theta(x(t),t)\Big|_{t_0}^{t_f} - \xi^T M(x(t_0),t_0) + \nu^T N(x(t_f),t_f) + \int_{t_0}^{t_f} (H(x,u,\lambda,t) - \lambda^T \dot{x})\,dt .$$

The necessary conditions can be found by finding the variation in J due to small perturbations about the optimal control and corresponding trajectory.

TABLE 7.4 THE PONTRYAGIN MINIMUM PRINCIPLE

State equations $\dot{x} = f(x,u,t)$ $(x(t) \in R^k,\ u(t) \in R^m)$

minimize $J = \theta(x(t),t)\Big|_{t_0}^{t_f} + \int_{t_0}^{t_f} \phi(x,u,t)dt$

Hamiltonian $H = \phi + \lambda^T f$ $(\lambda(t) \in R^k)$

	initial constraint	terminal constraint	auxiliary constraint	Necessary Conditions: minimization of H	Necessary Conditions: canonical equations	Necessary Conditions: transversality
1	t_0 fixed and $x(t_0)=x_0$ or none	t_f fixed and $x(t_f)=x_f$ or none	none	$\frac{\partial H}{\partial u}=0$	$\dot{x}=\frac{\partial H}{\partial \lambda}$ $\dot{\lambda}=-\frac{\partial H}{\partial x}$	$\delta x^T[\frac{\partial \theta}{\partial x}-\lambda]=0$ $t=t_0,t_f$ varies according to constraints. See text
2	t_0 fixed $x(t_0)=x_0$	t_f fixed $\lVert x(t_f)\rVert^2=1$	$\theta \equiv 0$	as in 1	as in 1	$\delta x^T(t_f)x(t_f)=0$ $\delta x^T(t_f)\lambda(t_f)=0$
3	t_0 fixed $M(x(t_0),t_0)=0$ where $M:R^{k+1}\to R^r$	t_f fixed $N(x(t_f),t_f)=0$ where $N:R^{k+1}\to R^q$	none	as in 1	as in 1	$\lambda(t_0)=\frac{\partial \theta}{\partial x_0}+(\frac{\partial M^T}{\partial x_0})\xi$ $\lambda(t_f)=\frac{\partial \theta}{\partial x_{t_f}}+(\frac{\partial N^T}{\partial x_{t_f}})\nu$
4	t_0 fixed and $x(t_0)=x_0$	t_f unspecified and determined when $N(x(t_f),t_f)=0$	$\theta(x(t_0),t_0)\equiv 0$	as in 1	as in 1	$\lambda(t_f)=\frac{\partial \theta}{\partial x_{t_f}}+(\frac{\partial N^T}{\partial x_{t_f}})\nu$ $[H+\frac{\partial \theta}{\partial t}+(\frac{\partial N^T}{\partial t})]_{t=t_f}=0$
5	as in 4	as in 4	none	optimal u satisfies $H(u)\le H(v)$ for any $v \in U$	as in 1	as in 4

The interpretation and use of the results listed in Table 7. will be illustrated by examples bearing the same number as the principle used in the example.

Example 1

Consider the system $\dot{x} = u$. Minimize the performance index

$$J = \int_0^{t_f} (u^2+x^2)dt.$$

The Hamiltonian is

$$H = u^2+x^2+\lambda u.$$

The canonical equations are

$$\dot{x} = \frac{\partial H}{\partial \lambda} = u \tag{68}$$

$$\dot{\lambda} = -\frac{\partial H}{\partial x} = -2x \tag{69}$$

To obtain a system of equations in x and λ, the minimization of H is used to express u as a function of x and λ. Here

$$0 = \frac{\partial H}{\partial u} = 2u+\lambda$$

and so $u = -\lambda/2$. Hence, (68) can be written as

$$\dot{x} = -\lambda/2. \tag{70}$$

The transversality conditions provide the boundary conditions required to solve the system (69) and (70) as shown in (a) and (b) below. In general, a two-point boundary value problem is obtained.

The general solution of the system (69) and (70) can be obtained by differentiating (70) with respect to t and substituting for $\dot{\lambda}$ from (69). This yields

$$\ddot{x} = x$$

whose solution is

$$x = ae^{-t} + be^{t} \tag{71}$$

where a and b are constants. This determines $u = \dot{x}$ and $\lambda = -2u$.

(a) Suppose $x(0) = 1$, $x(t_f) = 0$ and $t_f = 1$. Since $\delta x(t_0) = \delta x(t_f) = 0$ the transversality condition yields $x(0) = 1$ and $x(1) = 0$. Substituting these values in (71) for $t = 0$ and $t = 1$ yields

$$1 = a + b$$

$$0 = ae^{-1} + be.$$

The solutions of these equations are

$$a = -e^2(1-e^2)^{-1}, \qquad b = (1-e^2)^{-1}$$

(b) Suppose $x(0) = 1$, $x(t_f)$ is unspecified and $t_f = 1$. Since $\delta x_{t_f} \neq 0$ and $\theta \equiv 0$, the transversality condition gives $[\frac{\partial\theta}{\partial x} - \lambda]_{t=t_f} = 0$ and so $\lambda(1) = 0$. The condition $x(0) = 1$ once again yields $a+b = 1$. From (70) and (71) it follows that

$$\lambda = -2(-ae^{-t}+be^{t}).$$

Hence, the condition $\lambda(1) = 0$ gives

$$0 = -ae^{-1}+be.$$

The solution of the last equation and $a+b = 1$ are

$$a = e^2(1+e^2), \qquad b = (1+e^2)^{-1}.$$

Example 2

The state equations of a system consisting of tww cascaded integrators is

$$\dot{x}_1 = x_2 \qquad x_1(0) = 0$$

$$\dot{x}_2 = 0 \qquad x_2(0) = 0.$$

The system is controlled so that when $t = 1$

(72) $$x_1^2(1) + x_2^2(1) = 1.$$

Find the control u which minimizes the performance index

$$J = \frac{1}{2} \int_0^1 u^2 dt$$

The Hamiltonian

$$H = \frac{1}{2} u^2+\lambda_1 x_2+\lambda_2 u$$

The optimal control is found from the equation $\frac{\partial H}{\partial u} = 0$ and is

(73) $$u = -\lambda_2$$

The transversality conditions reduce to the equations

$$\lambda_1\delta x_1+\lambda_2\delta x_2 = 0 \ ; \qquad x_1\delta x_1+x_2\delta x_2 = 0$$

at the final time t = 1. Hence

$$(74) \qquad \frac{\lambda_1(1)}{x_1(1)} = \frac{\lambda_2(1)}{x_2(1)} = k$$

where k is a constant. Using the canonical equations it follows from (72), (73) and (74) the problem of finding the optimal control is completely resolved by solving the two-point boundary value problem

$$(75) \qquad \begin{array}{ll} \dot{x}_1 = x_2 & x_1(0) = 0 \\ \dot{x}_2 = -\lambda_2 & x_2(0) = 0 \\ \dot{\lambda}_1 = 0 & \lambda_1(1) = kx_1(1) \\ & x_1^2(1)+x_2^2(1) = 1 \\ \dot{\lambda}_2 = -\lambda_1 & \lambda_2(1) = kx_2(1) \end{array}$$

The four first order differential equations are linear and time invariant. Howevery, the explicit solution of this problem is difficult to determine because of the nonlinear terminal constraint (72). In general, nonlinearity complicates many problems and necessitates the use of numerical methods as described in Chapter 10 of Sage and White [57].

Example 3

Consider the system described in Example 2 with the new initial condition that the trajectory intersects the manifold

$$(76) \qquad M(x,t) = x_1-x_2$$

at t = 0 and the new terminal condition that the trajectory lies on the manifold

$$(77) \qquad N(x,t) = \frac{1}{2}x_1^2-x_2 + 1$$

at t = 1. Apply the third transversality condition in Table 7.4 to the functions and N yields

$$(78) \qquad \lambda_1(0) = \xi \quad ; \quad \lambda_2(0) = -\xi$$

$$(79) \qquad \lambda_1(1) = \nu x_1(0) \quad ; \quad \lambda_2(1) = -\nu$$

The fact that the trajectory intersects the manifolds M = 0 and N = 0 gives

(80) $$x_1(0) = x_2(0)$$

(81) $$x_2(1) = \frac{1}{2} x_1^2(1) + 1$$

The new boundary conditions (78) - (81) can be used to determine the four unknown constants in the general solution of the differential system (75).

Example 4

The system in Example 1 is reexamined with the initial conditions x(0) = 0 and the terminal time determined from the intersection of the trajectory with the manifold

$$N(x,t) = x-t-1.$$

Using the initial condition, the general solution (71) specializes to

(82) $$x(t) = a(e^{-t}-e^{t})$$

Since $u = \dot{x}$, it follows from (82) that

(83) $$u(t) = -a(e^{-t}+e^{t}).$$

The fourth transverality conditions in Table 7.4 yield

(84) $$\lambda(t_f) = \nu$$

(85) $$[u^2+x^2+\lambda u-\nu]_{t=t_f} = 0.$$

Recalling that $\lambda = -2u$ and using (82), (83) and (84), it follows that (85) reduces to

(86) $$2a-e^{-t_f}+e^{t_f} = 0$$

provided that $a \neq 0$. Using the fact that the trajectory intersects the manifold N = 0 gives

(87) $$a(e^{-t_f}-e^{t_f}) -t_f-1 = 0.$$

Equations (86) and (87) can be used to determine a and t_f.

Example 5

A point initially at x(0) = 1 is to be moved to

$$N(x_t, t_f) \equiv x_{t_f} = 0$$

in the shortest possible time. Its trajectory satisfies the differential equation $\dot{x} = u$. The control variable u satisfies $|u| \leq 1$.

To solve this problem the performance index $J = t_f$ must be minimized. Note that $\theta \equiv 0$, $t_0 = 0$ and $\phi = 1$. The Hamiltonian

$$H = 1+\lambda u$$

is minimized when

$$u(t) = -\text{sign}\ \lambda(t).$$

The canonical equations are

$$\dot{x} = -\text{sign}\ \lambda; \qquad \dot{\lambda} = 0$$

The given boundary conditions are

$$x(0) = 1; \qquad x(t_f) = 0$$

The transversality conditions yield

$$\lambda(t_f) = \nu \quad ; \ 1+\lambda(t_f)u(t_f) = 0$$

The solution of the system satisfying the auxilliary conditions is

$$x = -t+1, \quad u = -1, \ \lambda = 1; \ t_f = 1.$$

<u>Notes on Table 7.4.</u>

1. In row 5, the condition $u(t) \in U$ for $t \in [t_0, t_f]$ can often be described by a set of inequalities of the form

   $$g(u,t) \geq 0$$

   where $g: R^{m+1} \to R^r$. This inequality constraint can be converted to an equality constraint by writing

(88) $$\dot{z}_i^2 = g_i(u,t) \qquad z_i(t_0) = 0, \qquad i = 1,2,\ldots,r$$

   for each component of g. The constraints (88) are adjoined to the performance index under the integral sign by using Lagrange multipliers $\Gamma(t)$ in the form $-\Gamma^T(t)\ (g-\dot{z}^2)$. The necessary conditions for minimizing the performance index are obtained by studying its variation as previously.

The necessary conditions are the same as those listed in row 5 of Table 7.4 except that the minimal u is determined from

(89) $$\frac{\partial H}{\partial u} - \Gamma^T(t)\,\frac{\partial g}{\partial u} = 0$$

and the fact that Lagrange multipliers Γ must satisfy

(90) $$\Gamma_i \dot{z}_i = 0 \qquad (t \in [t_0, t_f])$$

To apply the above method to the system described in Example 5 the constraint $|u| \leq 1$ is replaced by

$$g_1(u,t) = (1-u)(1+u) \geq 0 \quad .$$

Thus (68) becomes

(91) $$\dot{z}_1^2 = (1-u)(1+u) \qquad z_1(0) = 0$$

while (89) is

(92) $$\lambda \ + \Gamma_1 \ 2u = 0.$$

Since the transversality condition implies that $\lambda \neq 0$, (92) shows that $\Gamma_1 \neq 0$. Hence, it follows from (90) that $\dot{z}_1 = 0$. Therefore, $u = \pm 1$ from (91). The solution $u = 1$ is incompatible with the equations and initial and terminal conditions in Example 5. Hence, as previously, $u = -1$.

An alternative approach is to replace (88) by

(93) $$y_i^2 = g_i(u,t) \qquad i = 1,2,\ldots,r.$$

The conversion of an inequality constraint to an equality constraint was proposed by Valentine [64] and extended by Berkovitz [6].

2. Besides the control variable constraints appearing in row 5, the state variables are frequently constrained by inequalities of the form

$$h(x,t) \geq 0$$

where $h: R^{k+1} \to R^s$. The inequality constraints are eliminated by introducing a new variable x_{k+1}, where

(94) $$\dot{x}_{k+1} = \frac{dx_{k+1}(t)}{dt} = f_{k+1} = \sum_{i=1}^{s} (h_i(x,t))^2 H(h_i)$$

and

$$H(h_i(x,t)) = \begin{cases} 0 & h_i \geq 0 \\ K_i > 0 & h_i < 0 \end{cases}$$

By introducing the corresponding Lagrange multiplier λ_{n+1}, it can be shown that the only modifications in the necessary conditions are the addition of (94) to the state equations with the boundary conditions

(95) $$x_{n+1}(t_0) = x_{n+1}(t_f) = 0$$

and the replacement of the adjoint variable equations by

(96) $$\dot{\lambda} = \frac{d\lambda(t)}{dt} = -\frac{\partial H}{\partial x} - \frac{\partial f_{n+1}}{\partial x} \lambda_{n+1}$$

(97) $$\dot{\lambda}_{n+1} = \frac{d\lambda_{n+1}(t)}{dt} = 0.$$

Further details and sufficiency conditions for the state-space constraint problem appears in Funk and Gilbert [24].

The following heuristic proof of Theorem 1 is given to motivate Pontryagin's principle. The rigorous proof takes more than 50 pages in the original paper. The proof requires a fundamental lemma on the Hamiltonian. The following notation will be used

$$[\frac{\partial h}{\partial x}]_{ij} = [h_x]_{ij} = [\frac{\partial h_i}{\partial x_j}]$$

Note that $(h_x)^T = (h^T)_x$ and $(\lambda^T h)_x = h_x^T \lambda$.

Lemma 1

Assume that the admissible control u and the related trajectory x satisfy the system of differential equations in Theorem 1 with performance index J. The adjoint vector λ satisfies assumption (i) of Theorem 1. Then the variation δJ of the performance criterion due to a permissible[10] variation $\delta u(t)$ of the control of

[10] This means the control $u+\delta u$ is admissible and brings the system from the initial to the final state by following the trajectory $x+\delta x$.

(98) $$\delta J = \int_{t_0}^{t_f} \delta H(t)dt$$

where

$$\delta H = (\phi_u^T + \lambda^T h_u)\delta u.$$

Proof

Let $x(t) + \delta x(t)$ be the trajectory corresponding to the control $u(t)+\delta(j(t)$. As a first order approximation δx satisfies the differential equation

(99) $$\frac{d}{dt}(\delta x) = h_x \delta x + h_u \delta u.$$

Next the time derivative of $\lambda^T \delta x$ will be computed and the values of $\dot{\lambda}^T$ and $\frac{d}{dt}(\delta x)$ will be substituted from (65) and (99):

(100) $$\frac{d}{dt}(\lambda^T \delta x) = (-\phi_x^T - \lambda^T h_x)\delta x + \lambda^T (h_x \delta x + h_u \delta u)$$

After some simple algebra and the addition of $\phi_u^T \delta u$ to both sides of (100) the following result is obtained

(101) $$\frac{d}{dt}(\lambda^T \delta x) + \phi_x^T \delta x + \phi_u^T \delta u = (\lambda^T h_u + \phi_u^T)\delta u$$

Integrating both sides of (101) from t_0 to t_f and using the definition of δH, (101) can be transformed into.

(102) $$[\delta x^T (\lambda - \frac{\partial \theta}{\partial x})]_{t_0}^{t_f} + [\delta x^T \frac{\partial \theta}{\partial x}]_{t_0}^{t_f} + \int_{t_0}^{t_f} (\phi x^T \delta x + \phi_u^T \delta u)dt = \int_{t_0}^{t_f} \delta H dt$$

Using (66) and the definition of δJ, (101) reduces to (98).

Proof of Theorem 1

The optimal trajectory is characterized by $\delta J \geq 0$ for any permissible infinitesimal variation δu. Thus, from (98), a necessary condtion for the minimum is

(103) $$\int_{t_0}^{t_f} \delta H dt \geq 0$$

for any permissible variation δu.

$$\delta u(t) = \begin{cases} \delta \bar{u} & t_1 - e < t < t_1 + e \\ 0 & \text{elsewhere} \end{cases}$$

is a permissible variation. It follows from (103) that the necessary condition for a minimum is $\delta H(t_1) \geq 0$. In other words, at any time t_1, the function of u, $H(x(t_1),u,\lambda(t_1),t_1)$, reaches its minimum for $u = u(t)$. This proves that (67) holds and thus Theorem 1.

In the proof of the lemma

$$||\delta u|| = [(\delta u_1)^2 + (\delta u_2)^2 + \ldots + (\delta u_m)^2]^{1/2} \to 0.$$

However, in the proof of the theorem variations of the type $||\delta x|| \to 0$ and $||\delta u||$ bounded as $e \to 0$ were used. The given proof of Theorem 1 is <u>not</u> rigorous.

(c) <u>The Discrete Minimum Principle</u>

The following theorem is the discrete version of Pontryagin's minimum principle.

<u>Theorem 2</u>

For the control u_i and related trajectory x_i, which satisfies

(104) $$x_{i+1} = h(x_i,u_i,i) \qquad i = i_0,\ldots,i_f-1$$

and inequality constraints of the form

(105) $$u_i \in U \qquad i = i_0,\ldots,i_f-1$$

where $U \subset R^m$ is a given set and i_0 and i_f are fixed integers to minimize the performance criterion

(106) $$J = \theta(x_i,i)\Big|_{i=i_0}^{i=i_f} + \sum_{i=i_0}^{i=i_f-1} \phi(x_i,u_i,i)$$

it is necessary that a vector λ_i exist such that:

(i) λ_i satisfy the adjoint equation

(107) $$\lambda_i = \frac{\partial H}{\partial x_i}$$

where the Hamiltonian

(108) $$H(x_i,u_i,\lambda_{i+1},i) = \phi(x_i,u_i,i) + \lambda_{i+1}^T h(x_i,u_i,i);$$

the initial and final conditions (transversality conditions) are:

(109) $$\eta_{i_0}^T [\lambda_{i_0} - \frac{\partial\theta_{i_0}}{\partial x_{i_0}}] = 0;$$

(110) $$\eta_{i_f}^T [\lambda_{i_f} - \frac{\partial\theta_{i_f}}{\partial x_{i_f}}] = 0.$$

(ii) for all $i = i_0,\ldots,i_f-1$

(111) $$H(x_i,u_i,\lambda_{i+1},i) = \min_{v_i \in U} H(x_i,v_i,\lambda_{i+1},i).$$

The proof of these necessary conditions is similar to the proof of Theorem 1. The assumptions required for this proof are that for each $i = i_0,\ldots,i_f-1$, $\phi(\ ,\ ,\ i):R^k \times R^m \to R$ is continuously differentiable (c.d.) in both arguments, $\theta(.,i):R^k \to R$ is c.d. for $i = i_0,i_f$, $h(.,u_i,i):R^k \to R^k$ is c.d. for $i = i_0,\ldots,i_f-1$ and $u_i \in U$ and the set $\{(x_i,v_i,i)|v_i \in U\}$ is convex for $i = i_0,\ldots,i_f-1$ and $x_i \in R^k$. Hautus [30] has weakened these conditions.

A more general discrete minimum principle, involving state-space constraint, and its proof can be found in Hautus [30]. Sufficient conditions for a related problem appear in McClamrock [46].

Note that for the unconstrained situation in which $U = R^m$, equation (111) implies the necessary condition.

(112) $$\frac{\partial H}{\partial u_i} = 0 \qquad i = i_0,\ldots, i_f-1$$

A perturbation method can be used to develop necessary conditions for this situation in an analogous manner to the method used in (b).

In general, the solution of the above nonlinear, two-point boundary value problem is solved by reiterative techniques using a digital computer. The discrete maximum principle provides an optimization method which is the natural one to use. as an approximation for many continuous problems. Care must be used in determining a sequence of discrete optimal control problems which converge (in some sense) to the associated continuous control problem. Further discussion of this point can be found in Budak [15] and Cullum [20,21].

If the Hamiltonian or performance criterion is linear in the control variables Theorem 2 cannot be applied. A direct analysis in the (H,u_i) plane is carried out to find the control, which are called singular. An example will be given in the next chapter.

(d) Optimization of Several Performance Criteria

It would seem that optimal control theory could be applied in cancer chemotherapy to discover more efficient drug protocols. However, a major problem is the formulation of meaningful performance criteria. Even if this difficulty is overcome, there may be several criteria, $J_1,\dots,J_s$ to minimize or maximize simultaneously. Frequently these criteria are in opposition. For example, the maximum number of cancerous cells must be killed while white blood cells must be spared and other toxic effects of the drug minimized. Rarely will a control vector u* exists which minimizes all the criteria simultaneously. Therefore, the problem is to find the best compromise solution.

One approach to the problem is to form a single performance criterion

$$J = \sum_{i=1}^{s} c_i J_i \quad , \quad \sum_{i=1}^{s} c_i = 1; \qquad c_i \geq 0,$$

which would then be minimized. For example, a small weight might be assigned to the performance criterion corresponding to the number of white bloods cells if these cells are replaced. Then it would be permissible to destroy a large number of such cells. However, the choice of c_i requires a value judgement by the scientist which is often difficult to make.

Another approach, originating in economics, is due to Pareto [52]. It optimizes the criteria in the following sense.

Definition 6

A control u* is Pareto-optimal (or nondominated) if and only if for every other admissible control u

$$J_i(u) \leq J_i(u^*) \qquad i = 1,2,\dots,s$$

implies

$$J_i(u) = J_i(u^*) \qquad i = 1,2,\dots,s.$$

Hence, if u^* is Pareto-optimal there is at least one i such that $J_i(u) > J_i(u^*)$ for any admissible control or else $J_i(u) = J_i(u^*)$ for $i = 1,2,\ldots,s$.

A simple example of Pareto-optimality appears in Fig. 7.13.

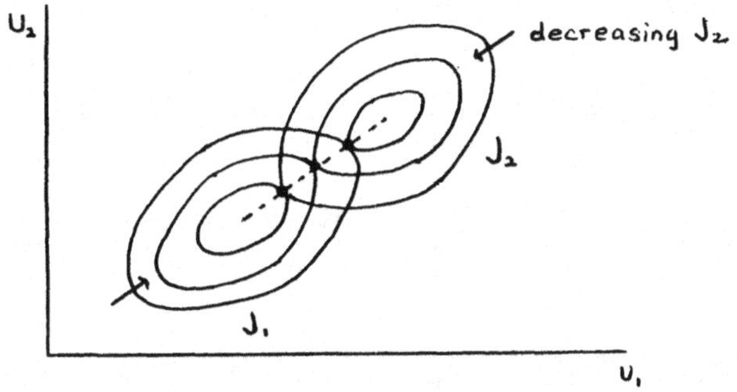

Figure 7.13 Pareto-optimality.

Contours of constant cost are plotted in R^m. The Pareto-optimal controls $u \in R^m$ are points of tangency of the contours.

The following lemma appears in Leitmann [42].

Lemma 2

If there is a $c \in R^s$, $c_i > 0$ for $i = 1,2,\ldots,s$ and $\sum_{i=1}^{s} c_i = 1$ such that the admissible control u^* satisfies

$$J(u^*) \leq J(u)$$

for all admissible controls u, where

$$J = \sum_{i=1}^{s} c_i J_i$$

then u^* is a Pareto-optimal control with respect to the performance criterion J_i.

Note that an optimal control found by the first approach is a Pareto-optimal control by Lemma 2. Hence, once all Pareto-optimal controls have been found the most satisfactory solution can be selected.

The Pareto-optimal controls are among the solutions of an optimal control problem with a scalar performance criterion which is a linear combination of the given performance criterion as shown in Theorem 3 due to Yu and Leitmann [70].

Consider the system

(113) $$\dot{x} = h(x(t),u(t)),\ x(t_0) = x_0$$

where $x(t) \in R^k$, $u(t) \in R^m$ and $f{:}R^k \times R^m \to R^k$ is C'.

A control $u{:}[t_0,t_f) \to U$ a given subset of R^m, is admissible if it is bounded and measurable and if it generates a solution $x{:}[t_0,t_f] \to R^k$ of (113) such that $x(t_0) = x_0$ and

(114) $$N(x(t_f)) = 0$$

for some $t_f \in [t_0,\infty)$ where $N{:}R^k \to R^q$ is a given C' function and $\frac{\partial N}{\partial x}(x)$ has rank q on $\{x|N(x) = 0\}$.

The performance criterion column vector J has components

(115) $$J_i(u,x) = \int_{t_0}^{t_f} \phi_i(x(t),u(t))dt \qquad i = 1,2,\ldots,s.$$

where $\phi_i{:}R^k \times R^m \to R$ is C'.

Remark: Measurable functions include most functions that are studied in an elementary calculus course, such as, continuous and piecewise continuous functions.

Theorem 3

If u^* is Pareto-optimal with respect to J and x^* the corresponding trajectory satisfying (113) - (115) there exists an $\alpha \in R^s$, $\alpha \geq 0$, a continuous $\lambda{:}[t_0,t_f^*] \to R^k$ with $(\lambda(t),\alpha) \neq 0$ on $[t_0,t_f^*]$, a constant q vector ν and a Hamiltonian function

(116) $$H(x,u,\lambda,\alpha) = \alpha^T\phi$$

where $\phi = (\phi_1,\ldots,\phi_s)^T$ such that

(i) λ is a solution of the adjoint equation

(117) $$\dot{\lambda}_i(t) = -\frac{\partial H(x^*,u^*,\lambda,\alpha)}{\partial x_i} \qquad i = 1,2,\ldots,k$$

(118) $$\lambda_i(t_f) = \sum_{j=1}^{q} \nu_j \frac{\partial N_j(x^*(t_f^*))}{\partial x_i} \qquad i = 1,2,\ldots,k$$

(ii) $H(x^*,u^*,\lambda,\alpha) \leq H(x^*,u,\lambda,\alpha)$

for all admissible u.

A relatively simple proof of Theorem 3 has been given by Schmitendorf and Leitmann [59].

The classic book on optimal control theory is by Pontryagin et al. [53]. Further details on optimal control theory and its application can be found in the books by Athans and Falb [2], Intriligator [34] and Leitmann [41]. Derivations of the continuous and discrete Pontryagin minimum principle by removing constraints using Lagrange multipliers and then applying perturbation techniques are given by Sage and White [57]. Berkovitz [8] is a modern rigorous treatment of optimal control theory. A short history of optimal control theory is contained in a paper by Berkovitz [7].

REFERENCES

1. Adolph, E.F., Origins of Physiological Regulations, Academic Press, New York, 1968.

2. Athans, M. and Falb, P.L., Optimal Control Theory: An Introduction to the Theory and Its Applications, McGraw-Hill, New York, 1966.

3. Banks, H.T., Modeling and Control in the Biomedical Sciences, Vol. 6 of Lecture Notes in Biomathematics, Springer-Verlag, New York, 197 .

4. Bayliss, L.E., Living Control Systems, English Univ. Press, London, 1966.

5. Bellman, R.,Adaptive Control Processes: A Guided Tour, Princeton University Press, Princeton, 1961.

6. Berkovitz, L.D., Variational methods in problems of control and programming, J. Math. Anal. Appl., 3, 145-69,1961.

7. Berkovitz, L.D., Optimal Control Theory, Am. Math. Monthly, 83, 225-239, 1976.

8. Berkvotiz, L.D., Optimal Control Theory, Springer-Verlag, New York,1974.

9. Bernard, C., Les Phenomenes de la Vie, Paris, 1878.

10. Bode, H.W., Feedback Amplifier Design, Bell System Tech. J., 19, 42, 1940.

11. Bode, H.W., Network Analysis and Feedback Amplifier Design, Van Nostrand, Princeton, 1945.

12. Brogan, W.L., Modern Control Theory, Quantum Press, New York, 1974.

13. Brown, J.H.U. and Grann, D., Engineering Principles in Physiology, Vol. 1-3, Academic Press, New York, 1973.

14. Brown, J.H.U., Jacobs, J.E. and Stark, L. (editors). Biomedical Engineering, F.A. Davis, Philadelphia, 1971.

15. Budak, B.M., Difference Approximations in Optimal Control Problems, SIAM J. Control, 7, 18-31, 1969.

16. Clark, C.W., Mathematical Bioeconomics. The Optimal Management of Renewable Resources, J. Wiley, New York, 1976.

17. Clynes, M. and Milsum, J.H. (editors). Biomedical Engineering Systems, McGraw-Hill, New York, 1970.

18. Cannon, W.B., Organization for physiological homeostasis, Physiol. Revs., 9, 399,1929.

19. Coddington, E.A. and Levinson, N., Theory of Ordinary Differential Equations, McGraw-Hill, New York, 1955.

20. Cullum, J., Discrete approximations to continuous optimal control problems, SIAM J. Control, 7, 32-49, 1969.

21. Cullum, J., An explicit method for discretizing continuous optimal control problems, J. O. T. A., 5, 1970.

22. Dorf, R.C., Modern Control Systems, Addison-Wesley, Reading, 1967.

23. Fredericq, L., Influences du milieu ambiant sur la composition du sang des animaux aquatiques, Arch. de Zool. Exper. et. Gen., 3, 35, 1885.

24. Funk, J.E. and Gilbert, E.G., Some sufficient conditions for optimality in control problems with state space constraints, SIAM J. Control, 8, 498-504, 1970.

25. Gilbert, E.G., Controllability and observability in multivariable control systems, J. Soc. Ind. Appl. Math. - Control Series, Series A, Vol. 1, No. 2, 128-151, 1963.

26. Goodwin, B.C., Temporal Organization in Cells, A Dynamic Theory of Cellular Control Processes, Academic Press, New York, 1963.

27. Grodins, F.S., Control Theory and Biological Systems, Columbia Univ. Press, New York, 1963.

28. Guyton, A.C. and Crowell, J.W., Cardiac deterioration in shock: I. Its progressive nature, in Hershey, S.G. (editor). Shock, Little, Brown and Co., Boston, 1-12, 1964.

29. Guyton, A.C., Taylor, A.E. and Granger, H.J., Circulatory Physiology. I Cardiac Output and Its Regulation, Saunders, Philadelphia, 1973. II Dynamics and Control of the Body Fluids, ibid, 1975.

30. Hautus, M.L.J., Necessary conditions for multiple constraint optimization problems, SIAM J. Control, 11, No. 4, 653-69, 1973.

31. Hazen, H.L., Theory of servomechanisms, J. Franklin Instl., 218, 279, 1934.

32. Hurwitz, A., On the conditions under which an equation has only roots with negative real parts, Math. Annalen, 46, 273-284, 1895. In Selected Papers on Mathematical Trends in Control Theory, Dover, New York, 70-82, 1964.

33. Iberall, A.S. and Guyton, A.C. (editors). Regulation and Control in Physiological Systems, Instrument Soc. of America, Pittsburgh, 1973.

34. Intriligator, M.D., Mathematical Optimization and Economic Theory, Prentice Hall, Englewood Cliffs, 1971.

35. Kalman, R.E., On the general theory of control systems, in Automatic and Remote Control (Proc. IFAC Moscow 1960). Vol. I, Butterworth, London, 481-492, 1961.

36. Kalman, R.E., Mathematical description of linear dynamical systems, J. Soc. Ind. App. Math. - Control Series, Series A, Vol. 1, No. 2, 152-192, 1963.

37. Kalmus, H., Regulation and Control in Living Systems, Wiley, New York, 1966.

38. Kernevez, J.P., Control, optimization and parameter identification in immobilized enzyme systems, in Analysis and Control of Immobilized Enzyme Systems, edited by Thomas, D. and Kernevez, J.P., North-Holland, Amsterdam, 199-225, 1976.

39. Kernevez, J.P. and Thomas, D., Numerical analysis and control of some biochemical systems, Appl. Math and Optimization, 1, 222-285, 1975.

40. Kline, J., Biological Foundations of Biomedical Engineering, Little, Brown and Co.., Boston, 1976.

41. Leitmann, G., An Introduction to Optimal Control, McGraw-Hill, New York, 1966.

42. Leitmann, G., Cooperative and Non-Cooperative Many Player Differential Games, Springer-Verlag, Vienna, 1974.

43. Longini, Jr., I.M., Ackerman, E. and Elveback, L.R., An optimization model for influenza epidemics, Math. Biosc., 38, 141-157, 1978.

44. Mayr, O., The Origins of Feedback Control, M.I.T.Press, Cambridge, 1970.

45. Maxwell, J.C., On governors, Proc. of the Roy. Soc. of London, 16, 1868, in Selected Papers on Mathematical Trends in Control Theory, Dover, New York, 270-283, 1964.

46. McClamroch, N.H., A sufficiency condition for discrete optimal processes, Int. J. Control, 12, 157-61, 1970.

47. Milsum, J.H., Biological Control System Analysis, McGraw-Hill, New York, 1966.

48. Milhorn, H.T., Applications of Control Theory to Physiological Systems, Saunders, Philadelphia, 1966.

49. Minorsky, N., Directional stability of automatically steered bodies, J. Am. Soc. Naval Eng., 34, 280, 1922.

50. Nyquist, H., Regeneration theory, Bell System Tech. Journal, 11, 126-147, 1932.

51. Papoulis, A., The Fourier Integral and Its Applications, McGraw-Hill, New York, 1962.

52. Pareto, V., Cours d'Economie Politique, Lausanne, Rouge, 1896.

53. Pontryagin, L.S., Boltyanskii, V.G.; Gamkrelidge, R.V. and Mischenko, E.F. The Mathematical Theory of Optimal Processes, Interscience (Wiley), New York, 1962.

54. Riggs, D.S., Control Theory and Physiological Feedback Mechanisms, William and Wilkins, Baltimore, 1970.

55. Rosenbleuth, A., Wiener, N. and Biglow, J.H., Behavior, purpose, and teleology, Phil. Sci. 10, 18-24, 1943.

56. Routh, E.J., Dynamics of a System of Rigid Bodies, Macmillan, New York, 1892.

57. Sage, A.P. and White, C.C., Optimum Systems Control, Prentice Hall, Englewood Cliffs, N.J., 1977.

58. Saucedo, R. and Schiring, E., Introduction to Continuous and Digital Control Systems, MacMillan, New York, 1968.

59. Schmitendorf, W.E. and Leitman, G. A simple derivation of necessary conditions for Pareto optimality, IEEE Trans. on Automatic Control, AC - 19, 601, 1974.

60. Schwan, H.P. (editor). Biological Engineering, McGraw-Hill, New York, 1969.

61. Stear, E.B., Application of control theory to endocrine regulation and control, Annals of Biomed. Eng., 3, 439-455, 1975.

62. Swan, G.W., Mathematical Studies of the Harvesting and Control of Biologically Renewable Resources. Unpublished report to the National Science Foundation, Washington, D.C., 1974.

63. Talbot, S.A. and Gessner, Systems Physiology, Wiley, New York, 1973.

64. Valentine, F.A., The problem of Lagrange with differential inequalities as added side conditions, in Contributions to the Calculus of Variations, 1933-1937, University of Chicago Press, Chicago, 1937.

65. Van Valeknburg, M.E., Network Analysis, 2nd ed., Prentice-Hall, Englewood Cliffs, N.J., Chapter 10, 1964.

66. Vyshnegradskii, I.A., On controllers of direct action, Izv., SPB Tekhnolog. Inst., 1877.

67. Wiener, N., Cybernetics,MIT Press, Cambridge, Rev. ed., Wiley, New York, 1961.

68. Wickwire, K., Mathematical models for the control of pests and infectious diseases: a survey, Theor. Pop. Biol. 11, 182-238, 1977.

69. Yamamoto, W.S. and Brobeck, J.R. (editors). Physiological Controls and Regulations, Saunders, Philadelphia, 1965.

70. Yu, P.L. and Leitmann, G., Non dominated decisions and cone convexity dynamic multicriteria decision problems, JOTA, 14, No. 5, 573-84, 1974.

71. Zadeh, L.A. and Desoer, C.A., Linear System Theory, The State Space Approach, McGraw-Hill, New York, 1963.

VIII. TOWARDS MATHEMATICAL CHEMOTHERAPY

8.1 Introduction

Since the discoverer of a "wonder" drug for cancer chemotherapy is more likely to be a biological scientist than a mathematician, one may ask what contribution a mathematician can make to cancer chemotherapy. The answer to this question is that a mathematician can discover ways to use existing drugs more efficiently. Even an efficacious drug can appear worthless if it is not administered properly.

To illustrate the importance of properly designed drug protocols, consider the following highly simplied, discrete, model of the cell cycle. The generation time of each cell is 5 units and is divided into 5 states. A cell in state i (i = 1,2, 3,4) at time t enters state i+1 at time t+1. A cell in state 5 at time t divides into 2 cells and both cells enter state 1 at time t+1.

Suppose that at t = 0 there are x_i cells in state i for i = 1,2,...,5. Then at t = 0 the system can be described by the row vector

$$(1) \qquad (x_1,x_2,x_3,x_4,x_5).$$

When t = 1, the row vector (1) is transformed according to the above into the vector

$$(2) \qquad (2x_5,x_1,x_2,x_3,x_4).$$

After four similar transformations (at t = 5), the population of cells has doubled and is described by

$$(3) \qquad (2x_1,2x_2,2x_3,2x_4,2x_5).$$

Consider a drug which kills cells in all phases of the cell cycle (a cycle nonspecific drug). Assume that when the drug is applied, it instantly kills a certain fraction of all the cells (normal as well as cancerous). Moreover, suppose that the drug is instantly decomposed so that there is no effect due to the cumulative dosage. The same concentration of the drug is used for each dose. Suppose each dose reduces the population to a fraction f of its size ar the instant the drug was administered. For simplicity, assume that the drug only acts on two different cell populations. The first population is normal with generation time 5 and originated from n_0 cells at time $t = 0$. The second population is cancerous with generation time 10 and originated from c_0 cells at time $t = 0$. If the drug is administered at $t = 10k$, $k = 1,2,3,\ldots$, then the instant after the nth dose is administered, the size of the cancerous population is $f^n 2^n c_0$ while the size of the normal population is $f^n 2^{2n} n_0$. The object of the therapy is to reduce the number of cancerous cells to less than 1 while maintaining the normal population's size above rn_0 for some fixed positive $r < 1$. Translating these objectives in terms of the sizes of the populations after n doses of the drug, yields the following inequalities for the cancerous and normal cells, respectively,

$$f^n 2^n c_0 < 1 \tag{4}$$

$$f^n 2^{2n} n_0 > rn_0. \tag{5}$$

Solving inequalities (4) and (5) for f gives

$$\frac{1}{4} r^{1/n} < f < \frac{1}{2} c_0^{-\frac{1}{n}} \tag{6}$$

Inequality (6) has a solution[1] if and only if

$$\frac{1}{2} c_0^{-\frac{1}{n}} < 1 \tag{7}$$

and

$$\frac{1}{4} r^{\frac{1}{n}} < \frac{1}{2} c_0^{-\frac{1}{n}}$$

or

$$(c_0 r)^{1/n} < 2. \tag{8}$$

If n is sufficiently large, inequalities (7) and (8) hold. Hence, it is theoretically possible to eradicate the cancer cells while maintaining the normal popula-

[1] In clinical trials the inequalities may not be satisfied since the concentration of the tested drugs and the timing of the doses are determined empirically. Hence, a potentially useful drug may appear worthless.

tion at a functional level. More generally, the last assertion remains valid if the cycle time of normal cells is less than the cycle time of the cancerous cells.

On the other hand, if the cycle time of the normal cells is greater than or equal to the cycle time of the cancerous cells, the cancer cells cannot be eradicated without destroying the normal population using a cycle nonspecific drug. Any drug protocol which would eliminiate the cancerous cells would also destroy the slower growing normal population. However, in some situations, a drug which does not kill cells throughout the whole cell cycle (a cycle specific drug) can be effective. For example, assume that the generation times of a cancer and a normal cell are 5 and 9 units, respectively. Moreover, suppose that at $t = 0$ the cancerous population is described by the row vector (1) while the normal population is described by

$$(n_1, n_2, n_3, \ldots, n_9).$$

The same concentration of the cycle specific drug is administered at $t = 3k$, $k = 1,2,\ldots$. Assume that the drug acts in the same manner as the cycle nonspecific drug described above except that it only kills cancer cells in state 3 and normal cells in state 5. Once again f will denote the surviving fraction of cells in these states. When $t = 15n$ the number of surviving cancer cells is

$$f^{5n} 2^{3n} c_0$$

where $c_0 = x_1 + \ldots + x_5$. Hence, the number of cancer cells can be reduced to less than 1 provided that

$$f < 2^{-\frac{3}{5}} c_0^{\frac{1}{5n}} \tag{9}$$

In the normal population, only cells which are descendants of cells that were in states 2,5 and 8 at $t = 0$ will be eliminated. The remaining normal cells will continue to proliferate. Hence, by selecting f to satisfy inequality (9) it is possible to eliminate the cancerous cells without destroying the normal population.

A cycle specific drug can be effective even if the cycle times of both populations are the same. It is sufficient that at predictable times, a large percentage of the cancer cells and a small percentage of the normal cells are in the drug-susceptible parts of their cycle. This distribution of cells in their cycles may not occur naturally. However, the desired distribution may be obtained by using other drugs to synchronize the cells.

The above examples illustrate that the construction of efficient drug protocols is a complex mathematical problem. There is a subtle interplay between the cell kinetics of normal and cancerous cells, the scheduling of doses and the concentration of the employed drug. A more sophisticated analysis than presented in this section is required to describe the actual response of real populations of cells to drugs. Some early attempts at quantitative cencer chemotherapy are reviewed by Aroesty et al. [1]. The remainder of this chapter will present some recent research in this area. The work on leukemia will be deferred until the next chapter since it requires some discussion on the development of blood cells.

8.2 Growth Laws and Cycle Nonspecific Cancer Chemotherapy

The essence of the simple treatment of cycle nonspecific cancer chemotherapy, presented in the first section, was the growth law of the tumor. In this section, the introductory discussion will be made more realistic by using the more general growth laws in Chapter 2. First some intuitive approaches will be presented. Then it will be shown how optimal control theory can be used to devise programs for cancer chemotherapy.

Priore [56] formulated the following mathematical model for analyzing the short term response of various human tumors to 5-fluorouracil and 6-mercaptopurine. During intervals between doses of the chosen drug the tumor was assumed to follow an exponential growth process. Only a fraction of the tumor cells were assumed to be susceptible to the drug. Moreover, the proportion of the susceptible tumor cells killed by the drug was assumed to be a logistic function of the logarithm of the cumulative dosage of the drug. The model was fitted by a nonlinear least squares method. Estimates of the tumor growth rate and the drug effect were obtained for each tumor. An analogous simple approach was used by Skipper et al. [63] to describe the effect of cytoxan on certain solid tumors. Clinically relevant information on the length of treatment required for curing childhood acute lymphoid leukemia [34] and choriocarcinoma [3] have been determined by application of an exponential growth model in combination with the constant fraction cell kill principle.

A two-compartment pharmacodynamic model for the characterization of chemotherapeutic agents was proposed by Jusko [39]. The model is analogous to the one discussed in section 4 of Chapter 2, except that all the rate parameters are constants and the compartments have a different interpretation. One compartment is used for modeling the administration of the drug while the other, consisting of target cells, represents the site of the chemotherapeutic effect. A portion of the drug that reaches the pharmacologic site is involved in an irreversible reation (rate constant k) with the receptors of the target cells. This produces a cell loss rate kX_tC, where C is the number of target cells and X_t is the measure of the amount of drug at the receptor site. Differential equations representing the continuity of flow through each compartment can then be derived. These equations are used to devise a theoretical model for representing the treatment of mice with the drug cyclophosphamide. An interesting prediction of the model is that the subdivision of the drug into four fractional doses did not significantly change the dose-effect constants of the drug. Pharmocokinetic priciples in chemotherapy are further developed in Jusko [40].

Unfortunately, cycle nonspecific drugs ultimately damage normal cells such as bone marrow and intestinal epithelium cells. Hence, the above models must be generalized to include the deleterious effects of the antineoplastic drugs. This was done by Berenbaum [8]. Berenbaum considered two populations of cells (cancer cells and normal cells). The population of cancer cells was assumed to either grow exponentially or grow according to the Gampertz equation. It was assumed that the normal cells were maintained at a steady-state level by homeostasis and so their growth was approximated by the logistic equation.

The effect of the drug and the drug schedule were simulated in the following manner. Initially a dose D of the drug is given which reduces both normal and tumor cells by a fraction F. The cells now grow according to their original growth law from the new initial condition. After a time interval t, a second dose of the drug of the same magnitude of the first is administered. This process is then repeated. For example, if a tumor is in exponential growth, then the net surviving fraction after n doses is

$$N_0(F2^{t/T})^n \tag{10}$$

where N_0 is the number of tumor cells present initially and T is the doubling time. The surviving fraction of cells is related to the dose by the dose-response curve. Berenbaum studies three types of dose-response curves: exponential ($F = e^{-\alpha D}$), exponential with a shoulder ($F = 1-(1-e^{-\alpha D})^{\beta}$) and hyperbolic ($F = (\frac{D}{D_0})^{-\gamma}$).

The important features of a typical regrowth curve for the drug effect and schedule assumed by Berenbaum is shown in Fig. 8.1. After a single dose D of the

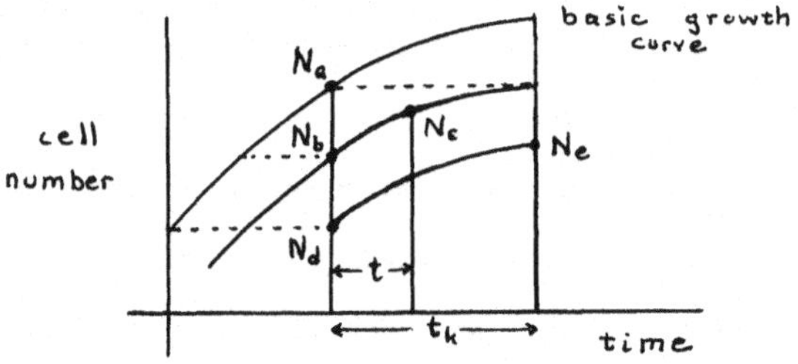

Figure 8.1 Tumor regrowth between two doses.

drug, the tumor population is instantly reduced to $N_b = N_aF$ where $F < 1$. Tumor regrowth continues along the basic growth curve translated to the right. After a time t the tumor reaches size N_c. If $t < t_*$, $t = t_*$, $t > t_*$ then $N_c < N_a$, $N_c = N_a$, $N_c > N_a$, respectively. On the other hand, suppose the time between doses is fixed and is t_*. Furthermore, assume the dose is increased so that the tumor population is instantly reduced to $N_d = GN_a$, where $G < F$. Tumor regrowth continues along the basic growth curve translated to the right as shown in Fig. 8.1. After a time t_* the tumor has increased to size $N_e < N_a$. Similarly, it can be shown that if $G > F$ the tumor will regrow to a size that is larger than its original size N_a.

The above reasoning can be repeated to study the effect of administering n fixed doses at intervals of time t. The tumor response for different values of the single dose surviving fraction F is illustrated in Fig. 8.2. If $F > F_e$, $F = F_e$, $F < F_e$, then the population size at diagnosis is increased, unchanged, reduced, respectively. These graphical results can be translated into analytical terms.

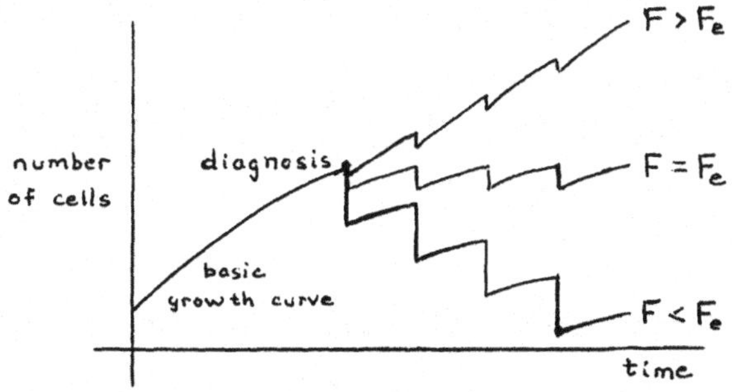

Figure 8.2 Tumor response to fixed doses at fixed time intervals.

For example, for exponential growth it follows from (10) that F_e satisfies $F_e\ 2^{t/T} = 1$ and so $F_e = 2^{-t/T}$.

Berenbaum applied the above reasoning to a population of tumor cells which grew according to either the exponential or Gompertz law. He obtained an inequality to be satisfied by F in order that all tumor cells be eradicated. Another inequality was obtained for F from the condition that the normal cells reach a new tolerable steady-state level. Finally, these inequalities for F were translated into inequalities for the dose D by using the dose-response curves.

These results are difficult to apply in vivo, since the response to therapy must be monitored after repeated doses of the drug. This requires a measurement of tumor size during the course of therapy. For nonsolid tumors it might be possible to estimate the number of tumor cells by direct sampling. For example, in acute leukemia the leukemic blasts could be counted. However, the size of a solid tumor is difficult to estimate even when the tumor is accessible. The exact extent of the tumor may not be obvious. Moreover, tumor nodules may contain variable but significant numbers of nonviable cells, normal tissue elements and interstitial fluid which also make sizing difficult. Hence, the clinical applications of mathematical growth models require the development of new methods of determining tumor size - for example, by radioactive labelling or by the use of biochemical markers or substances characteristic of the tumor.

The use of marker substances to determine tumor size is not just idle speculation. The excess synthesis of a gamma globulin, IgG_{mp} (or M-component), was used as a quantitative measure of the number of "live" tumor cells in multiple myeloma by Salmon and Smith [59]. Multiple myeloma is a plasma cell tumor characterized by very aggressive infiltration of the marrow [43]. The anaplastic plasma cells are mutations of lymphocytes which exhibit invasive growth and produce atypical antibodies.

Sullivan and Salmon [66] used measurements of synthesis of M-components to determine the total number of tumor cells present in patients with multiple myeloma. For each patient, the total number of tumor cells, N, was shown to satisfy the Gompertz equation

$$N = N_0 e^{(A/\alpha)(1-e^{-\alpha t})} \tag{11}$$

with patient-specific parameters N_0, A and α.

Therapy consisted of fixed doses of a drug at a fixed time interval. The drugs employed were mephalan and predisone, both cycle-nonspecific drugs. From the clinical data for each patient they obtained tumor regression curves similar to those in Fig. 8.2. When a smooth curve was fitted to the points at the peaks of each sawtooth, the resulting curve was discovered to satisfy a Gompertz equation with the same retardation coefficient, α, as measured for the untreated tumor during growth.

The above experimental results were explained in [66] by means of the same model which led to Fig. 8.2. It was assumed that if N(t) is the number of viable tumor cells at time t of drug administration then N(t+) the remaining number of viable cells just after the drug is given satisfies

$$\ell n\left(\frac{N(t+)}{N(t)}\right) = R$$

where R is some constant depending on the dosage of the drug. Hence, the fraction of cells that survive a single dose is

$$F = e^R.$$

It follows from (11) that N(t) satisfies the differential equation

$$\frac{d}{dt} \ell n\left(\frac{N}{N_\infty}\right) = -\alpha \ \ell n\left(\frac{N}{N_\infty}\right) \tag{12}$$

where

$$N_\infty = N(\infty) = N_0 e^{A/\alpha} \tag{13}$$

Suppose that the (k+1)th dose is administered a time

$$t_k = kT$$

where T is the interval between two successive doses. Let N_k be the size of the tumor cell population immediately before the (k+1)th dose. Then from (12)

$$\int_{FN_k}^{N_{k+1}} \frac{d\, \ell n(N/N_\infty)}{\ell n(N/N_\infty)} = -\alpha \int_{t_k}^{t_{k+1}} dt$$

which yields the recursion formula

$$\ell n\, N_{k+1} = e^{-\alpha T}\, \ell n\, N_k + Re^{-\alpha T} + \ell n\, N_\infty (1-e^{-\alpha T}) \tag{14}$$

Substituting for N_∞ in (14) from (13) and defining the constant B by

$$Re^{-\alpha T} + \frac{A}{\alpha}(1-e^{-\alpha T}) = \frac{B}{\alpha}(1-e^{-\alpha T})$$

equation (14) can be rewritten as

$$\ell n\, N_{k+1} = e^{-\alpha T}\, \ell n\, N_k + \frac{B}{\alpha}(1-e^{-\alpha T}) + \ell n\, N_0 (1-e^{-\alpha T}) \tag{15}$$

It can be shown by induction from (15) that

$$N_k = N_0 e^{B/\alpha\, (1-e^{-\alpha k T})} \tag{16}$$

Hence, the drug-induced tumor regression also follows Gompertzian kinetics with the same retardation coefficient, α, as measured for the gorwing, untreated tumor.

Sullivan and Salmon used the Gompertzian model of tumor growth and regression to predict the course of therapy. The methodology also provided a guide, well in advance of the actual occurrence, to the ultimate failure of the therapy to reduce the tumor further. The predictive and analytic accuracy of the model was confirmed for those cases where both tumor growth and regression were measured.

The natural history of the myeloma was reconstructed from the gorwth curves. Sullivan and Salmon discovered that the duration of the disease is shorter than the previous estimates. The subclinical phase of the disease (up to 2.5×10^{11} cells) occurs from less than 4 months to approximately 18 months. However, most patients are not diagnosed at this threshold point. Recognition usually occurs after the tumor has undergone further but ever slowing growth for several additional years. In the average patient fewer than 5 years elapse from the initial tumor cell doubling to its clinical presentation with from 10^{11} to more than 10^{12} cells in the body.

The development of drug resistance is not the explanation for the failure of a cycle nonspecific drug to reduce the tumor further. Therapeutic equilibrium occurs when the number of cells destroyed by a single dose of the drug equals the number of cells born during the drug free period T. Therefore, at equilibrium

$$N_k = N_{k+1}. \tag{17}$$

The common value in (17) is just N_e, the number of tumor cells at equilibrium. From (14) and (17) it can be shown that

$$\ell n\ N_e = \frac{Re^{-\alpha T}}{1-e^{-\alpha T}} + \ell n\ N_\infty\ . \tag{18}$$

The ability to predict the maximum degree of regression and the time until therapeutic equilibrium is important clinically for drug scheduling. It is thought that the plateauing phenomena during therapy results from the expansion of the proliferative fraction of the tumor as the total population is reduced by a cycle nonspecific drug. Thus, when the point of equilibrium is reached, it is natural to use antimetabolites that attack tumor cells during the proliferative cycle. This strategy has been applied by Griswold et al. [28] in treating the phasmacytoma of the hamster, which also grows according to the Gompertizian law. The plasmacytoma cannot be cured with a potent cycle-active drug (cytosine arabinoside) when the tumor is extensive. However, it can be cured by the same drug if the total tumor mass is first reduced by several logs with cycle nonspecific drugs. A reversal of the sequence of drugs is ineffective. Burke and Owens [17] have used related reasoning in developing their experimental protocol for acute leukemia, where VCR and cytoxan (cycle nonspecific drugs) were used first to reduce tumor size and ara-C then employed.

The approach of Sullivan and Salmon is discussed further in Salmon and Durie [58]. A clinical application of Gompertzian kinetics to the chemotherapy of multiple mycloma appears in a recent unpublished paper of Cox [18].

A natural generalization of the approach of Sullivan and Salmon to therapy is to consider drug toxicity. For example, the effects of the drug on a normal population could be studied as in Berenbaum [8] Alternatively, large concentrations of the drug could be used initially when the tumor is growing relatively fast and smaller concentrations of the drug when the tumor is growing relatively slowly. The objective of this new therapy would be the reduction of the number of tumor cells to a preassigned level while minimizing the cumulative toxicity effects of the anti-cancer drug. Optimal control theory can be used to quantitate this new approach.

Petrovskii [54] introduced the use of optimal control theory for cancer problems. He presents a mathematical model of a tumor based on a system of ordinary differential equations. However, the biological meaning of some terms in the equation is not clear. Moreover, the model is not analyzed. Earlier papers by Petrovskii et al. [53] and Zakharova et al. [73] contain references to applications of optimal control theory to biomedical problems. The first detailed discussion of the application of methods of optimal control theory appears to be the paper of Bahrami and Kim [4]. This paper will be discussed in the following section on cycle specific therapy.

Swan and Vincent [67] applied optimal control theory analysis to obtain a program of chemotherapy for IgG multiple myeloma. The raw data on ten patients who had undergone treatment with a melphalan, cyclophosphamide and prednisone program (MCP) were supplied to the authors by Dr. Salmon. Swam and Vincent assumed that the cumulative effect of the three drugs could be represented by a single (fictitious) drug administered at the end of each time interval. They also argued that the continuous administration of the drug would shorten the regrowth phase of the tumor. Hence, the final level of cancer cells as well as the total amount of the drug used should be lower than for periodic drug administration.

To verify this heuristic argument they applied optimal control theory analysis using the assumption that the (fictitious) drug was administered continuously.

Swan and Vincent's mathematical model is based on the following assumptions:

(a) The untreated tumor grows according to the Gompertz equation (11).

(b) The size of the tumor at the time $t = 0$ is known from clinical observations.

(c)[2] The action of the drug a time t can be written in the form

$$f(v)N$$

where N is the population size. Here $f(v)$ is the rate of control per cell and is given by

$$f(v) = k_1 v/(k_2+v) \tag{19}$$

where k_1 and k_2 are constants and $v = v(t)$ represents the concentration of the drug at time t.

(d) After the administration of a given dose at a given time the anti-neoplastic drug is assumed to decay with a time constant γ. Accordingly,

$$\frac{dv}{dt} = -\gamma v. \tag{20}$$

It is assumed that the turnover time of the drug is about 1 day $(\gamma \simeq 1)$.

(e) For purposes of data reduction it was assumed that each patient had the same equilibrium number (N_∞) of tumor cells with $\ell n\, N_\infty = 29$ and an average value of $\alpha = .015$ per day. These values are reasonable according to the results obtained by Sullivan and Salmon [66].

Under the above assumptions the kinetics of the tumor are described by

$$\frac{dN}{dt} = -\alpha\, N\, \ell n(\frac{N}{N_\infty}) - f(v)N \qquad N(0) = N_0. \tag{21}$$

The differential equation for Gompertzian growth can be reduced to the differential equation for exponential growth by means of the transformation

$$y = \ell n(N/N_\infty). \tag{22}$$

---

2 These and similar assumptions will be discussed in Section 4.

For ease in computation Swan and Vincent defined a nondimensional time by

(23) $$\tau = \alpha t.$$

Using transformations (22) and (23) and the value of f(v) given in (19), equations (21) and (20) reduce to

(24) $$\frac{dy}{d\tau} = -y-pu/(1+u)$$

(25) $$\frac{du}{d\tau} = -\beta u$$

where

$$p = k_1/\alpha, \quad u = v/k_2; \quad \beta = \gamma/\alpha .$$

For a typical $\alpha = .015$ and $\gamma \approx 1$ then $\beta \approx \frac{1}{.015}$. It is assume that $\beta = 100$.

To apply the model represented by the differential equations (24) and (25) the parameter p must be estimated from the date for a given patient. Assume that at each of the nondimensional times $0,\tau_1,\tau_2,\ldots$ a single dose of the (fictitious) drug of dosage u(0) is administered. The growth of the tumor in the interval (τ_1,τ_2) is obtained by integrating (25) to yield

$$u(\tau) = u(0)\exp(-\beta(\tau-\tau_1))$$

and substituting this result into (24) which integrates to give

(26) $$y(\tau) = y(\tau_1)\exp(-(\tau-\tau_1))+ \frac{p}{\beta} \exp(\beta\tau_1-\tau) \int_{\tau_1}^{\tau} \frac{\exp [(1-\beta)s]ds}{1+u(0)\exp[-\beta(s-\tau_1)]}$$

The integral appearing in (26) can be approximated by using the fact that $\beta >> 1$. Hence, throughout (τ_1,τ_2)

(27) $$\exp(s-\beta s) \approx \exp(\beta_1- \beta s).$$

Using (27) and the identity

$$\exp(\beta\tau_1-\tau) = \exp(\beta\tau_1-\tau_1)\exp(\tau_1-\tau)$$

the integral in (26) can be integrated to give

(28) $$y(\tau) \approx [y(\tau_1)-k]\exp[-(\tau-\tau_1)]+ \frac{p}{\beta} \exp[-(\tau-\tau_1)] \ \ell n \ \{1+u(0)\exp[-\beta(\tau-\tau_1)]\}$$

where

(29) $$k = (p/\beta)\ \ln[1+u(0)].$$

The quantity k is called the drug effectiveness parameter. Once k is evaluated then p can be determined from (29).

To find k note that since $\beta >> 1$ the effect of the second term in (28) is negligible except over a very short time interval (τ_1,τ^*), where $\tau_1 < \tau^* << \tau_2$. Hence, for $\tau > \tau^*$, equation (28) can be approximated by only the first term on the right side. Setting $\tau = \tau_2$ the first term on the right side of the approximation (28) can be solved for k yielding

(30) $$k = y(\tau_1) - y(\tau_2)\exp(\tau_2-\tau_1).$$

The four quantities on the right of equation (30) are known from the raw data. Swan and Vincent discovered that the drug effectiveness parameter not only varies from patient to paitent, but may vary in time in an individual patient. A value k_α is assigned to each patient by averaging the computed values of k. Since k_α is known (29) can be used to define p:

(31) $$p = \beta k_\alpha/\ \ln\ [1+u(0)].$$

once a numerical value has been assigned to u(0). To determine u(0), Swan and Vincent assumed that a standard MCP program is one that is administered every four weeks. This time period and the amount of drug given to a patient results in $\frac{u(0)}{1+u(0)} = \frac{1}{2}$. Thus, u(0) = 1 for the standard MCP program of the fictitious drug. Equation (31) simply becomes

(32) $$p = (100/\ \ln\ 2)k_\alpha$$

from which p can be evaluated.

After calculating the model parameters, as described above, the results predicted by the model with the raw data from the ten patients. Half of them showed good agreement. In general, good agreement was obtained when the drug effectiveness parameter remains relatively constant with time for an individual patient.

Since the model has some degree of realism it seems reasonable to use it to predict the effect of alternate control programs. Swan and Vincent proposed minimizing the total amount of drug in the body[3]:

(33) $$\min J = \min \int_0^T u \, d\tau \qquad (0 \leq u < \infty)$$

from an initial introduction at $\tau = 0$ up to $\tau = T$ and reducing the number of the tumor cells at $\tau = T$ to a preassigned level $y(T) = y_f$. The nondimensional number of cancer cells at any time τ satisfies (24) with $y(0) = y_0$.

Equation (25) is not included in the optimal control problem since decay is assumed negligible because the drug is administered continuously.

Using the techniques for solving optimal control problems described in Chapter 7, the analysis begins by forming the Hamiltonian

$$H = u-\lambda[y+pu(1+u)^{-1}].$$

The necessary conditions $\partial H/\partial u = 0$ and the fact that $u > 0$ determine that

(34) $$u(\tau) = (x(\tau) + \sqrt{x(\tau)}/(1-x(\tau))$$

where $x = -y/p$. Note that the level of drug dose is given as a function of the nondimensional number of cancer cells present. In other words, the control program is in a closed loop form.

The form of the control function indicates that the tumor is reduced in a continuous manner. Moreover, the dose gradually increases as the cancer cells decrease.

When the control function defined by (34) is used in the original differential equation (21) the number of cancer cells

$$N(t) = N_\infty \exp[-\frac{k_1}{\alpha}\{1-[1-(\frac{\alpha N(0)}{k_1})^{1/2}]\exp(-\alpha t/2)\}^2].$$

The plateau level is $N(t) = N_\infty \exp(-k_1/\alpha)$. Using this last expression Swan and Vincent showed that the attainment of the plateau level is much swifter than with the standard program. To achieve this level the accumulated amount of

---

3 They also imposed the restriction $0 \leq u \leq u_{max}$; however, they never solved the optimal control problem with this constraint.

cytotoxic drug is about 1/40 of the accumulated amount with the standard MCP program.

Theoretically, the above protocol results in improvements in recovery time using a smaller cumulative drug dosage. However, this is true only if the drug can be administered in a continuous fashion and toxicity effects held to a tolerable level. It is not clear that this can be achieved, especially since the original MCP program required rests from the periodic administration of the drugs in order to avoid toxicity. Thus, clinical trials are required to determine if the optimal control problem is practical.

Another difficulty is that the drug effectiveness parameter may vary considerably with an individual patient. In this situation, the superiority of the optimal control program is not clear since the number of cancer cells may not be expressible in a simple closed form. However, the optimal control program can still be applied since u is a function of the drug effectiveness parameter.

A natural generalization of the above problem is to consider toxicity. Besides the population of tumor cells, referenced by the subscript 2, consider a population of normals cells, referenced by the subscript 1. Both populations grow according to the Gompertzian growth law with different parameters. The effect of the drug on both populations is described by (19). Assume that the parameter k_2 is the same for both populations[4]; however, the parameter k_1 can vary. Since both the number of normal cells and the number of tumor cells satisfy a differential equation of the form (21), transformation (22) reduces these equations to

(35) $$\frac{dy_1}{dt} = -\alpha_1 y_1 - \gamma_1 u/(1+u) \qquad y_{10} = y_1(0)$$

(36) $$\frac{dy_2}{dt} = -\alpha_2 y_2 - \gamma_2 u/(1+u) \qquad y_{20} = y_2(0)$$

where

$$u = \frac{v}{k_2} \; ;$$

[4] If the parameter k_2 depends on the population, the optimal control problem can be solved by the method described in Swan and Vincent's paper.

γ_1 and γ_2 are the values of the parameter k_1 appearing in (19) for population 1 and 2, respectively. It is assumed that the drug dosage cannot exceed a fixed maximum and so

(37) $$0 \leq u \leq u_m,$$

where u_m corresponds to the maximum dosage. One aim of therapy would be to maximize the number of normal cells and minimize the number of tumor cells at the end of the treatment period or equivalently to simultaneously minimize the following performance criteria:

(38) $$J_1 = -\int_o^T y_1 dt$$

(39) $$J_2 = \int_o^T y_2 dt.$$

The optimal control problem formulated in the preceding paragraph can be solved using the notion of Pareto-optimality. According to Theorem 3 of Chapter 7, necessary conditions for Pareto-optimality can be found by applying the standard techniques of control theory using the performance criterion $\sigma_1 J_1 + \sigma_2 J_2$ where $\sigma_1 > 0$ and $\sigma_2 > 0$. The performance criterion can be written as $\sigma_1(J_1+\sigma J_2)$ where $\sigma = \sigma_2/\sigma_1$. Since $\sigma_1 > 0$ it suffices to minimize $J_1 + \sigma J_2$. Accordingly, the optimal control u minimizes the Hamiltonian

(40) $$H = -\dot{y}_1+\sigma\dot{y}_2+\lambda_1(-\alpha_1 y_1-\alpha_1 u/(1+u)) + \lambda_2(-\alpha_2 y_2-\gamma_2 u/(1+u))$$

Using (35) and (36), H can be reduced to the form

(41) $$H = [\gamma_1-\sigma\gamma_2-\lambda_1\gamma_1-\lambda_2\gamma_2]\frac{u}{1+u} + [\alpha_1-\lambda_1\alpha_1]y_1-[\sigma\alpha_2+\lambda_2\alpha_2]y_2$$

The necessary condition $\frac{\partial H}{\partial u} = 0$ does not determine a value of u because of the form of H. Instead the optimal u is determined by considering the coefficient of $\frac{u}{1+u}$ in

(42) $$S(t) = -(\lambda_1\gamma_1+\lambda_2\gamma_2) + \gamma_1-\sigma\gamma_2,$$

the so-called, <u>switching function</u>. Noting that, $\frac{u}{1+u}$ is an increasing function of u, it follows that H will be minimized when

(43) $$u = \begin{cases} 0 & S(t) > 0 \\ u_m & S(t) < 0 \end{cases}$$

However, when $S(t) \equiv 0$ on $[t_1,t_2]$ the value of u cannot be determined.

In order to determine the sign of S(t) the functions λ_1 and λ_2 must be found from the adjoint equations

(44) $$\dot{\lambda}_1 = \alpha_1\lambda_1 - \alpha_1 \qquad \lambda_1(T) = 0$$

(45) $$\dot{\lambda}_2 = \alpha_2\lambda_2 + \alpha_2\sigma \qquad \lambda_2(T) = 0.$$

Substituting the solutions of (44) and (45) for λ_1 and λ_2 in (42) yields

(46) $$S(t) = \gamma_1 e^{\alpha_1(t-T)} - \sigma\gamma_2 e^{\alpha_2(t-T)}$$

The sign of S depends on the relative magnitude of the parameters appearing in (46), which in turn depend on whether the normal and tumor cells are (a) alike or (b) unalike.

(a) $\alpha = \alpha_1 = \alpha_2$; $\gamma = \gamma_1 = \gamma_2$ (tumor and normal cells are similar)

In this situation

(47) $$S(t) = \gamma(1-\sigma)e^{\alpha(t-T)}.$$

There are three possibilities

(i) $\sigma = 1$. Here, $S = 0$ and u is undetermined.

(ii) $1 > \sigma$. In this situation maximizing the number of normal cells is considered to be of paramount importance. Here, $S(t) > 0$ and so $u = 0$ and no drug is given.

(iii) $\sigma > 1$. Minimizing the number of cancer cells is most desirable. From (47), $S(t) < 0$ and the maximum dosage of drug ($u = u_m$) is given continuously.

(b) $\alpha_1 > \alpha_2$ (tumor and normal cells are different)[5]

Note that S(t) is not identically zero, since $\alpha_1 > \alpha_2$ and so can have at most one zero because the exponential functions are monotone functions. Thus, there are four possible forms of the control function:

(i) $u \equiv 0$ if $\gamma_1 > \sigma\gamma_2\, e^{(\alpha_2-\alpha_1)(t-T)}$ $\qquad (0 \leq t \leq T)$

To see that this possibility can actually occur let $\gamma_1 = \gamma_2$, $\alpha_1 = 2\alpha_2$; $1 > \sigma e^{\alpha_2 T}$.

---

[5] The plateau value of the tumor cell population is greater than that of the normal cell population.

It follows from (46) that $S(t) > 0$ for $0 \leq t \leq T$.

(ii) $u \equiv u_m$ if $\gamma_1 < e^{(\alpha_2-\alpha_1)(t-T)}$ $\quad (0 \leq t \leq T)$

This possibility will occur when $S(t) < 0$ on $[\sigma,T]$, for example, if $\sigma > 1$, $\gamma_1 = \gamma_2$ and $\alpha_1 = 2\alpha_2$.

(iii) u switches from 0 to u_m at $t_s \in [0,T]$.

This situation is impossible as can be seen by rewriting (46) in the form

$$S(t) = e^{\alpha_1 t}(\gamma_1 e^{-\alpha_1 T} - \sigma\gamma_2 e^{-\alpha_2 T} e^{(\alpha_2-\alpha_1)t}).$$

The sign of S(t) is determined by the expression in parenthesis. If $S(t) > 0$ for some $t_1 \in [0,T]$ then $S(t) > 0$ for every $t \geq t_1$ since the quantity in parenthesis increases as t increases.

(iv) u switches from u_m to 0 at $t_s \in [0,T]$ if and only if $\sigma\gamma_2 \leq \gamma_1 \leq e^{(\alpha_1-\alpha_2)T}$.

To show that this possibility can actually occur the <u>switching time</u>, t_s, will be determined by solving the equation $S(t_s) = 0$, where S is defined by (46). This yields

$$t_s = T - \frac{1}{(\alpha_1-\alpha_2)} \ln \frac{\gamma_1}{\sigma\gamma_2} .$$

The switching time $t_s \in [0,T]$ if and only if

$$1 \leq \frac{\gamma_1}{\sigma\gamma_2} \leq e^{(\alpha_1-\alpha_2)T}$$

Note that if $\frac{\gamma_1}{\sigma\gamma} = 1$ then $t_s = T$ and $u = u_m$ while if

$$\frac{\gamma_1}{\sigma\gamma_2} = e^{(\alpha_1-\alpha_2)T}$$

then $t_s = 0$ and $u \equiv 0$.

Once the form of u is determined the final size of the normal population ($y_1(T)$) and the cancerous population ($y_2(T)$) can be determined by solving the differential equations (35) and (36). An analogous problem with the action of the drug described by $f(v) = kv$ was solved by Zietz [74].

Another criterion for optimizing drug therapy is to try and keep the number of normal cells above a certain level L_1 and the number of tumor cells below a certain level L_2 as long as possible. A possible analytic interpretation of the last proposal is

(48) $$\text{maximize} \int_0^T (y_1-L_1)dt;$$

(49) $$\text{minimize} \int_0^T (y_2-L_2)dt.$$

The new optimal control problem defined by (35),(36),(37),(48) and (49) can be solved by the same technique that was used for the previous problem with performance criteria (38) and (39). It can be shown that the switching time in the situation that u switches from u_m to 0 at t_s satisfies

$$\frac{\alpha_1\gamma_2\sigma}{\alpha_2\gamma_1} = \frac{e^{\alpha_1(t_s-T)}-1}{e^{\alpha_2(t_s-t)}-1} .$$

Zietz [74] studied an analogous problem using a linear drug law. He was unable to find an analytic solution for the switching time and proposed a numerical solution for his problem which was not very efficient. No proof was given that the solution converged.

Still another criterion which can be used to optimize cancer chemotherapy is to replace (38) by

$$y_1 \geq L_1$$

for $0 \leq t \leq T$. Using the technique described in Chapter 7 for handling a state variable constraint a system of differential equations can be obtain for the state and adjoint variables. However, the equations must be solved numerically since an analytic solution is not apparent.

Many other equally plausible performance criteria using different laws for the effect of drugs and growth of cells can be formulated. Some examples appear in Swan [68]. Each criterion has a corresponding "optimal" solution. Which of these solutions is the best? Although theoretically one solution may be better than another in actuality the reverse may be true. The theoretically superior solution may lead to toxicity while the theorically inferior solution does not. Furthermore, most solutions will involve some approximations to reality and so may not actually behave as predicted theoretically. Thus, optimal drug protocols must be tested by clinical trials.

One of the major difficulties in applying optimal control theory is formulating realistic performance criteria. Much work remains to be done by experimentalists to determine which population of normal cells are affected by a particular drug and to quantitate these effects. The effects of the administration of multiple drugs should be investigated and theoretical models to describe their action should be formulated. At present drugs are not administered continuously. Therefore, discrete versions of the above continuous models should be studied using the discrete optimal principle.

8.3 Cycle Specific Chemotherapy

One of the first attempts to mathematically analyze specific chemotherapy was the report by Merkle et al. [46]. The Stuart-Merkle cell model, described in Chapter 3, was used to show how treatment schedules for an idealized S-phase specific agent could be found. The parameters of the cell model are defined in terms of the mean cycle time, T_c, and the variance of the cycle time. In addition, for computational simplicity, it was assumed that T_s/T_c is a constant, where T_s is the mean duration of the S phase. This last condition along with the duration of drug exposure determines the effect of the drug. When the frequency of exposure to the drug and the parameters described above are fixed, the tumor growth rate exhibits erratically damped oscillations as T_c increases. This behavior is contrary to expectations and may be related to the assumption T_s/T_c = constant. Even if the more realistic assumption T_s = constant were adopted and the time spent in G_0 taken into consideration, the results might still not be realistic. Recall that the cell model suffers from the inherent defect that the cells, on the average, grow younger as time increases.

Valeriote et al. [69] formulated a deterministic mathematical model using the exponential growth law to interpret the in vivo experimental results on the surviving fraction of AK lymphoma colony-forming units, after exposure to the mitotic-acting drug vinblastine, whose activity decreases exponentially with time.

The fraction of cells which survived a given dose after exposure was recorded up to the generation time. There was good agreement between the predicted and experimental results except for cells with very long generation times. An interesting suggestion is that the above cell-killing method could be used for the determination of generation times. This approach would be useful when classic autoradiographic methods are inappropriate.

Bagshawe [2] discusses many factors which enter into chemotherapy in a qualitative and graphical manner. He proposes that more realistic models of tumors be formulated. In addition, he suggests that the effects of drug combinations, cell loss and acquired resistance be incorporated into existing models. In the quantitative section of the paper Bagshawe studied the time required to reduce a given population of tumor cells to less than 1 cell under the following assumptions:

(a) The growth law is exponential

(b) A fixed dose of the drug instantaneously kills a certain fraction of the population.

An interesting consequence of the above calculations is that the duration of uninterrupted action for chemotherapy is longer than that used in practise.

Another mathematical model of cycle specific therapy by Kuzma et al. [42] is one which uses an exponential birth process. The number of cells in each phase is obtained by multiplying by a constant which depends on the phase. Both the immediate kill effect which is allowed to decrease with repeated doses and the delayed kill effect are taken into account. This last assumption is hypothetical since the biological laws are not known and are difficult to verify. Different doses of drugs are allowed but the doses must be administered at a fixed interval of time. Toxicity is also taken into account. The authors theoretically determine the required number of drug administrations, the total time to reduce the tumor to a preassigned level and the minimum initial immediate kill effect of the chemotherapy to insure regression of the tumor.

Wilson and Gehan [72] developed a program to simulate the normal cell cycle using Monte-Carlo techniques. They postulated that the duration of each state of the cell cycle is a random variable. The only departure from the usual description of the cell cycle was that once a cell entered the G_o state the cell could not return to the proliferating cycle. Each drug treatment was assumed to have a specified probability of killing a cell in the various states. This probability remains applicable for the period of activity of the drug. The program was able to simulate the proliferation of L-1210 cells in mice. Under the assumption that:

(a) the duration of each state follows a gamma distribution;

(b) the logarithm of the number of cells required to kill a mouse follows an exponential distribution,

the mean and variability of time to death of the cancerous mice were computed. The computed results were in reasonable agreement with the experimental results obtained by Skipper et al. [61].

Shackney [62] developed an elaborate discrete computer model for studying the treatment of L1210 leukemia with cytosine arabinoside. The multiple model parameters were found by trial and error by fitting computed curves to experimental data. Verification of the program's predictive capacity was obtained by comparing survival results to sets of experimental schedules which were not used in the original determination of parameters. There was good agreement between observed and predicted results, especially for small inoculum sizes and dosages.

The above approach is appropriate for predicting survival rates for a new untried drug protocol, when a considerable quantity of such data using related schedules for the same system has already been determined experimentally. It is difficult to see how the model can be generalized to guide work on other systems when survival rates and schedule dependency have not been so carefully measured. In particular the above computer model is inapplicable in a clinical setting.

A computerized version of the Hahn model for calculating the survival of normal and malignant cell populations after chemotherapy or radiotherapy is presented by Hahn and Steward [29]. The response of the cell to the agent is assumed to be a function of the cell's age or position in the cell cycle. This age-response function is calculated from experimental results as follows. Cultured, exponentially growing cells are synchronized. The agent is applied at various times after cell synchrony. After each administration cells are gathered. Each of these cells is grown individually and the fraction of surviving colonies is recorded. The plot of surviving fraction versus time of administration defines the experimental age-response function. Since the cells are only partially synchronized the experimental age-response function is used to infer the age-response function (ARF). The age-response function can then be used to find the fraction of cells which are killed and so removed from a given compartment (age interval) of the Hahn type model. Thus, the therapeutic response to the agent can be simulated.

The authors suggest their model is directly applicable to murine leukemia and possibly some human leukemias. The last assertion seems doubtful at the present time because of the complicated experimental procedures required to find the ARF and other parameters of the model.

A pharmacodynamic model for the quantitative analysis of dose-time-cell-survival curves produced by the administration of cell-cycle-specific-chemotherapeutic agents was proposed by Jusko [41]. The model consists of a proliferating compartment containing cells sensitive to the drug and a resting compartment containing cells resistant to the drug. The drug effect is modelled by a loss of cells from the proliferating compartment. This loss is proportional to the number of proliferating cells and the transient amount of drug to which these cells are explosed. By employing certain pharmacokinetic approximations, the model is used to predict the behavior of the drugs vincristine, vinblastine, aribinosylcytosine and cyclophosphamide on lymphoma and hematopoietic cells in the mouse femur. The model agrees well with the experimental data.

In the first part of the paper by Jansson [76] the growth of a population containing one type of cell is modelled using a linear differential equation with constant coefficients. This method is extended to study the growth of mixed populations where competition between different types of cells take place. The relative number of diploid, endomitotic tetraploid and marked tetraploid Ehrlich ascites cells predicted by the model was in agreement with the experimental data.

In the second part of the paper there is a discussion on how the cell cycle changes with the size of the population. The biological technique for analysis is the precentage labelled mitoses curve.

In the third part of the paper a discrete multicompartmental model is used to simulate the growth of two kinds of cells - malignant and healthy. The G_0 phase is included in the model. It is assumed that the therapeutic agent kills all cells in the S phase. Parameters were chosen to approximately simulate the behavior of diploid and tetraploid Ehrlich ascites cells. The resulting curves of the number of malignant and normal cells in the S phase showed that these two populations reached their local maxima at different times. Jansson introduced the natural protocol of administering the drug at the times when the number of malignant cells in the S phase reached their local maxima. Provided the activity of the drug lasted for a sufficiently small interval of time, the above protocol greatly reduced the number of malignant cells while preserving the normal cells. The disastrous consequences of the wrong choice in time for treatment is also illustrated.

More refined models for simulation of chemotherapy are described in Jansson [77]. These models not only include competition between different cell populations but also consider differentiation of cells.

The CELLSIM program, described in Donaghey and Drewinko [23] is another computer program for the simulation of cell kinetics and chemotherapy. It has the advantage that it employs a vocabulary similar to that used by those engaged in cell kinetics research. Thus, the learning time for the language is minimal. Ten different probability distributions can be used to describe the time spent in each phase. The user can describe how cells are to be distributed in the various phases, the precentage of cells killed in each phase, the time the cells are to be killed and so on. A user's manual has been written by Donaghey [24].

Many of the computer models have only been used for theoretical studies. Moreover, there is no indication of how to apply these models to quantitate the effects of a particular drug. Other models have been applied by varying their parameters, by trial and error, until the simulated results agree with the experimental results. A model of the cell cycle based on difference equations, was designed by Eisen and Macri [25] to remedy these defects. The model is similar to that presented in section 3 of Chapter 6 except that there is a single delay line. To apply this model the cell cycle parameters must first be determined - for example, by autoradiography or microfluorometry. Then, the drug parameters are varied by a computerized optimal search procedure until the theoretically simulated results agree with the experimental results. For instance, until the generated and experimental labelling indices or DNA histograms are close to each other according to some criterion, such as least squares.

One of the few analytic approaches to cancer chemotherapy involving cycle specific drugs, is presented in the paper by Himmelstein and Bischoff [32]. The population of cells is described by Rubinow's maturity-time equation

$$\frac{\partial n}{\partial t} + \frac{1}{\tau}\frac{\partial n}{\partial \mu} = -\lambda n,$$

where $n(\mu,t)$ is the cell density function, τ is the constant population generation time and $\lambda(\mu,t)$ is the death or loss rate of cells. The initial condition is

$$n(\mu,0) = N_0\ \delta(\mu)$$

where $\delta(\mu)$ is the unit impulse function. In other words, it is assumed that a time zero N_0 cells are present, each with maturity $\mu = 0$. The boundary condition is given by binary mitosis:

$$n(0,t) = 2n(1,t)$$

Suppose that the drug is effective only during the interval (a,b) of the cell cycle. To describe this cell cycle specificity the loss function is assumed to be of the form

$$\lambda(\mu,t) = \begin{cases} 0 & \mu < a \\ \lambda(t) & a \leq \mu \leq b \\ 0 & \mu > b \end{cases}$$

Using the above conditions Himmelstein and Bischoff solved the partial differential equation by the method of generations described in Chapter 3. However, the resulting solution was not simple and had to be evaluated numerically. For simplicity, the effect of the drug was considered to be a function of the local concentration, $C(t)$, in the host organism. The drug-cell interaction was described by various loss functions such as

$$\lambda = (k_1 C(t)/(k_2+C(t)).$$

Several different functional forms of C were given. The extensions of this work to actual animal experiments including the various possible factors that influence the action of the drug (such as drug uptake and exposure time) and the determination of the local concentration of the drug will be described in the next section.

The paper of Bahrami and Kim [4] applies optimal control theory to cycle-specific chemotherapy in order to curtail the growth of cancer cells while minimizing host toxicity. Their discrete time model for describing the dynamics of cellular proliferation is a generalization of Takahaskis or Hahn's model described in Chapter 3. The mean generation time of the cell cycle of the cancer cells is divided into n equal intervals or age compartments. The transfer of cells from one age compartment to other age compartments in a unit of time is described by

(50) $$x(k+1) = \Phi_{pp}(k+1,k)\, x(k)+\Phi_{pm}(k+1,k)w(k)$$

(51) $$w(k+1) = \Phi_{mp}(k+1,k)\, x(k)+\Phi_{mm}(k+1,k)w(k)$$

where $x(k)$ is the n-dimensional cell age-state vector at time k for the proliferating cells and $w(k)$ is the state vector of the nonproliferating cells. The jth component of $x(k)$ is the number of proliferating cells in the jth age compartment. The state transition matrices Φ_{pp}, Φ_{pm}, Φ_{mp} and Φ_{mm} are assumed to be independent of the state vectors and time.

A cycle specific drug, administered at time k, is assumed to kill a fraction c_i of the number of proliferating cells in the ith age compartment. Thus, the vector $x_d(k)$, whose components are the number of cells in each compartment that killed by the antitumor drug from time k to k+1 can be expressed as

(52) $$x_d(k) = Sx(k)u(k).$$

In (52) S is an nxn diagonal matrix with elements c_i and $u(k) = 0$ indicates that at time k no drug is present and $0 < u(k) \leq 1$ indicates that at time k the drug is present with the maximum concentration when $u = 1$. The dynamics of the cell population affected by the drug can be obtained from (50) and (51) by removing $x_d(k)$ from the proliferating group, including $x_d(k)$ in the nonproliferating group and using a matrix M to assign all the dead cells (due to the drug) to the compartment for dead cells in the nonproliferating group. The resulting equations have the form

(53) $$y(k+1) = (A+Bu(k)\; y(k)$$

where

$$y = \begin{bmatrix} x \\ \\ w \end{bmatrix} \qquad A = \begin{bmatrix} \Phi_{pp} & \Phi_{pm} \\ \\ \Phi_{mp} & \Phi_{mm} \end{bmatrix} \qquad B = \begin{bmatrix} -\Phi_{pp}S & 0 \\ \\ MS-\Phi_{mp}S & 0 \end{bmatrix}$$

The performance criterion is

$$J = \alpha^T(N)x(N) + \sum_{k=0}^{N-1} \beta_k u(k) \tag{54}$$

where $\alpha(N)$ is a weighting vector with constant positive entries, N is the fixed terminal time and $\beta_k \geq 0$ for $k = 0,1,\ldots,N-1$ weighs the control effort $u(k)$. The purpose of an optimal drug protocol is to minimize J. This corresponds to achieving a low final population of cancer cells while keeping the amount of drug used small. The tradeoff between the final population and the amount of drug used is governed by the relative values of the $\alpha^T(N)$ and the β_k's and has to be based upon physical considerations.

The necessary conditions for minimizing the performance criterion in (54) subject to the constraints (53) and $u(k) \in [0,1]$ can be obtained from the discrete maximum principle described in Chapter 7. The Hamiltonian

$$H_k = \beta_k\, u(k) + \lambda^T(k+1)(A+Bu)y\,. \tag{55}$$

The adjoint equations are obtained from the relation

$$\lambda(k) = \frac{\partial H_k}{\partial y(k)}$$

and are

$$\lambda(k) = (A+Bu)\lambda(k+1)\,. \tag{56}$$

The optimal solution must satisfy the state equations (53) with initial conditions

$$y(o) = y_o \tag{57}$$

and the adjoint equations with final conditions

$$\lambda(N) = \alpha(N) \tag{58}$$

where the control $u(k)$ is selected to minimize H_l for $h = 0,1,2,\ldots,N-1$.

Bahrami and Kim used the following algorithm for solving the above problem.

(i) Select an initial control vector

$$u' = (u'(0),\ldots,u'(N-1))^T$$

where $u'(i) \in [0,1]$. Let the control vector at iteration j be denoted by u^j.

(ii) Solve in "Forward Time" and store

$$y(k+1) = (A+Bu^j(k))y(k) \quad y(0) = y_o \qquad (k = 0,1,\ldots,N-1)$$

(iii) Solve in "Backward Time" and store

$$\lambda(k) = (A+Bu^j(k))\lambda(k+1) \quad \lambda(N) = \alpha(N) \qquad (k = N-1,\ N-2,\ldots,2,1,0)$$

(iv) Find an index k^* such that

$$H_{k^*u(k^*)} \operatorname{sgn} \ (u^j(k^*)-1/2) = \max_{0<k<N-1} (H_{ku(k)} \operatorname{sgn}(u^j(k)-1/2))$$

$$\text{where } H_{ku(k)} = \frac{\partial H_k}{\partial u(k)} = \beta_k + \lambda^T(k+1)\ By(k)$$

If

(59) $$H_{k^*u(k^*)} \operatorname{sgn} \ (u^j(k^*)-1/2) > 0$$

then stop; u^j satisfies the necessary conditions of the maximum principle. If (59) is not satisfied, define

(60) $$u^{j+1} = \begin{cases} u_j(k) & k \neq k^* \\ 1-u^j(k) & k = k^* \end{cases}$$

and proceed to step (i) for another iteration.

The proof that the control vector iteration converges and leads to a solution is contained in the paper of Bahrami and Kim [4]. The optimal strategy is shown to be a bang-bang control and lies on the vertices of an N-dimensional cube contained in N-dimensional Euclidean space.

The application of their method to the treatment of spontaneous AKR leukemia did not yield accurate results. They used the Hahn model as described in Hahn's original paper. Unfortunately, some transition probabilities in a certain matrix in that paper are in error. These errors were corrected in a later paper of Hahn (1970) which appeared in Math. Biosciences.

Zietz [74] formulated a system of equations of the same form as (53) to describe the effect of chemotherapy on the growth of cancer and normal cells. The linear performance criterion was obtained from maximizing the number of normal cells while simultaneously minimizing the number of cancer cells in each age compartment by means of the Pareto Optimality principle. The resulting performance criterion had both positive and negative coefficients. The proof that the Bahrami and Kim algorithm leads to a solution is no longer valid since this proof requires that all coefficients in the performance criterion be positive. Zietz speculates that essentially the same algorithm applies to his more general problem. He provides several numerical examples to support this view.

The model of Eisen and Schiller described by equations (35) Chapter 2, can be used to illustrate the application of control theory to cycle specific chemotherapy. It will be assumed that the drug only acts on the proliferating cells and destroys a certain proportion of these cells. The kinetics of the tumor and the drug action can be described by

(61) $$\frac{dy_1}{dt} = (a-ku)y_1 + by_2$$

(62) $$\frac{dy_2}{dt} = cy_1 - dy_2$$

where y_1 and y_2 represent the number of proliferating and nonproliferating cells, respectively; a,b,c and d are constants which are all nonnegative except a which is arbitrary. The constant $k > 0$ depends on the maximum administered dosage of the drug. The variable amount of drug injected is represented by variations in the control variable $u = u(t)$, where

$$0 \leq u \leq 1.$$

Given the initial number of cells,

(63) $$y_1(0) = y_{10}; \quad y_2(0) = y_{20}$$

the object of the therapy is to reduce in a given time T, the number of cells to

(64) $$y_1(T) = y_{11}; \quad y_2(T) = y_{21}$$

while using the minimum amount of the drug. Thus, the performance criterion

(65) $$J = \int_0^T u\, dt$$

must be minimized. As described previously, many other equally plausible strategies are possible.

To solve the above problem, the control function u must be chosen to minimize the Hamiltonian

(66) $$H = u+\lambda_1[(a-ku)y_1+by_2] + \lambda_2[cy_1-dy_2]$$

The Hamiltonian cannot be minimized by setting $\frac{\partial H}{\partial u} = 0$, since it is linear in u. The solution of this singular problem requires studying the coefficient of u or the switching function

(67) $$S(t) = 1-\lambda_1 y_1 k$$

The Hamiltonian will be minimized if

$$u = \begin{cases} 1 & \text{whenever } S(t) < 0 \\ 0 & \text{whenever } S(t) > 0 \end{cases}$$

The control function is not determined when S(t) = 0 on some finite interval contained in [0,T].

To determine the behavior of the switching function requires the adjoint equations

(68) $$\dot{\lambda}_1 = -(a-ku)\lambda_1-c\lambda_2$$

(69) $$\dot{\lambda}_2 = -b\lambda_1 + d\lambda_2$$

and the derivative of S(t) or $(\dot{\lambda_1 y_1})$. The last derivative can be found by multiplying (61) by λ_1 and (69) by y_1 and adding the resulting equations. This procedure gives

(70) $$(\dot{\lambda_1 y_1}) = b\lambda_1 y_2 - c\lambda_2 y_1$$

Similarly, by using (62) and (69), it can be shown that

(71) $$-(\dot{\lambda_2 y_2}) = b\lambda_1 y_2-c\lambda_2 y_1$$

The above results will now be applied to study S(t) by considering the following possibilities.

(i) b = c = 0.

It follows from (67) and (70) that S(t) is constant on [0,T]. Hence S(t) = 0 or S(t) does not change sign on [0,T]. It can be shown by examples that both possibilities can occur. When $S(t) \equiv 0$, the control function u is not uniquely determined.

(ii) $b = 0;\ c \neq 0$

If $S \equiv 0$ on some interval in $[0,T]$, then $\dot{S} = 0$. Equation (70) implies that $\lambda_2 \equiv 0$ on this same interval or $y_1 \equiv 0$. It follows from (69) and (61) respectively, that $\lambda_2 \equiv$ or $y_1 \equiv 0$ on $[0,T]$. Since $y_1 \not\equiv 0$, because $y_1(0) = y_{10} \neq 0$, $\lambda_2 \equiv 0$ on $[0,T]$ which implies $S \equiv 0$ on $[C,T]$. Conversely if $\lambda_2 = 0$ at any point of $[0,T]$ then S is constant on $[0,T]$.

Next suppose that $S(t_0) = S(t_1) = 0$ for $0 < t_0 < t_1 < T$ and $S(t) \neq 0$ on (t_0,t_1). Assuming that λ_1 and y_1 are continuous it follows that there exists a point ξ such that $t_0 < \varepsilon < t_1$ and ξ is a relative extreme point of S. Hence, $\dot{S}(\xi) = 0$ and it follows that $y_1 \equiv 0$ or $\lambda_2 \equiv 0$. In either situation S is constant on (t_0,t_1) which is a contradiction. Summarizing, either S is constant on $[0,T]$ or at most 1 switch occurs.

(iii) $b \neq 0;\ c = 0$.

The conclusions are similar to (ii)

(iv) $b \neq 0;\ c \neq 0$

Suppose $S \equiv 0$ on some subinterval of $[0,T]$; then

$$\lambda_1 y_1 = \frac{1}{k} \tag{72}$$

and $\dot{S} \equiv 0$ on this same subinterval. Multiplying (70) by $y_1 y_2$ it follows that

$$b(\lambda_1 y_1)\, y_2^2 = c(\lambda_2 y_2) y_1^2 \tag{73}$$

Thus, $\lambda_1 y_1$ and $\lambda_2 y_2$ have the same sign. Equation (71) implies

$$\lambda_2 y_2 = K \tag{74}$$

where K must be a positive constant because of (72), (73) and $k > 0$. Equations (72), (73) and (74) imply $y_2 = m y_1$, where

$$m = \sqrt{ckK/b}$$

Under the assumption that the number of nonproliferating cells is not proportional to the proliferating cells, it follows that S is not identical to zero on any subinterval of $[0,T]$. A criterion for determining the number of switches in sign of S is not known. This problem was also considered by Swan [66].

All of the applications of optimal control theory to chemotherapy, described in this chapter, only satisfy the necessary conditions for an optimal solution to exist. Research is required to investigate whether the solutions of the above problems also satisfy a set of sufficient conditions.

8.4 Pharmacokinetics

The clinical application of any chemotherapy model requires that the concentration of the cytotoxic or cytostatic agent in the target cells be known. Although this concentration is not immediately evident, it can sometimes be determined from the concentration of the injected drug. This determination is complicated by the fact that the active agent may be a metabolite of the drug which need not have the same concentration as the drug. Thus, pharmacokinetics, which is concerned with the prediction of the events occurring during the process of drug disposition throughout the body, plays a crucial role in cancer chemotherapy.

In this section procedures needed to construct pharmacokinetic models based on physiologic, biochemical and pharmacologic principles will be described. These models are required to develop optimal dosage regimens for clinical applications. Another important reason for studying pharmacokinetics is to aid in scaling the results of drug experiments on lower animals to man. Finally, there is the intellectual satisfaction of having quantitative predictive models based on underlying knowledge, rather than the empirical, curve-fitting approaches.

Pharmacokinetic models are based on the fact that certain parts of the body can be lumped together. These lumped parts will be called body regions or compartments. The choice and number of compartments are based on the anatomy and physiology of the body, the properties of the drug and the purpose of the model. An organ is a familiar example in which cells have been lumped together.

The following examples illustrate how lumping is influenced by the proposed use of the model. The concentration of tracer in an indicator-dilution experiment is measured within a time scale which is the same as the mean circulation time (about 1 minute in man). Thus the details of the transit time in the arteries and veins must be taken into account and the circulatory system cannot be lumped into 1 compartment. Next, suppose the model is to be used for studying cancer chemotherapy. The time scale of interesting events for most cancer drugs are of the the order of hours or days. After an hour, the injected drug has circulated about sixty times and so its concentration in the blood is essentially uniform. Hence, the entire circulatory system can be lumped into one "blood pool."

The dependence of the choice of compartments is illustrated in Fig. 8.4 and Fig. 8.5 for methotrexate and ara-C. Note that a muscle compartment is required for methotrexate and not for ara-C, since negligible amounts of ara-C are present in the muscles. On the other hand, ara-C has a substantial concentration in the bone marrow while methotrexate does not, as reflected in their compartmental configurations. Extra compartments are required to describe the biliary secretion of methotrexate.

There is no one set of rules for automatically choosing approximate compartments. In fact, the definition of the various body regions is flexible and can depend on the situation. For example, the major barrier to the transport of many drugs is the cellular membrane and not the capillary membrane. In such a situation, the tissue compartment is defined to be the intracellular space while the blood compartment is the vascular space plus the interstitial space. A crucial test of a compartmental model is whether the results predicted by the model agree with the experimental data. However, it is possible that an entirely different model can also be in close agreement with the experimental data. In such a situation, the model whose compartments have the firmest anatomic basis and is most compatible with the available physiologic and pharmacologic facts is preferable.

Compartmental models are highly simplified versions of tissues or organs that they represent. A tissue is a complex arrangement of cells and interstitial material containing blood vessels. The arteries branch into smaller ones which branch into arterioles which lead to highly permeable capillaries. There are hundreds of capillaries per artery. The capillaries lead to venules, small veins and finally to large veins. There are thousands of capillaries per vein. Hence, the ultimate vessels in tissues consist of a large number of small, interconnected containers. Transport of material occurs through the capillaries. Schmidt [61] has considered some aspects of this type of flow and transport in this microcirculation.

Detailed movies of living tissue, made by Block [16], shows that the flows through the various small vessels seem to occur in spurts, first through one region and then through another. It is apparent that a detailed mathematical model of a tissue would be difficult to formulate and if formulated would be difficult to use as part of a larger overall model. Consequently, the arterioles, capillaries and vanules are represented by a single blood compartment which for simplicity also contains the interstitial fluids. All of the cells (other than red blood cells), assumed to be of the same type, are lumped into the intracellular compartment which is separated from the blood compartment by a (fictitious) cell membrane. The flow of blood into and out of the tissue by means of many vessels is represented by a single average flow into and out of the blood compartment. The resulting model is shown in Fig. 8.3.

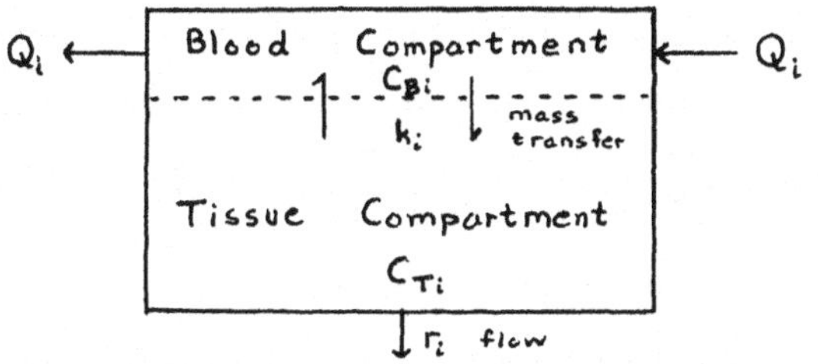

Figure 8.3. Compartmental model of ith organ (or tissue)
The rate r_i represents removal by secretion, metabolism, etc.

Many details of transcapillary, interstitial and transmembrane fluid flow are not known. For example, there is a small amount of bulk fluid movement through the capillary walls in addition to diffusion. Another complicating factor is the active transport of solutes through biological membranes. Active transport occurs when certain solutes are transferred in opposition to the concentration gradient in conflict with the usual laws of diffusion. As a first approximation these effects will be considered part of the overall diffusion and correlated by altering the permeability coefficients. Further details on the biology and physics of the transport of fluids can be found in Harris [33] and Bird et al. [9].

The volume of each compartment must be determined in order to obtain quantitative results. The blood volumes in the various organs (except the lungs) are obtained from the unverified postulate that the flow rate (v) through the arterioles, capillaries and vanules is about the same for every body region. Let ℓ, A and V_b be the average length, cross-sectional area and volume of the vessels being studied. It is assumed that ℓ must be related to the total size of the body region; in particular $\ell \sim M^{1/3}$. Since the volume flow rate Q = vA it follows that

$$v = \frac{Q}{A} = \frac{\ell Q}{V_b} \sim \frac{M^{1/3}Q}{V_b}$$

Hence, the volume of blood in i is

$$V_{B_i} = (\text{constant})\, M_i^{1/3} Q_i$$

Taking ratios between the ith and jth organs gives

$$\frac{V_{B_i}}{V_{B_j}} = \left(\frac{M_i}{M_j}\right)^{1/3} \frac{Q_i}{Q_j} \tag{75}$$

From various handbooks the organ flow rates, the organ weights and the total volume (500 ml.) of the arterioles, capillaries and venules can be found. Therefore

$$\sum_i V_{B_i} = 500 \tag{76}$$

The volumes of the blood compartments can be found from equations (75) and (76).

The volume of the tissue compartment (V_t) can be found from the assumption in Bischoff and Brown [11] that the ratio of the volume of interstitial fluid (V_{int}) to the total volume of cells is a constant - namely,

$$\frac{V_{int}}{V_T} = \frac{12}{23} = .521 \tag{77}$$

The weight of an organ is assumed to be the same as the total volume (V_{tot}) of the organ (density $\approx$ 1.0 g/cc.).

$$V_{tot} = V_{int} + V_T + V_B \tag{78}$$

Hence, V_t can be obtained by solving equations (77) and (78). This gives

$$V_T = \frac{V_{tot} - V_B}{1.52}$$

Further details can be found in Bischoff and Brown [11].

The mathematical description of the compartmental model of an organ (Fig. 8.3) is based on the principle of conservation of mass applied to each compartment in the following form:

$$(79)\quad \begin{pmatrix}\text{rate of change of}\\ \text{mass of free drug}\end{pmatrix} + \begin{pmatrix}\text{rate of change of}\\ \text{mass of bound drug}\end{pmatrix}$$

$$= \begin{pmatrix}\text{rate of}\\ \text{injection of absorption}\\ \text{of drug}\end{pmatrix} + \begin{pmatrix}\text{rate of inflow}\\ \text{of drug in the}\\ \text{blood}\end{pmatrix} + \begin{pmatrix}\text{rate of formation}\\ \text{of the drug}\end{pmatrix}$$

$$- \begin{pmatrix}\text{rate of outflow}\\ \text{of drug with blood}\end{pmatrix} - \begin{pmatrix}\text{rate of conversion}\\ \text{of the drug by}\\ \text{chemical reaction}\end{pmatrix} - \begin{pmatrix}\text{rate of elimination}\\ \text{of the drug by}\\ \text{other processes}\end{pmatrix}$$

All ionized forms of the drugs which will be study are only present in very small concentrations assumed to be zero.

The following notation will be used to write the mass balance equations for the various compartments.

NOTATION

C = drug concentration, mcg. /mℓ.

k_i = permeability coefficient of membrane i

k_F = reciprocal of nominal transit time in small intestine, min^{-1}.

k_G = saturable rate of intestinal absorption, mcg. / min.

k_k = clearance by kidney, ml. / min.

k_L = saturable rate of drug transport into bile, mcg./min.

K = Michaelis constant for intestinal absorption or bile formation, mcg./ml.

Q = plasma flow rate, ml./min.

r = drug-transport rate in bile, mcg./min.

R = tissue-to-plasma equilibrium distribution ratio for linear binding, dimensionless

t = time, min.

V = volume of compartment, ml.

w = fraction of water in a tissue

x = bound concentration per unit amount of protein

τ = nominal residence time in bile subcompartment, min.

Subscripts

B = blood

G = gastrointestinal tract

GL = gut lumey

K = kidney

L = liver

Le = lean

M = muscle

P = plasma

T = tissue

1,2,.. = bile or gut-lumen subcompartments

Bi etc. = blood in compartment i

Using the mass balance principle and the notation stated above,the mass balance equation for the ith blood compartment is

$$w_{Bi}\ V_{Bi}\ \frac{dC_{Bi}}{dt} + P_{Bi}\ \frac{dx_{Bi}}{dt} \tag{80}$$

$$= \quad Q_i(W_B\ C_B+p_Bx_B-W_{Bi}C_{Bi}-P_{Bi}x_{Bi}) + k_i(C_{Ti}-C_{Bi})$$

where the last term represents the membrane transfer of the free drug. The mass balance equation for the tissue compartment is

$$w_{Ti}\ V_{Ti}\ \frac{dC_{Ti}}{dt} + P_{Ti}V_{Ti}\ \frac{dx_{Ti}}{dt} = K_i(C_{Bi}-C_{Ti}) - r_i(C_{Ti},x_{Ti}) \tag{81}$$

where the last term represents removal by secretion, metabolism, etc.

To actually solve the mass balance equations numericals values of all the parameters must be known. As seen above, the various organ flows and volumes can be estimated. Plasma protein binding can be measured in vitro experiments. The most difficult to obtain are the specific membrane permeabilities and not much information is available at the present time. To overcome this difficulty, the assumption of flow-limited conditions is often introduced in pharmacokinetics. According to this assumption the blood leaving a tissue is in diffusion equilibrium with the tissue. Hence, the free concentration of the drug in the blood and tissue are the same, that is,

(82) $$C_i = C_{Bi} = C_{Ti}$$

A mathematical criterion for flow-limited conditions was derived by Dedrick and Bischoff [21] by comparing the solutions of (80) and (81) with the solution of the simplied equation (83) resulting from using (82) and some algebra. The criterion is $k_i >> Q_i$. However, some knowledge of the permeability is still required to use this criterion and such data are scarce. Hence, flow limited-conditions are assumptions which cannot often be verified.

Using (82) and adding (80) and (81) yields

(83) $$(w_{Bi}V_{Bi}+w_{Ti}V_{Ti})\frac{dC_i}{dt} + p_{Bi}V_{Bi}\frac{dx_{Bi}}{dt} + p_{Ti}V_{Ti}\frac{dx_{Ti}}{dt} =$$

$$Q_i(w_BC_B+p_Bx_B-w_{Bi}C_i-p_{Bi}x_{Bi})-r_i(C_i)$$

Note that the mass transfer term has been eliminated and the exact value of k_i need not be known, just that it is very large.

Dedrick and Bischoff [21] introduced the concept of effective protein binding in the tissues mainly because of the lack of much rigorous data. The basic assumption is that the binding isotherm is the same for all tissues and plasma so that

$$x_B = \psi C(B) \quad \text{implies} \quad x_{Ti} = \psi(C_{Ti}).$$

The differences in various types of tissue binding are entirely accounted for by the various effective protein amounts, p_{Ti}. These are determined from tissue binding experiments.

Under the further assumption that binding is linear

$$x = \psi(C) = BC$$

equation (83) reduces to

(84) $$(w_{Bi}\,V_{Bi} + w_{Ti}\,V_{Ti} + p_{Bi}\,BV_{Bi} + p_{Ti}BV_{Ti})\frac{dC_i}{dt}$$

$$= Q_i(w_BC_B + p_BBC_B-w_{Bi}C_i-p_{Bi}BC_i) - r_i(C_i)$$

The concentrations actually measured in the laboratory are the total (free and bound) wet tissue values defined by

(85) $$\bar{C}_B = (w_B + p_B B) C_B$$

(86) $$\bar{C}_{Bi} = (w_{Bi} + p_{Bi} B) C_i$$

(87) $$C_i = \frac{(w_{Bi} V_{Bi} + p_{Bi} BV_{Bi} + w_{Ti} V_{Ti} + p_{Ti} BV_{Ti}) C_i}{V_{Bi} + V_{Ti}}$$

The basis of equation (87) can be stated in words

$$\bar{C}_i = \frac{\begin{pmatrix}\text{amount in}\\ \text{equilibrium blood}\end{pmatrix} + \begin{pmatrix}\text{amount in}\\ \text{actual tissue}\end{pmatrix}}{\text{total volume of blood and tissue}}$$

Note that by using C_i the effective protein amounts need not be found explicitly.

The total concentration of a tissue is frequently reported as a tissue/blood distribution ratio

(88) $$R_i = \left(\frac{\bar{C}_i}{\bar{C}_{Bi}}\right) = \left(\frac{\bar{C}_i}{\bar{C}_B}\right) \quad \text{equilibrium, } C_i = C_B$$

Note that for linear binding R_i is a constant. The second equality in (88) indicates that the tissue venous outflow concentration can be replaced by the blood pool concentration only under equilibrium conditions, i.e. long-time constant infusion. Bischoff and Dedrick [12] showed that R_i could be obtained by finding the long-time slopes ratio from a bolus injection under certain conditions. In particular, Dedrick et al. [22b] have shown these ratios are essentially equivalent for the linear bound drug, methotrexate.

Using the total concentrations defined in (85), (86) and (87) and the tissue/blood distribution ratio (88) the mass balance equation (84) simplifies to

(89) $$V_i \frac{dC_i}{dt} = Q_i \left(\bar{C}_B - \frac{\bar{C}_i}{R_i}\right. - r_i(\bar{C}_i)$$

where the total volume of the organ is represented by

$$V_i = V_{Bi} + V_{Ti}.$$

As another example of the use of (79), the mass balance equation for the blood pool compartment, representing the lumped vascular system, can be written as

(90) $$w_B V_b \frac{dC_B}{dt} + p_B V_B \frac{dx_B}{dt} = -Q_B(w_B C_B + p_B x_B)$$
$$+ \sum_i Q_i (w_{Bi} C_{Bi} + p_{Bi} x_{Bi} \quad + \text{(injections)}$$

The summation in (90) is over all organs i whose venous outflow feeds directly into the blood compartment. Equation (90) is valid for whole blood for a solute that immediately equilibrates with erythrocytes or for plasma when the solute is excluded from the erythrocytes. Using total concentrations and tissue/blood distribution ratios, equation (90) can be reduced to

$$V_B \frac{d\bar{C}_B}{dt} = -Q_B \bar{C}_B + \sum_i Q_i \frac{\bar{C}_i}{R_i} + \text{injections.} \tag{91}$$

The above concepts will be applied to study in some detail the pharmacokinetics of methotrexate as developed by Bischoff et al [14]. A compartmental model must take into account the following major physiological facts. Following an intravenous injection of methotrexate there is a very rapid drop in plasma concentration and an increase in gut-lumen concentration. This concentration drop results from rapid localization in the tissues and excretion. The increase in gut-lumen concentration indicates the significant role of the biliary route. Chemical analysis and radioactivity counting indicate that methotrexate is not conjugated or metabolized by the liver. However, several minutes are required to attain maximum concentration in the liver. This effect is probably also influenced by biliary clearance and bile formation in the liver. The excretion of the drug was by the renal or fecal route. There is linear binding of methotrexate by tissues at plasma concentrations above .1 meg./ml. At low concentration levels, the binding becomes nonlinear.

The diagram of the basic model, including the enterohepatic circulation, appears in Fig. 8.4.

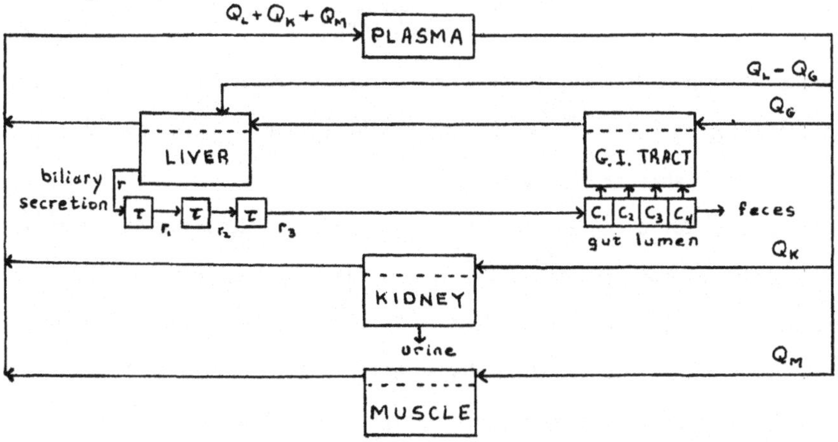

Figure 8.4 Important body compartments in methotrexate distribution.

Specializing (80) and (81) the mass balance equations for the plasma and muscle compartments can be written as follows. To simplify the notation, the total concentration will be denoted by C instead of $\bar{C}$.

Plasma:

$$V_P \frac{dC_P}{dt} = \text{(injection function)} + Q_L \frac{C_L}{R_L} + Q_K \frac{C_k}{R} + Q_M \frac{C_M}{R_M} - (Q_L + Q_K + Q_M)C_P \tag{92}$$

Muscle:

$$V_M \frac{dC_M}{dt} = Q_M(C_P - \frac{C_M}{R_M}) \tag{93}$$

The compartmental equation for the kidney is similar to that for muscle except for a term $-k_K\ C_K/R_K$ which is required to represent the renal excretion of methotrexate. The renal clearance constant, k_K was obtained by comparing integrated plasma concentration data with cumulative urinary excretion. The equation for the kidney is

Kidney:

$$V_K \frac{dC_K}{dt} = Q_K(C_P - \frac{C_K}{R_K}) - k_K \frac{C_K}{R_K} \tag{94}$$

The mass balance equation for the liver can be deduced from (79).

Liver:

$$V_L \frac{dC_L}{dt} = (Q_L - Q_G)(C_P - \frac{C_L}{R_L}) + Q_G(\frac{C_G}{R_G} - \frac{C_L}{R_L}) - r_o \tag{95}$$

where

$$r_o = \frac{k_L(C_L/R_L)}{K_L + (C_L/R_L)} \tag{96}$$

is a Michaelis-Menten type of expression for the secretion rate (mcg. / min.) of methotrexate out of the liver cells into the bile ducts. Note that the liver secretion is saturable at high doses.

The reason for using a multicompartment model to describe bile formation and secretion is to account for the delay before the drug in the bile reaches the gut-lumen. If a step function is used to introduce a delay, the theoretical curve does not agree with the observed S-shaped efflux bile curve. It is known from chemical engineering that a series of discrete compartments with the inflow rate greater than the outflow rate provides both a delay and a smooth response. The following equations describe such compartments.

Bile ducts:

$$\tau \frac{dr_i}{dt} = r_{i-1} - r_i \qquad (i = 1,2,3) \tag{97}$$

The model parameter τ is called the holding time. Three equations were adequate to model separate experiments where the bile duct was cannulated and concentrations of methotrexate in the bile were measured directly. An extensive description of this scheme appears in the paper by Bischoff et al. [13].

The transit of the drug down the gut is also handled by using a multicompartment model. This approach provides a smooth response and also accounts for different concentrations in different portions of the lumen. Gut absorption from the lumen to the tissues can occur in each compartment. The possibility of saturable and nonsaturable effects is accounted for by the introduction of a Michaelis-Menten term $(k_G C_i / CK_G + C_i)$ and the term bC_i, respectively. The mass balance equations are:

Gut tissue:

$$V_G \frac{dC_G}{dt} = Q_G\left(C_P - \frac{C_G}{R_G}\right) + \sum_{i=1}^{4} \frac{1}{4}\left(\frac{k_G C_i}{K_G + C_i} + bC_i\right) \tag{98}$$

Gut lumen:

$$\frac{dC_{GL}}{dt} = \frac{1}{4} \sum_{i=1}^{4} \frac{dC_i}{dt} \tag{99}$$

$$\frac{V_{GL}}{4} \frac{dC_1}{dt} = r_3 - k_F V_{GL} C_1 - \frac{1}{4}\left(\frac{k_G C_1}{k_G + C_1} + bC_1\right)$$

$$\frac{V_{GL}}{4} \frac{dC_i}{dt} = k_F V_{GL}(C_{i-1} - C_i) - \frac{1}{4}\left(\frac{k_G C_i}{k_G + C_i} + bC_i\right) \quad , i = 2,3,4. \tag{100}$$

Using the above equations, which describe the compartmental model for methotrexate distribution, the pharmacokinetic behavior of methotrexate was able to predicted in mice, rats, dogs, monkeys and man over a wide dose range. The predicted results agreed closely with the experimental results. Since the same model with parameters scaled for the size of the animals was used for the different species, the model must be a reasonable approximation of the actual physiologic and events. Thus, the model should not only be of aid in interpreting experiments but also useful in predicting results in lieu of performing experiments. The values of the model parameters, the sources for these values and information on scaling these parameters can be found in Bischoff et al. [14] and Bischoff [10].

The pharmacokinetic behavior of the drug cytosine arabinoside (ARA-C) is quite different than methotrexate. Its distribution is dominated by rapid metabolism in the liver and organs of man. Fig. 8.5 shows the model devised by Dedrick et al. [22a] for studying the detailed disposition of ARA-C. The volumes of the compartments are proportional to the depicted size in Fig. 8.5

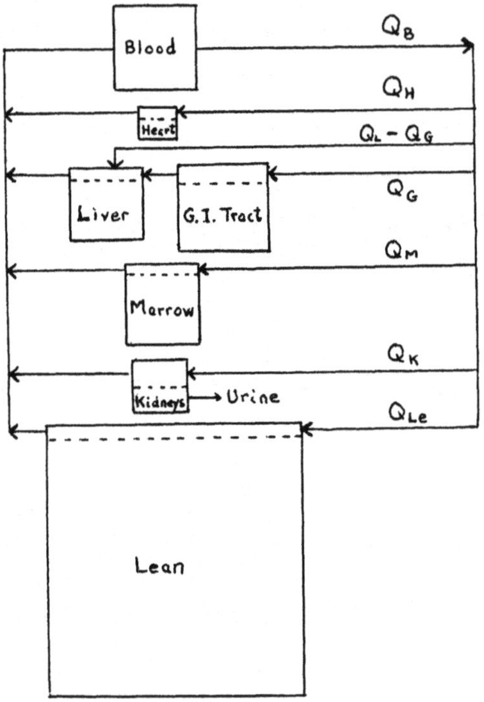

Figure 8.5 Compartmental model for ARA-C distribution.

The choice of compartments is based on the following physical facts. ARA-C disappears from the blood more rapidly than can be accounted for by urinary excretion, which is important. ARA-C is deaminated to ARA-U by homogenates of human liver. ARA-U is a significant inhibitor of ARA-C decimation only when the molar ratio of ARA-U to ARA-C is large. ARA-C is very soluble in water and is not bound significantly to plasma protein. Thus lean tissue should be an important resevoir for the drug and its metabolite. The marrow is represented because it is a critical site of toxicity. The gastrointestinal tract is included to allow simulation of oral doses or metabolic activity.

The compartmental models is described mathematically by a set of mass-balance equations on each compartment for ARA-C and ARA-U. Once again, experimental evidence suggests that flow-limited conditions hold and that the tissue to blood or plasma distribution is essentially unity. The possibility of metabolism by a Michaelis-Menten process, $(\frac{vC}{K+C})V$, is included in each compartment, even though there may be negligible activity in certain tissues. Most of the details in setting up and simplifying the mass balance equations for ARA-C will be omitted, since the procedure is similar to that used previously for methotrexate.

The following are the complete set of equations for ARA-C. Only the tissue volume is used as a basis for metabolism by the enzyme located in the tissue. For example, the kidney balance equation is

Kidney:

$$(102) \qquad V_K \frac{DC_K}{dt} = Q_K C_B - Q_K C_K - K_K C_B - \left(\frac{v_{max,K} C_K}{K_{m,K} + C_K}\right) V_{KT}$$

where

$v_{max,K}$ = enzyme activity at saturation, μg/g min

$K_{m,K}$ = Michaelis constant, μg|m|.

The numerical values for $v_{max,K}$ and $K_{m,K}$ are obtained from in vitro reaction rate studies using measurements of the enzyme (pyrimidine nucleoside deaminase) levels in the tissue.

Blood:

$$(103) \qquad V_B \frac{DC_B}{dt} = (Q_H C_H + Q_L C_L + Q_M C_M + Q_K C_K + Q_{Le} + C_{Le}) - Q_B C_B - \left(\frac{v_{max,B} C_B}{K_{m,B} + C_B}\right) V_B + \text{(injection function)}.$$

Heart:

$$(104) \qquad V_H \frac{dC_H}{dt} = Q_H C_B - Q_H C_H - \left(\frac{v_{max,H} C_H}{K_{m,H} + C_H}\right) V_{HT}$$

Liver:

$$(105) \qquad V_L \frac{dC_L}{dt} = (Q_L - Q_G) C_B + Q_G C_G - Q_L C_L - \left(\frac{v_{max,L} C_L}{K_{m,L} + C_L}\right) V_{LT}$$

G.I. Tract:

$$(106) \qquad V_G \frac{dC_G}{dt} = Q_G C_B - Q_G C_G - \left(\frac{v_{max,G} C_G}{K_{m,G} + C_G}\right) V_{GT}$$

Marrow:

$$(107) \qquad V_M \frac{DC_M}{dt} = Q_M C_B - Q_M C_M - \left(\frac{v_{max,M} C_M}{K_{m,M} + C_M}\right) V_{MT}$$

Lean Tissue:

$$(108) \qquad V_{Le} \frac{dC_{Le}}{dt} = Q_{Le} C_B - Q_{Le} C_{Le} - \left(\frac{v_{max,Le} C_{Le}}{K_{m,Le} + C_{Le}}\right) V_{LeT}$$

Cumulative excretion in urine:

$$(109) \qquad U = \int_0^t K_K C_B dt.$$

Analogous equations can be written for the metabolite, ARA-U. These resemble (102) through (109) except that the metabolism terms are positive, since ARA-U is being formed from ARA-C by deamination.

The resulting set of equations can be solved numerically. The values of the model parameters can be found in Dedrick et al. [22a]. The solutions give the concentration of ARA-C and ARA-U in each of the seven compartments and the cumulative urinary excretion of both. From these solutions it is easy to predict quantities of interest such as the ratio of ARA-C to ARA-U in the urine on a cumulative basis or for any time interval. The predicted results agreed with the experimental results. For instance, the model predicted a rapid drop in ARA-C plasma plasma concentration followed by a slower exponential decrease.

Bischoff et al. [15] combined Rubinow's cell model with the results of the above pharmacokinetic studies to predict the effects of ARA-C scheduling on mouse L-1210 leukemia cells. These cells exhibit essentially exponential growth during the main portion of cell growth. The total cell density,

$$N(t) = N_o\ 2^{-t/\tau}\ e^{-\Lambda(t)} \tag{110}$$

can be found by assuming that the asymptotic solution of Rubinow's equation is of the form

$$n(\mu,t) = N_o e^{\beta t}(h/\mu)\ \exp(-\Lambda(t))$$

where the unknown constants and functions are determined from the boundary conditions, the initial conditions and the cell loss function.

The mean loss rate of cells by drug action is modeled by

$$\mu(\alpha,t) = \frac{K_1(\alpha,t)\ C(t)}{K_2(\alpha,t) + C(t)}$$

where $C(t)$ is the local concentration of ARA-C and K_1 and K_2 are functions which can be chosen to reflect the cycle-specificity of the drug. For instance, if $a \le \alpha \le b$ corresponds to the S phase, then $K_1(\alpha,t) = 0$ outside that interval. For L-1210 cells, it is known that the S phase is approximately 80 or 90% of the entire cell cycle time. The region in α, outside the range $a \le \alpha \le b$ is so small that cycle specificity can be ignored. Hence, it is assumed that both K_1 and K_2 are functions of t only.

Recall that the pharmacokinetic model of ARA-C predicted that after a short time the plasma concentration of ARA-C decreased exponentially. Hence, it is assumed that

$$C = C_o e^{-t/t_d}$$

where

$$t_d = (\text{half-life})/(\ell n2)$$

Then the loss function can be determined as

$$\Lambda(t) = K_1 t_d \left[\frac{K_2+C_o}{K_2+C_o e^{-t/t_d}}\right]$$

The functions K_1 and K_2 are approximated by constant values and can be evaluated from a single schedule. The function Λ can then be substituted in (110). The model can then be used to predict the results of other types of experiments. The predicted results approximated the experimental results quite closely.

In the future, as more biochemical, physiological and pharmacological details are discovered the above pharmacokinetic models will be refined. Fro example, it ha has recently been discovered that the active intracellular form of ARA-C is the triphosphate, ARA-CTP, not ARA-C itself. ARA-CTP acts by blocking DNA synthesis and possibly by introducing ARA-C moieties into the DNA coding sequences. Therefore, it is important to determine the ARA-CTP concentration within target cells.

The paper by Morrison et al. [50] extends the model of ARA-C described above. The new model includes the simulation and prediction of intracellular metabolites and enzyme kinetics within those target cells which are the site of the ultimate cytotoxic and cytostatic effect. It is shown that the kinetics of ARA-C and ARA-CTP in vivo are quite different. Thus, the use of ARA-C concentrations to predict states of the ara-CTP-dependent properties of toxicity can lead to large errors.

The refined model of ARA-C pharmacokinetics is applied by Lincoln et al. [44] to simulate the treatment of L-1210 leukemia. Cell proliferation is described by a generalization of Rubinow's equation which includes cell-to-cell variability in maturation rates. The pharmacokinetic model and cell model are related by means of a loss function which relates cell lethality and progression delay with the time course of intracellular ARA-CTP concentration. Simulations of in vitro and in vivo experiments performed in a number of laboratories are described. The predicted results approximate the laboratory data.

A challenging problem would be the application of optimal control theory with a detailed pharmacokinetic model in order to design more efficient drug protocols.

8.5 Remarks

There are many theoretical as well as practical problems which must be overcome before some of the ideas in this chapter become clinically applicable. These difficulties will be discussed below and possible avenues of research will be suggested.

Any form of cycle-specific chemotherapy requires a rapid method of determining the percentage of the population of tumor cells (and also certain normal cells) in each phase of the cell cycle. Improvements in the techniques of flowmicrofluorometry may increase the clinical applicability of this intrument. However, the complicated treatment required for preparing suspensions of solid tumors has led some investigators to explore the possibility of using pattern recognition for determining the phases of the cell cycle. The advantage of this method is that cells require relativity little preparatory work. A recent approach involves the use of a computerized optical instrument called the quantimet [60]. This machine automatically measures certain cell parameters such as the diameter and perimeter of the cells and rapidly constructs a frequency histogram of these quantities. Although the rapid determination of cell kinetic parameters may eventually be possible by this approach, research in this area is still in its infancy.

Another practical problem is obtaining tissues for analysis since chemotherapy may alter the prevalent type or phase frequency of tumor cells. New surgical techniques for obtaining repeated samples in a rapid, innocuous and painless manner will have to be developed. Moreover, improved mathematical models of cells may reduce the required number of samples.

A related problem is the determination of the actual size of the tumor at various instants of time in order to evaluate the effectiveness of the chemotherapy. If the tumor is large and localized, palpation may be helpful. However, if the tumor is small, metastasized, or unpalpable, new methods for determining its size must be developed. One avenue of approach is computerized tomography. Another approach is the development of quantitative relationships between the size of the tumor and (a) certain characteristic antigens, or (b) the heat the tumor omits, or (c) the pressure the tumor exerts. Other possibilities are ultrasonagraphy and the use of radioactive isotopes.

Complex cell models, such as the one of Eisen and Macri [25], should be combined with one of the sophisticated pharmacokinetic models described in the previous section. The results can then be compared with various simplified models in order to discover which parameters are important. A reduction in the number of parameters seems necessary in order to model the intersection of several different drugs.

Models for blocking cells in some phase of the cell cycle and killing the blocked cells with another drug should be developed. The development of such a mocel would have to be guided by experimental results, since the exact mechanism for the action of many anticancer drugs is not known.

Drug interaction is intimately connected with the quantitative kinetics of complete biochemical pathways. Little is known about these pathways since mosst studies have been qualitative or have only considered a single enzyme. A possible approach is to condense the extremely complicated pathway into a few important rate controlling steps. Some examples of this tactic appear in the works of Raughi et al.[55] and Garfinkel et al. [25]. Some progress towards a complete cell model has been made by Heinmets [30] and Davison [20]. However, further investigation and verification of these models is still required.

Werkheiser and coworkers [70,71] have considered the effects of combinations of drugs on cell growth by formulating a model of deoxyribonucleic acid (DNA) biosynthesis. Velocity equations of the simple Michaelis-Menten form were used to obtain an open steady-state system regulated by a network of feedback controls. Good qualitative agreement was obtained for most of the combination of drugs studied. To explain why such simple models with unknown kinetic parameters can give insights into complex chemotherapeutic systems, they postulated that the topology of the network was the dominant factor. This intuitive notion of network form might be formalized by using the ideas in the papers of Glass [27], Hill and Simmons [31] and Oster et al. [51,52].

Werkheiser's approach has been extended to analyze the transient and steady state effects of combination of some antimetabolites. The Michaelis-Menten system of differential equations presented by Milgram were recast in a standard form suitable for numerical solution by Schiller and Eisen. The calculations were performed by Milgram and are detailed in Milgram and Nicolini [49].

Further research is required in order to refine pharmacokinetic models. For example, flows across membranes should be treated in a more realistic manner. Some progress in this direction has been made by Mikulecky and coworkers [47,48] using n nonequilibrium thermodynamics.

The problem of achieving an ideally sought level of therapeutic response without exceeding predetermined safe and tolerable levels of adverse drug reactions has received little attention. Control system analysis has been applied to this problem by Smolen et al. [64]. This approach should be extended to models incorporating both pharmacokinetics and control theory.

None of the main theoretical methods (the calculus of variations, the Pontryagin minimum principle, dynamic programming, linear and quadratic programming) are particularly well suited for finding optimal therapeutic regimens. The difficulties with these approaches is that they lead to problems which are presently unsolvable both analytically and numerically as pointed out in the paper by Creasey et al. [19]. Thus, in many situations, one is forced into using simulation in conjunction with optimal search methods for determining optimal drug protocols. One difficulty with this last method is that there is, in general, no assurance that the result is not just a local optimum rather than a global optimum. Another major difficulty is that there is no optimal search technique that converges rapidly for all problems, all data sets and all starting values. Finally, since some of the models involve partial differential equations, optimal control theory should be applied to such systems.

Most of the models for chemotherapy use average values of parameters and predict the average behavior of the tumor. Although such models are suitable for experimental work, they are not suitable for individual chemotherapy. It might be possible to modify some of these models by determining and using individual physiologic parameters. Not only will a large number of biological experiments be required but new methods of identifying and estimating the relevant parameters must be developed along the lines described in Bekey et al. [5] and Ismaile et al. [35].

A new approach for customizing cancer chemotherapy to the individual seems necessary since tumors of the same type often exhibit different behavior in different patients. Moreover, the same drug affects different people in different ways. Methods for determining which of several possible drugs is the most effective in treating a particular tumor should be developed. A possible approach might involve testing the given drugs on cultured tumor cells. In vivo, the size of the tumor and selected population of normal cells should be continuously monitored after administering the chosen drug. The concentration of the next dose and the time that it should be administered should depend upon the values of these measured parameters. The possibility of neutralizing a given drug when too large a dose has been administered should also be considered. In other words, a sampled-data feedback type of approach should be used. Some parameters of the model which is used would be known while others would have to be estimated. Some ideas for constructing such sampled-data feedback models are contained in Lincoln et al. [44] and Prasad and Ibidapo-Obe [55].

Some promising new drugs are produced normally by the human body. These are less destructive to normal cells than the conventional anti-cancer drugs. The best known natural cancer fighter is interferon, which was first isolated in 1957. Tests in Sweden have shown that interferon doubled the survivor rate of children with bone cancer. Large-scale trials on its use for cancers of the lymph system, bone marrow, skin and breast have just begun.

Dr. K. Isselbacher of Massachusetts General Hospital has found another promising anti-cancer agent called CAGA. This drug has caused a 90 percent reduction in growth rates for human breast and pancreatic cancers transplanted to mice. Unlike interferon, which is species specific, CAGA from humans works equally well against hamster tumors.

Dr. S. Green of the Memorial-Sloan-Kettering Cancer Center in New York, has discovered a substance which he calls nHG. This agent has killed some human tumors in mice as quickly as 24 hours after injection into their blood.

These three new drugs are expensive and difficult to obtain in a pure form. Their exact chemical composition is a mystery and it is not known how they work.

A major problem in chemotherapy is that the drug does not act solely at the required sites but also affects other organs. The development of local acting drugs would alleviate this problem. One approach would be the addition of moieties to the drug which would concentrate the cytotoxic agent at the desired site. For instance, the addition of iodine might concentrate the cytotoxan in the thyroid gland. Combining a ferromagnetic substance with the drug and using magnetic fields might increase the concentration of the drug at nearly any desired site. Thus smaller doses of the drug would be required. Another approach would be to implant the drug plus some inert moiety directly into the tumor. The inert chemical would allow the drug to be released at a rate sufficient for controlling the cancer but not rapidly enough to cause toxic effects in other parts of the body.

The difficulty in applying optimal control theory to cancer chemotherapy increases with the number of different normal cell populations which must not be harmed by the drug. Experimentalists should rank these populations according to the order in which toxic effects appear. In some situations, only the toxic effects for the most sensitive populations would have to be considered. This would greatly simplify the associated optimal control problem.

Finally, models of cancer chemotherapy should be combined with models of the immune system in order to study the joint and dependent effects of the drug and the immune response on the tumor.

REFERENCES

1. Aroesty, J., Lincoln, T., Shapiro, N. and Boccia, G. Tumor growth and chemotherapy: mathematical methods, computer simulations, and experimental foundations, Math. Biosciences, 17, 243-300, 1973.

2. Bagshawe, K.D. Tumor growth and anti-mitotic action; the role of spontaneous cell losses, Brit. J. Cancer, 22, 698-713, 1968.

3. Bagshawe, K. Choriocarcinoma: The Clinical Biology of the Trophoblast and its Tumors, Edward Arnold Ltd., London, 180-220, 1969.

4. Bahrami, K. and Kim, M. Optimal Control of multiplicative control systems arising from cancer therapy, IEEE Trans. Auto. Cont., Vol. AC-20, 537-542, 1975.

5. Bekey, G. and Benken, J. Identification of biological systems: a survey, Automatica, 14, 41-47, 1978.

6. BBellman, R., Jacquez, A. and Kalaba, R. Some mathematical aspects of chemotherapy, Proc. Fourth Berkeley Symposium of Math. Stat. and Prob., 4, 57-66, Univ. of Cal. Press, Berkeley, 1961.

7. Bellman, R. and Dreyfys, S. Applied Dynamic Programming, Princeton University Press, Princeton, 1962.

8. Berenbaum, M.C. Dose-response curves for agents that impair cell reproductive integrity, Brit. J. Cancer, 24, 434-445, 1969.

9. Bird, R.B., Stewart, E.W. and Lightfoot, E.W. Transport Phenoma, Wiley, New York, 1960.

10. Bischoff, K.B. Some fundamental consideration of the applications of Pharmacokinetics to cancer chemotherapy. Cancer Chemotherapy Reports, 59, 777-93, 1975.

11. Bischoff, K.B. and Brown, R.G. Drug distribution in mammals, Chem. Eng. Prog. Symp. Series, 84, 64, 33-45, 1968.

12. Bischoff, K.B. and Dedrick, R.L. Generalized solution to linear, two-compartment, open model for drug distribution, J. Theor. Biol. 29, 63-83, 1970.

13. Bischoff, K.B., Dedrick, R.L., Zaharko, D.S. and Slater, S. A model to represent bile transport of drugs, Proc. Ann. Conf. Eng. Med. Biol., 12, 89, 1970.

14. Bischoff, K.B. Dedrick, R.L., Zaharko, D.S. and Longstreth, J.A. Methotrexate Pharmacokinetics, J. Pharmac. Science, Vol. 60, No. 8, 1128-33, 1971.

15. Bischoff, K.B., Himmelstein, K.J., Dedrick, R.L. and Zaharko, D.S. Pharmacokinetics and cell population growth models in cancer chemotherapy, in Chemical Engineering in Medicine and Biology: Advances in Chemistry, American Chemical Society, No. 118, 47-64, 1973.

16. Bloch, E.H. A quantitative study of hemodynamics in the living microvascular system, Am. J. Anat., 110, 125-145, 1962.

17. Burke, P.J. and Owens, A.H. Attempted recruitment of leukemic myeloblasts to proliferative activity by sequential drug treatment, Cancer, 28, 830-836, 1971.

18. Cox, E.B. Determination of possibility of cure, time to cure, and cell kill fraction in the Gompertz growth model, Duke University Medical Center, Durham, North Carolina, 1978.

19. Creasey, W., Fegley, K., Karreman, G. and Long, V. Designing optimal cancer chemotherapy regimens, in Modelling and Simulation, Vol. 9, Proceedings of the Ninth Annual Pittsburgh Conference, Instrument Society of America, Pittsburgh, 379-385, 1978.

20. Davison, E.J. Simulation of cell behavior: normal and abnormal growth, Bull. Math. Biol., 37, 427-58, 1975.

21. Dedrick, R.L. and Bischoff, K.B. Pharmacokinetics in applications of the artificial kidney, Chem. Eng. Prog. Symp. Series No. 84, 64, 32-44, 1968.

22a. Dedrick, R.L., Forrester, D.D. and Ho. D.H.W. In vitro-in vivo correlation of drug metabolism-deamination of 1-β-D-Arabinofuranosy cytosine, Biochemical Pharmacology, 21, 1-16, 1972.

22b. Dedrick, R.L., Zaharko, D.S. and Lutz, R.J. Transport binding of metrotrexate in vivo, J. Pharm. Sci., 62, 882-890, 1973.

23. Donaghey, C. and Drewinko, B. A computer simulation program for the study of cellular growth kinetics and its application to the analysis of human lymphoma cells in vitro, Comput. Biomed. Rès., 8, 118-128, 1975.

24. Donaghey, C. CELLISM II User's Manual, Industrial Engineering Department, University of Houston, 1975.

25. Eisen, M. and Macri, N. A model for drug action at the cellular level in Modelling and Simulation, Vol. 9,Proceedings of the Ninth Annual Pittsburgh Conference, Instrument Society of America, Pittsburgh, 393-99, 1978.

26. Garfinkel, D. et al. Simulation of the Krebs Cycle, Computers and Biomed, Res. Res., 4, 1-125, 1971.

27. Glass, L. Classification of biological networks by their qualitative dynamics, J. Theor. Biol., 54, 85-107, 1975.

28. Griswold, D.P., Jr., Simpson-Herren, L. and Schabel, F.M., Jr. Altered sensitivity of a hamster plasmacytoma to cytosine arabinoside (NSC-63878). Cancer Chemother. Rep. 54, 338, 1970.

29. Hahn, G.M. and Steward, P.C. The application of age response functions to the optimization of treatment schedules, Cell Tissue Kinet., 4, 279-291, 1971.

30. Heinmets, P. Analysis of Norman and Abnormal Cells, Plenum Press, New York, 1966.

31. Hill, T.L. and Simmons, R.M. Free energy levels and entropy production associated with biochemical kinetic diagrams. Proc. Nat. Acad. Sci., 73, 95-99, 1976.

32. Himmelstein, K.J. and Bischoff, K.B. Mathematical representations of cancer chemotherapy effects, J. of Pharmacokinetics and Biopharmaceutics, Vol. 1, No. 1, 51-68, 1973.

33. Harris, E.J. Transport and Accumulation in Biological Systems, Butterworths, England, 1960.

34. Holland, J.G. Clinical studies of unmaintained remissions in acute lymphocytic leukemia, in The Proliferation and Spread of Neoplastic Cells (The University of Texas M.D. Anderson Hospital and Tumor Institute, 21st Annual Symposium on Fundamental Cancer Research, 1967) Williams and Wilkins, Baltimore, 453-62, 1969.

35. Ismail, M., Prasad, T. and Quintana, V. A methodology for modeling and simulation of biomedical systems and drug kinetics using nonlinear stochastic compartmental analysis, in Modeling and Simulation, Vol. 9, Proceedings of the Ninth Annual Pittsburgh Conference, Instrument Society of America, Pittsburgh, 373-378, 1978.

36. Jacquez, J.A., Bellman, R. and Kalaba, R. Some mathematical aspects of chemotherapy II. The distribution of a drug in the body. Bull Math. Biophys., 22, 309-322, 1960.

37. Jansson, B. Competition within and between cell populations. In Fraction Size in Radiobiology and Radiotherapy, 51-72, editors Sugahara, T., Revesz, L. and O. Scott, Williams and Wilkins, Baltimore, 1974.

38. Jansson, B. Simulation of cell-cycle kinetics based on a multicompartmental model, Simulation, 25, 99-108, 1975.

39. Jusko, W.J. Pharmacodynamics of chemotherapeutic effects: dose-time-response relationships for phase-nonspecific agents, J. of Pharmaceutical Sciences, 60, 892-895, 1971.

40. Jusko. W.J. Pharmacokinetic principles in pediatric pharmacology, Pediat. Clin. N. Am., 19, 81-100, 1972.

41. Jusko, W.J. A pharmacodynamic model for cell-cycle-specific chemotherapeutic agents, J. of Pharmacokinetics and Biopharmaceutics, Vol. 1, No. 3, 175-200, 1973.

42. Kuzma, J.W., Valand, I. and Bateman, J. A tumor cell model for the determination of drug schedules and drug effect in tumor reduction, Bull. of Math. Biophysics, 31, 637-650, 1969.

43. Lewis, A.E. Principles of Hematology, Appleton Century Croft, New York, 1970.

44. Lincoln, T., Morrison, P., Aroesty, J. and Carter, G. The computer simulation of leukemia therapy: combined pharmacokinetics, intracellular enzyme kinetics, and cell kinetics of the treatment of L-1210 leukemia by ARA-C, Cancer Chemotherapy Reports, 60, 1723-1739, 1976.

45. Lincoln, Th., Aroesty, J., Meier, G. and Gross, J.F. Computer simulation in the service of chemotherapy, Biomedicine, 20, 9-16, 1974.

46. Merkle, T.C. Stuart, R.N. and Gofman, J.W. The Calculation of Treatment Schedules for Cancer Chemotherapy, UCRL-14505, Lawrence Radiation Laboratory, Livermore, California, 1965.

47. Mikulecky, D.C. A network thermodynamic two-port element to represent the coupled flow of salt and current. Improved alternative for the equivalent circuit, Biophysical J., 25, 323-340, 1979.

48. Mikulecky, D.C., Huf, E.G. and Thomas, R.S. A network thermodynamic approach to compartmental analysis of Na^+ transients in frog skin, Biophysical J., 25, 87-106, 1979.

49. Milgram, E. and Nicolini, C.A. A preliminary report on the mathematical-numerical considerations of enzyme kinetics, Biophysics Division, Temple University Internal Report 4/75, 1-54, 1975.

50. Morrison, P.F., Lincoln. T.L. and Aroesty, J. Disposition of cytosine arabinoside (NSC-63878) and its metabolites: a pharmacokinetic simulation, Cancer Chemotherapy Reports, Part 1, Vol. 59, 861-75, 1975.

51. Oster, G., Perelson, A. and Katchalsky, A. Network thermodynamics, Nature, 234, 393-399, 1971.

52. Oster, G.S., Perelson, A.S. and Katchalsky, A. Network thermodynamics: dynamic modelling of biophysical systems, Quart. Rev. Biophys., 6, 1-134, 1973.

53. Petrovskii, A.M., Suchkov, V.V. and Shkhvatsabaya, I.K. Cure management of disease as a problem in modern control theory, Automatika I Telemekhanika, 34, 99-105 (English translation - Automation and Remote Control, 34, 767-771) 1973.

54. Petrovskii, A.M. Systems analysis of some medicobiological problems connected with treatment control, Automatika I Telemekhanika, 35, 54-62 (English tran translation - Automation and Remote Control, 35, 219-225) 1974.

55. Prasad, T. and Ibidapo-Obe, O. Stochastic analysis and control of physiologic systems: cancer detection and therapy, Int. J. Systems Sci., 8, 1233-42, 1977.

56. Priore, R.L. Using a mathematical model in the evaluation of human tumor response to chemotherapy, J. Natl. Cancer Inst., 37, 635-47, 1966.

57. Raughi, G.J., Liang, T. and Blum, J.J. A quantative analysis of metabolite fluxes along some of the pathways of intermediary metabolism in tetrahymena pyriformis, J. Biol. Chem., 250, 5866-76, 1975.

58. Salmon, S.E. and Durie, B.G.M. Applications of kinetics to chemotherapy for multiple myeloma, in Growth Kinetics and Biochemical Regulation of Normal and Malignant Cells, Williams and Wilkins (Drewinko, B. and Humphrey, R. M. ed.), William and Wilkins, Baltimore, 815-77, 1977.

59. Salmon, S.E. and Smith, B.A. Immunoglobulin synthesis and total body tumor cell number in IgG multiple myeloma, J. Clin, Invest., 49, 1114- 1121, 1970.

60. Sawicki, W., Rowinski, J. and Swenson, R. Change of chromatin morphology during the cell cycle detected by means of automated image analysis, J. of Cellular Physiol., 84, 423-428, 1974.

61. Schmidt, G.W. A mathematical theory of capillary exchange as a function of tissue structure, Bull. Math. Biophys., 14, 229-63, 1952.

62. Schackney, S.E. A computer model for tumor growth and chemotherapy and its applications to L-1210 leukemia treated with cytosine arabinoside (NSC-63878), Cancer Chemother. Rep. Part I, 54, 399-429, 1970.

63. Skipper, H.E., Schabel, F.M. and Wilcox, W.O. Experimental evaluation of potential anticancer agents, XIII. On the criteria and kinetics associated with "curability" of experimental leukemias, Cancer Chemother. Rep., 35, 111, 1964.

64. Smolen, V.F., Turrie, B.D. and Weigard, W.A. Drug input optimization: bioavailability-effected time-optimal control of multiple, simultaneous, pharmacological effects and their interrelationships, J. of Pharmaceutical Sci., 61, 1941-52, 1972.

65. Stuart, R.M. and Merkle, T.C. The Calculation of Treatment Schedules for Cancer Chemotherapy, Part II, UCRL-14505, Univ. of Cal. Lawrence Laboratory, Livermore, California, 1965.

66. Sullivan, P.W. and Salmon, S.E. Kinetics of tumor growth and regression in IgG multiple myeloma, J. Clin. Invest., 51, 1697-1708, 1972.

67. Swan, G.W. and Vincent, T.L. Optimal control analysis in the chemotherapy of IgG multiple myeloma, Bul. Math. Biol., 39, 317-337, 1977.

68. Swan, G.W. Optimal control in some cancer chemotherapy problems, unpublished report.

69. Valeriote, F.A., Bruce, W.R. and Meeker, B.E. A model for the action of vinblastine in vivo, Biophys. J., 6, 145-152, 1966.

70. Werkheiser, W.C. Mathematical simulation in chemotherapy, Ann. New York Acad. Sci., 186, 343-58, 1971.

71. Werkheiser, W.C., Gridney, G.B., Moran, R.G. and Nichol, C.A. Mathematical simulation of the interaction of drugs that inhibit deoxyribonucleic acid biosynthesis, Mol. Pharmacol., 9, 320-29, 1973.

72. Wilson, R.L. and Gehan, E.A., A digital simulation of cell kinetics with application to L-1210 cells, Computer Programs in Biomedicine, 1, 65-73, 1970.

73. Zakharova, L.M., Petrovskii, A.M. and Shtabtsov, V.I. Matrix model for selection of pharmacological treatment, Automatika I Telemkhanika, 34, 58-61 (English translation - Automation and Remote Control, 34, 1763-1764), 1973.

74. Zietz, S. Mathematical Modeling of Cellular Kinetics and Optimal Control Theory in the Service of Cancer Chemotherapy, Ph.D. Thesis, Dept. of Math., Univ. of California, Berkeley, California, 1977.

IX. MATHEMATICAL MODELS OF LEUKOPOIESIS AND LEUKEMIA

9.1 Introduction

In this chapter, some quantitative models of populations of normal and abnormal blood-related cells are presented. The steady state behavior of neutrophils (certain white blood cells) and their precursors is modeled via difference equations. Next, the time-dependent behavior of some blood-related cell populations are considered. Rubinow's partial differential equation is used to describe normal neutrophil production. Under the assumption that a neoplastic neutrophil precursor is subject to abnormal control conditions, Rubinow's model predicts the occurence of acute myeloblastic leukemia. The neutrophil model can also be used to simulate chemotherapy for acute myeloblastic leukemia. The resulting simulation suggests improvements in the drug protocol. Next a model of chronic granulocytic leukemia and a discrete model of acute lymphoblastic leukemia are discussed. Finally, a comprehensive computer model for granulopoiesis and its application to cancer chemotherapy is presented.

Judging the realism and significance of the above models requires some knowledge of the biology of blood-related cells and their neoplasms. A brief outline will be presented in the next section. In addition, particular cell populations will be discussed in sufficient detail to describe the construction of the corresponding mathematical models. Readers interested in building a firmer biological foundation can consult Cline [11], Lewis [26] and Wickramsinghe [49].

9.2 The Hemopoietic System and its Neoplasms

The hemopoietic system consists of:

(a) blood cells

(b) reticuloendothelial cells (defined below)

(c) organs and specialized tissue (spleen, liver, lymph nodes, thymus, bone marrow) containing the cells described in (a) and (b)

The phrase <u>reticuloendothelial cell</u> was coined by Aschoff and Kiyono because these cells were originally discovered in juxtaposition to the <u>reticulum</u> of connective tissues and to the membranous lining (<u>endothelium</u>) of blood vessels. However, the adjective "reticuloendothelial" was not applied to a cell solely because of its location. In addition, the cell had to be phagocytic and stainable by vital dyes.

One method of explaining the cellular composition of the hemopoietic system in a unified way is to consider this system to be part of a larger system, the reticuloendothelial system. The cells of the reticuloendothelial system are assumed to be derived from a primitive stem cell called the "undifferentiated reticulum cell." This cell is neither phagocytic nor stainable with vital dyes. As shown in Fig. 9.1 it is capable of differentiating into various types of

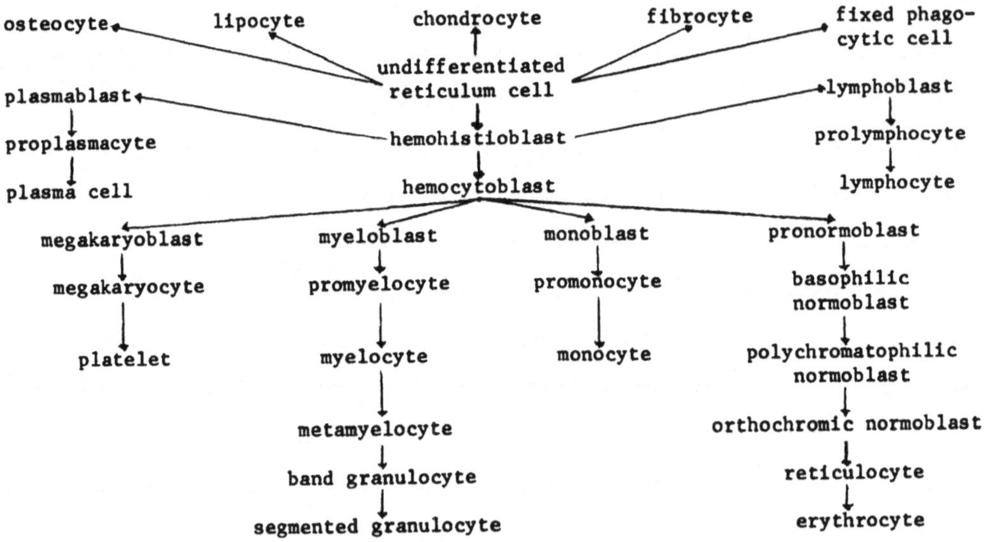

Figure 9.1 Cells of the reticuloendothelial system.

connective tissue cells (osteocytes, lipocytes, chondrocytes, and fibrocytes) as well as into the "classic" reticuloendothelial cell - the fixed phagocytic cell. The reticulum cell is also able to differentiate into hemopoietic cells are shown in Fig. 9.1. The mature form of blood cells are presented in greater detail in TABLE 9.1. At the present time, the primitive stem cell has not been identified. The undifferentiated reticulum cell, hemocytoblast and hemohistioblast cannot be distinguished from each other by morphological criteria.

Figure 9.1 is not a complete summary of the reticuloendothelial system. For example, the mature granulocytes can be further categorized as shown in TABLE 9.1.

TABLE 9.1 MATURE BLOOD CELLS IN HUMAN ADULTS

Type	platelets	red blood cells	white blood cells (leukocytes)				
			granulocytes			non granular leukocytes	
		erythrocytes	neutrophil	eosinophil	basophil	monocytes	lymphocytes
Average Number/c.c.	250,000	5 x 10^6	4,000	200	25	375	2,100
Functions	1. coagulation 2. release (a) serotonin (b) 5-hydroxytryptamine	respiratory gas exchange	1. appear in acute inflammation 2. phagocytosis of: (a) bacteria (b) Ag-Ab complexes 3. normally migrate to tissues 4. contain 17% of blood histamine	1. prominent in allergic reactions (attracted to histamine and increase release from bone marrow) 2. phagocytes for Ag-Ab complexes 3. large increase in parasitic infections 4. contain 17% of blood histamine	1. may participate in allergic reactions 2. accumulate at foreign protein 3. contain (a) 50% of blood histamine (b) heparin	1. phagocytic when changed to macrophage 2. pincytosis of protein 3. process Ag for immune response	1. can enter connective tissue and slightly phagocytic 2. immune response 3. form plasma

Furthermore, some of the "end-line" cells can undergo further development when exposed to suitable stimuli. For instance, resting lymphocytes can be transformed in several ways, as described in Appendix D, upon encountering an antigen. There is also some controversy about the number of stem cells and the number of maturation steps in the development of a particular series of cells. For example, Clarkson and Rubinow [14] have proposed the granulopoiesis model shown in Fig. 9.2 in order to explain the various types of leukemia which can occur. Thus, there

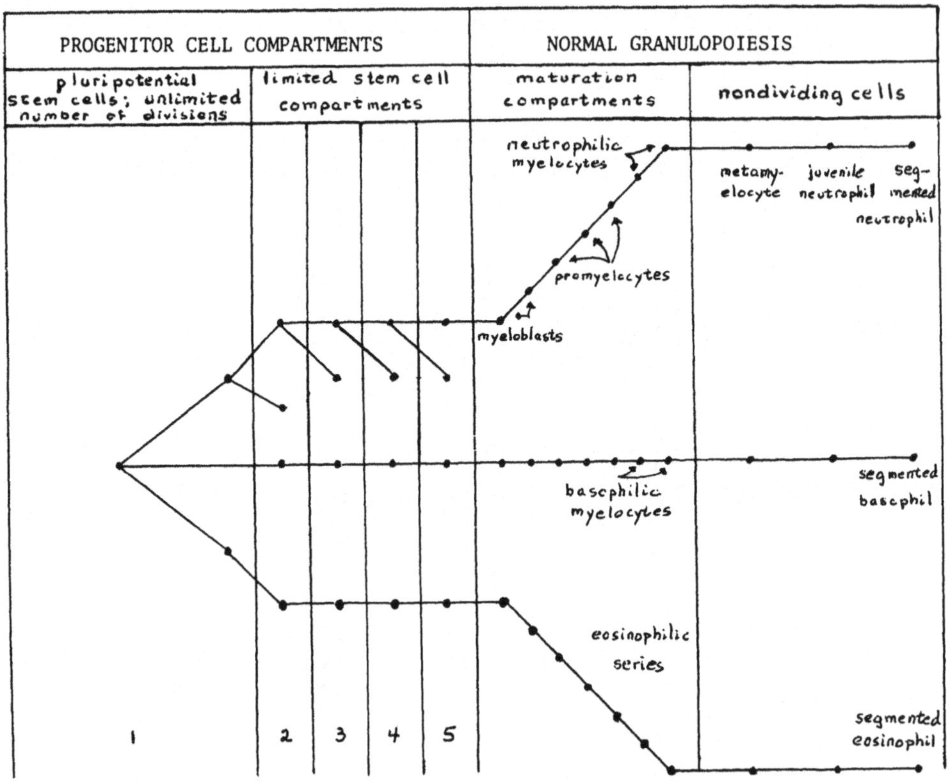

Figure 9.2 Hypothetical sequential scheme of normal granulopoiesis.

may be more stem cells than shown in Fig. 9.1 between the reticulum cell and the myeloblast. Furthermore, the exact stem cell from which a cell line originates is not certain. For example, megakaryoblasts may originate from myeloblasts.

There is a constant turnover of cells in the reticuloendothelial system. The total number of cells present at any given time represents the resultant of the rate of new cell formation and the rate of loss of old cells. In a healthy individual, the rate of new cell formation is regulated by a variety of feedback control mechanisms. These regulatory mechanisms keep the number of cells of each type within certain normal limits.

Pathologic conditions may stimulate primitive and resting reticuloendothelial cells to produce blood cells without any regard to overall requirements. This proliferative reaction may be either benign or malignant. Benign proliferation reactions often involve single cell types. Some examples are:

(a) lymphocytes and lymphoid tissue in viral infections,

(b) granulocytes in the bone marrow and spleen in bacterial infection;

(c) erythrocyte precursors in the bone marrow in hypoxia and hemolytic anemia.

The same is true in malignant proliferations. The lymphomas[1] and lymphocytic leukemias[2] are derived from lymphocytes or from their lymphatic precursors. The granulocytic leukemias affecting the bone marrow and spleen originate from granulocytes and their precursors.

Cells which have undergone malignant transformations have more virulent characteristics than merely increasing their number. They have the ability to infiltrate adjacent structures. Thus, they can penetrate lymph channels or capillaries. After being transported to a site different from the tissue of origin, they can form new colonies or metastases.

It is interesting to contrast the behavior of malignant cells with normal leukocytes. When there is myeloid hyperplasia the increased myeloid tissue may extend into adjacent lobules of marrow fat. However, the growth is confined to the marrow cavity and occurs only in response to appropriate stimuli. Granulocytes transported to distant sites may enter various tissues, but do not form colonies since they are end stage cells incapable of mitotic division. Similarly, lymph

---

[1] The major involvement in lymphoid tissue outside the bone marrow.

[2] The major involvement is the bone marrow.

nodes may become hyperplastic in response to appropriate stimuli. Normally, the new lymphocytes enter the lymphatics and do not invade the adjacent perilymphatic tissue.

At this point the reader might review the section on the development of cancer in Chapter 1 noting similarities with the following facts. There is good evidence that most forms of human cancer are monoclonal in origin. No consistent qualitative biochemical differences have been found between normal and leukemic cells at comparable levels of maturity. When a relapse occurs, the original abnormal leukemic line reappears. However, there may be some secondary abnormal lines as a result of clonal evolution. Hematopoietic tumors frequently progress from a benign, well-differentiated state to a malignant less differentiated state with greater autonomy of growth.

The etiology of leukemia is unknown. Some possible causes are the following:

(a) viruses

There is no definitive proof that there is a viral etiology for human leukemia. However, evidence accumulated from studies of leukemia in lower animals strongly supports the viral hypothesis for human leukemia. The viral etiology may be masked by other interacting factors such as genetic predisposition, ionizing radiation or the latency of the virus until activated by intrinsic or extrinsic factors.

(b) ionizing radiation

An increased incidence of leukemia has been reported in groups exposed to ionizing radiation such as:

(i) Japanese A-bomb survivors

(ii) radiologists

(iii) groups with medical radiation exposure for treatment or diagnosis

(iv) soldiers participating in simulated combat during A bomb tests

(v) residents in the vicinity of A-bomb tests.

(c) chemical agents

Certain chemicals may be leukemogenic since they affect the bone marrow. Exposure to benzene is leukemogenic. It has been suggested that phenylbutazone therapy and chloramphenicol therapy may cause leukemia.

(d) chromosomal abnormalities

The Philadelphia chromosome (or Ph'chromosome) is formed by translocation or deletions of portions of the normal 21 chromosome. It is found in most cases of chronic myelocytic leukemia. This abnormal chromosome is not found in normal persons or in those with benign myeloproliferative syndromes. mongoloids have trisomy 21 and a higher incidence of leukemia than normal children. This suggests that abnormalities of chromosome 21 are related to the development of leukemia.

Leukemia is classified according to:

(1) clinical duration of the disease

A leukemia having a relatively short duration is called acute; otherwise it is called chronic. The average untreated patient with acute leukemia survives about 4 to 6 months; some die within a few days. The acute diseases seems to have a relatively abrupt onset and are characterized by the appearance of young poorly differentiated, cell forms. The chronic diseases are characterized by predominance of mature cells, insidious onset and survival for months or years.

(2) leukocyte count in the peripheral blood

Leukemia is classified as leukemic (high leukocyte count), subleukemic (low leukocyte count but abnormal cells present), and aleukemic (low leukocyte count with no abnormal cells). It is improbable that truly aleukemic cases exist except as a stage in the development of leukemia.

(3) cell type in the blood and bone marrow

This classification of hematopoietic tumors is based chiefly on the predominant cell type. Classically it was believed that there were as many varieties of malignant hematopoietic diseases as there were lines of cells in the bone marrow. Outside the bone marrow, it was thought that there were two varieties of lymphomas corresponding to the two lines of cells in the lymphoid organs, lymphocytes (lymphosarcomas) and macrophages (reticulosarcomas).

Recently progress in morphological, cytophysiological and immunological classification of lymphocytes and their differentiated forms has led to the discovery of new types of these cells. Some classical hypotheses about certain types of lymphocytic cells has had to be changed. For example, the classical form of acute lymphocytic leukemia (ALL) was thought to be a proliferation of lymphocytic stem cells. Since these stem cells were found in the bone marrow, it was thought that the neoplastic element of ALL would be B lymphoid cells. However, investigations with immunological markers showed that more than 25% of all ALL cells carry at least one T marker. The origin of leukemia with T lymphocyte characteristics might be thymic rather than bone marrow. In light of these recent discoveries Mathé et al. [29] have proposed that the nomenclature and classification of hemopoietic neoplasms should be revised.

In order to present a unified concept of the various forms granulocytic leukemias, Clarkson and Rubinow [14] proposed the sequential scheme of granulopoiesis shown in Fig. 9.2. At the present time this scheme is purely theoretical since the stem cells in compartments 1-5 have not been identified. However, the scheme is useful for a logical classification of granulocytic forms of leukemia as shown in TABLE 9.2.

TABLE 9.2 TABLE SOME FORMS OF GRANULOCYTIC LEUKEMIAS

majority of transformed cells arrested in	type of leukemia
compartment 1	undifferentiated leukemia
compartment 2	chronic myelogenous leukemia (Ph' chromosome)
compartment 3	erythroleukemia
compartment 4	myelomonocytic (Naegli) leukemia
compartment 5	myeloblastic leukemia
promyelocytic stage	promyelocytic leukemia
neutrophilic stage	neutrophilic leukemia
eosinophilic stage	eosinophilic leukemia
basophilic stage	basophilic leukemia

Clarkson and Rubinow [14] also suggests that the simple compartmental classification of leukemia must be refined. Transitional forms may occur between any of the conventional diagnostic categories. Moreover, within any one category, there may be considerable variability in the direction of differentiation and in the degree of maturation. Both may change during the natural course of the disease or be altered by therapy. For example, the proportion of basophils, eosinophils, monocytes and megakaryocytes may vary greatly among different patients with chronic myeologenous leukemia or in any one patient during the course of his disease. Erythroleukemia may evolve into myeoblastic leukemia. There are also variable forms of leukemia that do not fit neatly into the scheme shown in Fig. 9.2. Clarkson and Rubinow [14] have pointed out that patients with forms of acute granulocytic leukemia (compartment 4 or 5) frequently have abnormalities of erythroid precursors. It seems likely that these abnormal cells arose from leukemic cell lines. The leukemic stem cell may originate in an earlier compartment (3 or 2). However, its progeny has a greater tendency for granulocyte development. Therefore, the characteristic hyperplasia found in typical erythroleukemia is not present at diagnosis. In some undifferentiated or stem cell leukemias it is, at present, difficult to determine to which precursor line these cells belong since they may lack any distinctive biochemical, ultrastructural or immunological features.

The pathology caused by hemapoietic neoplasms is a consequence of direct effects, indirect or secondary effects, effects of distant spread and metabolic effects. Consequences of some of these effects are summarized in Fig. 9.3

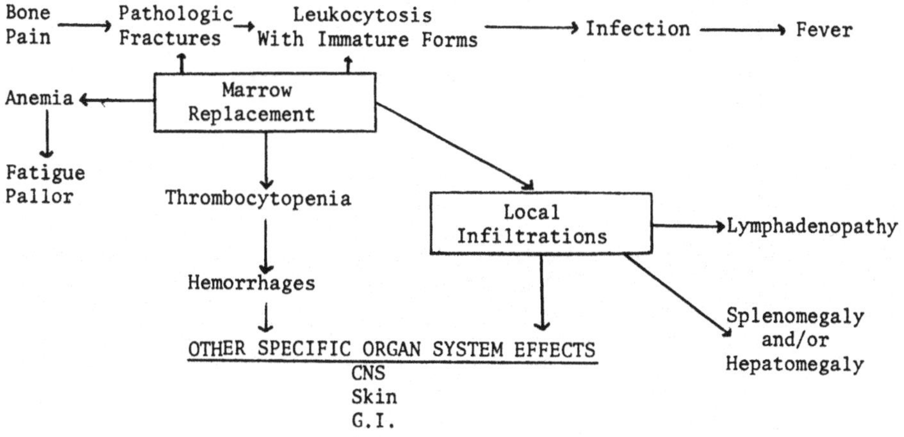

Figure 9.3 Some effects of hemapoietic neoplasms.

Distinguishing features of some common forms of leukemia appear in TABLE 9.3. Further details on the biology of blood cells and leukemia can be found in Cline

TABLE 9.3 DIAGNOSTIC FEATURES OF 5 COMMON LEUKEMIAS

The percentages in the table refer to the number of diseased patients exhibiting a particular disease characteristic.

	Acute Lymphoblastic	Acute Myeloblastic	Acute Monocytic	Chronic Myelocytic	Chronic Lymphocytic
Peak age incidence	children (3-4)	centered around teens	none	20-45	> 45%
leukocytic concentration	high in 50% normal or low in 50%	high in 50% normal to high	upper limits of normal to high	high in 100%	high in 100%
Differential WBC count	many lymphoblasts	many myeloblasts	many monocytes and monoblasts	entire myeloid series	small lymphocytes
Anemia	severe in > 90%	severe in > 90%	no but pallor	in 80% but mild	in 50% but mild
Platelets	low in > 90%	low in > 80% late in disease	small change	high in 60% low in 10%	low in 40%
sternal tenderness	65%	65%	rare	usual	rare
lymphadenopathy	moderate	minimal to moderate	cervical nodes not generalized	minimal	marked
splenomegaly	780%	60%	variable	usual and severe	usual and moderate
other features			soreness, ulceration of gums and buccal mucosa	1. granulocyte alkaline phosphatase low 2. Ph' chromosome positive in 85%	1. hemolytic anemia 2. hypogamma-globulinemia

[11], Hirsch [21], Lewis [26] and Wickramasinghe [45].

9.3 Steady State Models of the Hemopoietic System

Quantitative data has been accumulated on the kinetics and life cycle of some common blood cells such as erythrocytes and neutrophils. Therefore, it is feasible to study these data from a theoretical point of view as presented in Rubinow [35]. This approach uses the cell density function which depends on two variables, the time and a cell maturity measure. This concept of a cell density function has already been introduced in Chapter 3 Section 3, but will be described again in this section.

The system modelled is known as a differentiating cell system and is characterized by

(i) morphologically distinguishable cell classes,

(ii) the capability of cells in at least one of these classes to undergo mitosis;

(iii) daughter cells being like or unlike their parents.

An example of a differentiating cell system, which will be studied below, is the production of neutrophils. Six morphological classes of cells will be considered beginning with the myeloblast as shown in Fig. 9.1. It is believed that the myeloblast originates from another stem cell; however, such a cell has not been positively identified. To allow for this possibility a steady flux of newborn myeloblasts will be assumed to exist. Of the six classes only the first three (myeloblast, promyelocyte, myelocyte) are proliferative. The final three classes (metamyelocyte, band, segmented) are nonproliferative and are merely different maturation stages exhibited by a nonproliferative cell. Normally only the band and segmented forms leave the bone marrow to enter the blood. To simplify the model, the final three classes will be combined into a single class called the nonproliferating granulocyte class.

When a cell of a proliferative class reaches mitotic age it may or may not undergo mitosis. The three postulated types of mitosis are shown in TABLE 9.4 along with their probability of occurrence. Each of the probabilities are constant for all cells in class $i (i = 1,2,3)$. If a daughter cell is distinguishable from its mother, it is assumed that it belongs to the succeeding cell class. The cells of mitotic age which do not undergo mitosis are called <u>resting cells</u>. It is assumed that a constant fraction t_i of cells of mitotic age in the proliferating class i do not undergo mitosis. Furthermore, it is assumed that all resting cells in class i have the same life-time as the generation time T_i of dividing cell. After aging to T_i the resting cells enter the next class at age 0.

TABLE 9.4 TYPES OF MITOSIS

type	daughter cells	resemblance to mother	probability
heteromorphogenic	alike	2 dissimilar	q_i
homomorphogenic	alike	2 similar	r_i
asymmetric	unlike	1 dissimilar 1 similar	p_i

There is no loss in generality if asymmetric mitosis is not allowed ($p_i = 0$) for the following reason. If one cell undergoes hetermophogenic mitosis and another cell undergoes homomorphogenic mitosis, the result is indistinguishable from that of two cells undergoing assymetric mitosis. Mathematically, the transformation $r_i' = r_i+p_i/2$ and $q_i' = q_i+p/2$ would eliminate p if it appeared in the equations.

The cell density per unit maturation interval or cell density function will be denoted by $n_i(\mu,t)$, where the first three proliferative classes correspond to $i = 1,2,3$ and $i = 4$ denotes the nonproliferative class. The quantity $n_i(\mu,t)d\mu$ represents the number of cells in class i in the maturation interval μ to μ+du at time t. Here μ is the age of the cell measured from the moment when birth takes place. Thus, in class i, the maturity levels of newborn cells and mitotic age cells are $\mu = 0$ and $\mu = T_i$, respectively. The neutrophil production system is assumed to be in a steady state, so that

$$n_i(\mu,t) = n_i(\mu)$$

is time independent. Finally, assume that no cell death takes place during the maturation process. Then

$$n_i(\mu) = n_i \qquad 0 < \mu < T_i \tag{1}$$

which is a constant within the maturation interval of a given generation. Note that there will be a discontinuity in n_i when cell division occurs.

The cell density function at age zero, or the rate at which cells are being born at age 0 in a given class, equals the flux of cells from the preceding cell. As discussed above this flux of newborn cells can arise from:

(a) aging,

(b) heteromorphogenic mitosis,

(c) homomorphogenic mitosis;

(d) newborn cell flux from an exterior source designated s_o.

In particular only fluxes (c) and (d) are possible for the myeloblast class since a myeloblast can only originate from a stem cell or another myeloblast. Hence,

(2) $$n_1(0) = s_o+2r_1n_1(T_1-)$$

since $2r_1n_1(T_1-)$ represents the birth rate of cells of age zero at time T_1-.[3] Using (1) reduces (2) to the form

(3) $$n_1 = s_o+2r_1n_1.$$

The number of new promyelocytes per unit time, $n_2(o)$, can arise from fluxes (a), (b) or (c) and so

$$n_2(o) = 2q_1n_1(T_1-)+t_1n_1(T_1-)+2r_2n_2(T_2-)$$

which can be rewritten as

(4) $$n_2 = (2q_1+t_1)n_1+2r_2n_2$$

Similarly,

(5) $$n_3 = (2q_2+t_2)n_2+2r_3n_3.$$

and

(6) $$n_4 = (2q_3+t_3)n_3$$

since only fluxes (a) and (b) can occur in the nonproliferative cell class. Since each cell in a proliferating class is either a resting cell or destined to divide, the probabilities

(7) $$q_i+r_i+t_i = 1 \qquad i = 1,2,3.$$

[3] A - or + after T_1 signifies the value of μ just before or just after T_1, respectively. This notation is required because n_1 is discontinuous at T.

The 14 unknown parameters in (4)-(7) cannot be uniquely determined since there are only 7 equations. However, certain restrictions are placed upon the parameters. For example, (3) may be rewritten as $r_1 = 1/2 - 1/2\ s_o/n_1$. Since s_o and n_1 cannot be negative numbers, $r_1 \leq \frac{1}{2}$. Analogously it can be shown that $r_2 \leq \frac{1}{2}$ and $r_3 \leq \frac{1}{2}$; $r_i > \frac{1}{2}$ implies that the steady state hypothesis is not valid and that the class is in an exponential growth phase.

Three additional relations can be obtained as follows. The total population of the ith cell class

(8) $$N_i = n_i T_i \qquad i = 1,2,3,4$$

where T_i for i = 1,2,3 is the mean generation time of the ith cell class and T_4 is the mean life-time or transit time of nonproliferating granulocytes. The fraction of cells of the ith class in mitosis or the mitotic index

(9) $$f_{M_i} = \frac{n_i(1-t_i)T_{M_i}}{N_i} \qquad i = 1,2,3$$

where T_{M_i} is the mean length of the time for mitosis in class i. Finally, the fraction of cells of the ith class in the S phase or the labelling index

(10) $$f_{L_i} = \frac{n_i(1-T_i)T_{S_i}}{N_i} \qquad i = 1,2,3$$

where T_{S_i} is the mean length of the DNA synthesis period in the ith cell class.

An alternative form of (3)-(6) which are useful for computations may be obtained by eliminating q_i by using (7):

(11) $$n_1(1-2r_1) = s_o$$

(12) $$n_2(1-2r_2) = s_o+n_1(1-t_1)$$

(13) $$n_3(1-2r_3) = s_o+n_1(1-t_1)+n_2(1-t_2)$$

(14) $$n_4 = s_o+n_1(1-t_1)+n_2(1-t_2)+n_3(1-t_3)$$

From experimentally determined values of N_i (i = 1,2,3,4) and f_{M_i} (i = 1,2,3), Rubinow [35] was able to calculate q_i, r_i, T_{M_i} and T_i (i = 1,2,3) by utilizing (7)-(9) and (11)-(14) under the following assumptions:

(a) There is no stem cell influx into the myeloblast class ($s_o = 0$).

(b) There are no resting cells ($t_1 = t_2 = t_3 = 0$).

(c) The life time of nonproliferating granulocytes in the bone marrow is 89 hr. This value is consistent with experimental data.

(d) The mitotic time for all mitotically capable cells is the same ($T_{M_1} = T_{M_2} = T_{M_3}$).

The agreement between the experimental measurements of T_i $(i = 1,2,3)$ and the derived values of T_i was good. However, the calculated mitotic time, T_{M_i}, did not agree with the observed value.

Alternatively, dropping assumption (b), N_4, r_i and t_i $(i = 1,2,3)$ can be calculated from experimentally determined values of N_i, T_i and f_{M_i} $(i = 1,2,3)$. The calculated value of N_4 does not agree with the observed value.

Rubinow and Lebowitz [38], in an appendix, applied the above model to study other experimentally obtained data. They found that the mitotic index data was not consistent with either the labelling index data, or the generally accepted generation time estimates for myelocytes. They also studied a model suggested by other investigators and obtained mitotic times which disagreed with those calculated previously. Rubinow and Lebowitz concluded that the available estimates of the kinetic parameters of neutrophil production are not consistent with any given theretical scheme of proliferation.

Further experiments are needed to obtain corroborative information about mitotic times and other kinetic parameters of the neutrophil production process. Once this experimental data is obtained it can be used for testing more sophisticated models.

An example of another type of model for describing hemopoiesis appears in Clarkson and Rubinow [14]. This hypothetical model of a catenary compartment system is illustrated in Fig. 9.4. Here, pools or compartments (represented by circles in the figure) are used to denote the various cytological states of the hemapoietic system. The directed line segments represent allowable cell fluxes between the pools which they connect. Following division of a cell in compartment i a fraction r_i $(i = 1,2,\ldots,5)$ of the daughter cells remain in compartment i. Compartment 1 contains stem cells of unlimited proliferative capacity. The maintenance of a steady state requires that for every daughter cell that leaves the

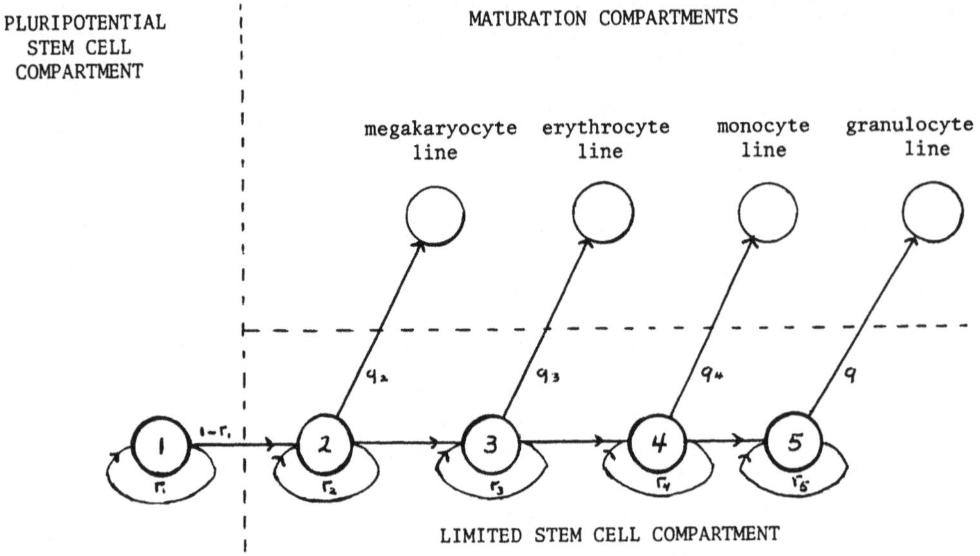

Figure 9.4 Catenary compartment system depicting hemopoietic stem cells.

compartment, one daughter cell remains in the compartment. Hence, $r_1 = 1/2$ in the steady state. Compartment 2 through 5 are called limited stem cell compartments since they have a limited proliferative capacity to maintain themselves for a fixed number of generations only. These compartments will be described in more detail in the next paragraph. The limited stem cell compartment $i(i = 1,2,3,4,5)$ has $r_i < 1/2$. However, in the steady state, their population size is maintained at a constant level by the influx of new cells from the preceding compartment. Note that from Fig. 9.4 a fraction q_i $(i = 1,2,3,4,5)$ of daughter cells in the ith compartment leave it to enter the associated committed maturation compartment, while the remaining fraction, $1-r_i-q_i$ $(i = 2,3,4)$, enters the succeeding limited stem cell compartment. The cells that enter the committed maturation compartment are called committed precursor cells. They are committed to a specialized line of development, such as the pronormoblast that leads to the erythrocyte or the myeloblast that leads to the granulocyte.

A cell in a limited stem cell compartment must undergo a change or age from one generation to the next. Its descendants can only remain in the compartment in which they originated for a fixed number of generations. This number is called the characteristic division number of the limited stem cell compartment and will be denoted by k_i. For example, suppose a cell has just entered compartment 2 from compartment 1. Its first division in compartment 2 is asymmetric, one daughter being type 2 and the other being a megakaryoblast. The daughter cell of type 2 divides in exactly the same manner as its mother and this type of division continues for k_2 generations. However, the daughter of the (k_2+1)st generation must produce one daughter of type 3 and a megakaryoblast. Note that for compartment 5 that after k divisions in it, the daughter cell undergoing the (k+1)st division must produce two committed precursor cells, which are myeloblasts. A postulated outcome of divisions in which both daughters are unlike and both different from the parent cell is called asymmetric heteromorphogenic division.

The consequence of the above scheme is that in the steady state, a fraction $k_i/(k_i+1)$ of dividing cells in compartment i undergo asymmetric division while a fraction $1/(k_i+1)$ of the dividing cells undergo asymmetric heteromorphogenic division. The fraction of cells that return to compartment i is $r_i = k_i/2(k_i+1)$ since two daughters are produced per cell division. The fraction that become committed precursor cells is $q_i = 1/2$ and the fraction that become cells of the succeeding limited stem progenitor type is $1-r_i-q_i = 1/2(k_i+1)$.

Using techniques analogous to those used in deriving (2)-(6), Clarkson and Rubinow [14] obtained equations for the steady state cell density function n_i of stem cell compartment i (i = 1,...,5). However, no experimental results are available to verify the theory.

An alternative to the catenary compartment of Fig. 9.4 is the mammillary or star shaped compartment system shown in Fig. 9.5. In this system, compartment 1, plays a central role in producing limited potential stem cells of types 2 through 5. Such a scheme appears to be simpler than that of Fig. 9.4 and may be the correct one. However, the complexity of the stem cell is increased since it must have the inherent ability to produce a variety of committee stem cells. Moreover, a corresponding regulatory mechanism is required for controlling the outcomes of division.

The frequency of occurrence of acute leukemias lends support the scheme shown in Fig. 9.4. Megakaryocytic leukemia and acute erythremia are the rarest of the

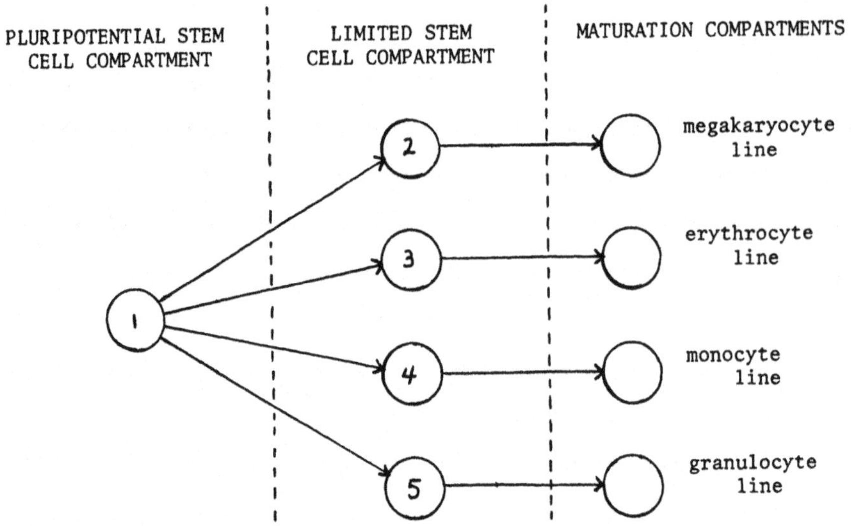

Figure 9.5 Mammillary compartment system representing hemopoietic stem cells.

acute nonlymphocytic types. Monocytic leukemia occurs with intermediate frequency and myelomonocytic and myeloblastic leukemias are the commonest. Suppose that the neoplastic lesion occurs in a stem cell and is equally likely to occur in any stem cell. Then megakaryocytic leukemia occurs when the lesion appears in compartment 1 and is transmitted to compartment 2 or when the lesion occurs in compartment 1. However, myeloblastic leukemia can occur if the lesion appears in any of the compartments 1 through 5. Therefore, myeloblastic leukemia should occur more frequently than megakaryocytic leukemia. The order of the other leukemias can be justified in a similar manner.

A combination of the catenary and mammillary schemes may be required to explain other facts. For example, the catenary compartment model does not predict the frequency of appearance of the various cell lines of chronic myelogenous leukemia and other myeloproliferative disorders. A scheme like that in Fig. 9.5 might have to be involved to explain how eosinophils, basophils and neutrophils can all be produced from compartment 5 in Fig. 9.4.

9.4 Kinetic Model of Neutrophil Production

The neutrophil production scheme is illustrated in Fig. 9.6.

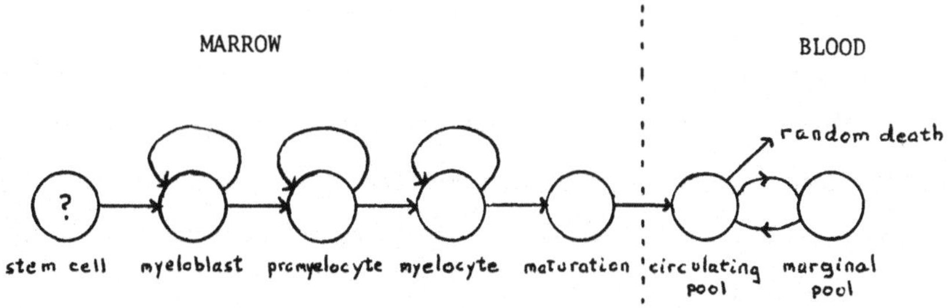

Figure 9.6 Production of normal neutrophils.

An unidentified stem cell is believed to exist in the marrow from which all granulocytes are derived. The earliest cytologically identifiable neutrophil precursor is the myeloblast which is proliferative and divides to produce more myeloblasts as well as promyelocytes. The latter divide and produce promyelocytes and myelocytes. Myelocytes are proliferative and give rise to a maturing and nonproliferating cell pool. This pool contains cytologically distinguishable cells which are called metamyelocytes, band forms and segmented neutrophils when ranked in order of increasing maturity. Eventually mature neutrophils are expelled into the blood from this maturation compartment by some unknown mechanism. The total blood neutrophil pool consists of two subcompartments, a circulating neutrophil compartment and a marginal neutrophil compartment. These pools contain approximately equal number of cells and equilibrate rapidly with each other.

There are several problems in formulating a detailed mathematical model of the neutrophil production scheme depicted in Fig. 9.6. One difficulty is the lack of precise knowledge of either the mean generation time or the mean transit time of cells in the proliferative compartments. Another difficulty is the lack of knowledge of the cell types of the daughters of a dividing cell. Models, such as those proposed in the last section, which try to circumvent this last difficulty, have not been satisfactory. The mean generation times and other parameters cannot be assigned to each of the proliferative compartments in a manner consistent with known observations regarding them.

In light of the above difficulties, Rubinow and Leibowitz [38] decided to accept the view that there is a single self-maintaining proliferative pool. The cells in this pool can be thought of as either representing myelocytes, the most common proliferative cell or as representing the mean behavior of all proliferating cells. This model, consisting of five compartments is depicted in Fig. 9.7. Rubinow and Leibowitz considered their model to be adequate for explaining virtually all known kinetic observations of neotrophils in the blood.

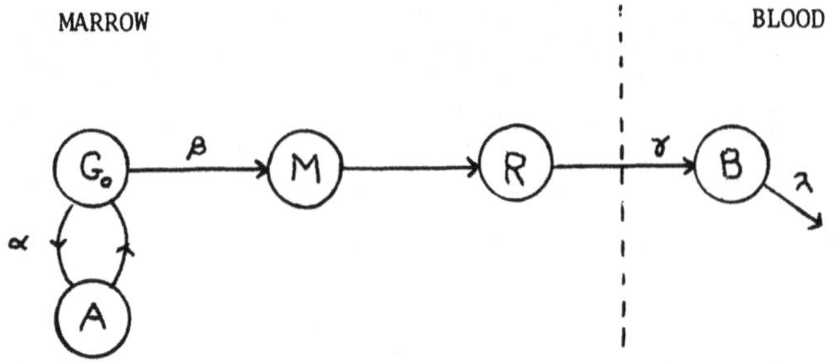

Figure 9.7 Simplified model of normal neutrophil production.

The pools or compartments in Fig. 9.7 are represented by labeled circles whose interpretation follows. The active compartment A and the resting compartment G_0 taken together comprise the proliferative pool. The compartments labeled M, R and B represent states of maturation, reserve and circulating blood, respectively. These three compartments are nonproliferative.

A cell which enters the active phase undergoes cell division a fixed time T_A later. This generation time T_A consists of the usual four cell-cycle phases G_1, S,G_2 and M. After mitosis the two daughter cells produced by division enter the G_0 state. In the G_0 phase a certain portion of the cells randomly leave to reenter the active compartment at a fractional rate α per unit time. Similarly, a certain portion of the cells randomly leave to enter the maturation compartment at a fractional rate β per unit time.

The analytic form of α and β is chosen to satisfy the following biological assumption. The control mechanism for cell production is able to recognize the total number, N(t), of neutrophils and their precursors and functions by trying to maintain N(t) at a fixed steady state level $\bar{N}$. Hence, when N(t) = $\bar{N}$ the rate of sending cells to the maturity compartment must equal the rate of sending cells to the active compartment. In other words $\alpha = \beta = \alpha_o$. If $N(t) < \bar{N}$, the resting compartment responds to the depletion in the total population by increasing production and by sending more cells to mature - that is, $\alpha > \alpha_o$ and $\beta > \beta_o$. On the other hand, if $N(t) > \bar{N}$ then $\alpha < \alpha_o$ and $\beta < \beta_o$. Specific functional forms of α and β which has the above properties are:

$$\alpha(t) = \alpha_o + \alpha_1[(\bar{N}/N)^{\nu} - 1] \tag{15}$$

$$\beta(t) = \alpha_o + \beta_1[(\bar{N}/N)^{\nu} - 1] \tag{16}$$

Here $\alpha_o, \alpha_1, \beta_1, \bar{N}$ and ν are positive constants, and $\alpha_1 > \beta_1$. The fact that $\alpha_1 > \beta_1$ insures stability about the value $N = \bar{N}$.

The biological or biochemical explanation of how the control mechanism functions is of no concern mathematically. However, the search for a biological explanation is being actively pursued. Hypotheses for the operation of such controls by a granulopoietic chalone, by colony stimulating factors (CSF) and CSF inhibitors are described in the papers by Bullough [4], Robinson and coworkers [3,34] and Chan and Metcalf [10], respectively.

Upon entering the maturation compartment all cells mature for a fixed time T_M and then enter the marrow reserve compartment R. The latter compartment is a "random" compartment like G_o because a certain proportion of the cells randomly leave the reserve compartment at a fractional rate γ per unit time and enter the blood. The blood compartment is also a random compartment from which cells randomly disappear at a constant fractional rate λ per unit time.

The blood compartment represents the combined neutrophil circulating and reserve pools. These two pools have been combined since the interaction between them is unknown and the rate of equilibration between them is rapid.

The analytic form of γ is chosen so that the following second biological assumption is satisfied. The release rate is increased whenever the total blood neutrophil population falls below its steady state value $\bar{N}_B$. However, if neutrophilia exists, then the marrow compartment releases cells at its minimum rate γ_o. These responses do not occur immediately; the existence of a time delay T_R is indicated by various experiments. A functional form of γ which has these properties is

$$(17) \qquad \gamma = \gamma(t) = \begin{cases} \gamma_0 + \gamma_1[(\bar{N}_B/N_B(t-t_R))^{\rho} - 1] & N_B(t-t_R) < \bar{N}_B \\ \gamma_0 & N_B(t-t_R) \geq \bar{N}_B \end{cases}$$

Here $\gamma_0, \gamma_1, \bar{N}_B, \rho$ and t_R are positive constants.

The neutrophil production model will be described by means of the cell density functions $n(\mu,t)$, $g(a,t)$, $m(\mu,t)$, $r(a,t)$ and $b(a,t)$ for the active, G_0, maturation, reserve and blood compartments respectively. The variables a,μ and t represent age, maturity and time, respectively. The density functions satisfy the following age-time equations of Chapter 3:

$$(18) \qquad \begin{aligned} &\frac{\partial n(\mu,t)}{\partial t} + \frac{\partial n(\mu,t)}{\partial \mu} = 0 && 0 < \mu < T_A \\ &\frac{\partial g(a,t)}{\partial t} + \frac{\partial g(a,t)}{\partial a} = -(\alpha+\beta)g(a,t) && 0 < a \\ &\frac{\partial m(\mu,t)}{\partial t} + \frac{\partial m(\mu,t)}{\partial \mu} = 0 && 0 < \mu \leqslant T \\ &\frac{\partial r(a,t)}{\partial t} + \frac{\partial r(a,t)}{\partial a} = -\gamma r(a,t) && 0 < a \\ &\frac{\partial b(a,t)}{\partial t} + \frac{\partial b(a,t)}{\partial a} = -\lambda b(a,t) && 0 < a. \end{aligned}$$

Here $a = 0$ or $\mu = 0$ characterizes cells which are entering a given compartment. The right hand sides of the equations (1) represent cell loss. Experiments in which which external agents are introduced which cause cell loss or disappearance from a given compartment require the introduction of a term $- D(\mu,t)$ (or $-D(a,t)$) to the right hand side of the corresponding compartmental equation. This term represents the rate of disappearance at time t of cells in the interval $\mu + d\mu$ (or a to $a+da$).

To complete the mathematical description of the model initial and boundary conditions are required. The initial conditions are

$$(19) \qquad n(\mu,0) = f_A(\mu) \qquad g(a,0) = f_G(a) \qquad m(\mu,0) = f_M(\mu)$$

$$r(a,0) = f_R(a) \qquad b(a,0) = f_B(a)$$

where each function is assumed to be prescribed. The manner in which new cells enter each compartment is described by the boundary conditions

$$(20) \qquad n(0,t) = \alpha N_0(t) \qquad g(0,t) = 2n(T_A,t) \qquad m(0,t) = \beta N_0(t)$$
$$r(0,t) = m(T_M,t) \qquad b(0,t) = \gamma N_R(t)$$

where $N_0(t)$, $N_R(t)$ respectively denote the total population of the resting and marrow compartments. The factor 2 appearing in the second equation in (20) signifies that two cells are produced by division of cells in the active compartment compartment when they reach the age T_A.

The total number of cells in each compartment are given by

$$(21) \qquad N_A(t) = \int^{T_A} n(\mu,t)du \qquad N_0(t) = \int_0^\infty g(a,t)da \qquad N_M(t) = \int_0^{T_M} m(\mu,t)du$$
$$N_R(t) = \int_0^\infty r(a,t)da \qquad N_B(t) = \int_0^\infty b(a,t)da$$

with the total population

$$(22) \qquad N(t) = N_A(t) + N_0(t) + N_M(t) + N_R(t) + N_B(t).$$

Since there is a time delay in the expression (17) for γ, the quantity $N_B(t-t_R)$ must be prescribed for $0 \le t \le t_R$:

$$(23) \qquad N_B(t-t_R) = F_B(t) \qquad 0 \le t \le t_R$$

In summary, the mathematical formulation of the simplified neutrophil production model consists of equations (18) in conjunction with the initial conditions (19), the boundary conditions (20) and the time delay condition (23).

In order to represent certain experiments involving the active compartment it is necessary to partition it into a number of subcompartments and write down the corresponding kinetic equations. For example, if cells are labelled with tritiated thymidine it is convenient to partition the active compartment into three compartments containing G_1, S and G_2+M cells. The first equation appearing in (18) is replaced by

$$\frac{\partial n_1(\mu,t)}{\partial t} + \frac{\partial n_1(\mu,t)}{\partial \mu} = 0 \qquad 0 < \mu \leqslant T_1$$

$$(24) \qquad \frac{\partial n_S(\mu,t)}{\partial t} + \frac{\partial n_S(\mu,t)}{\partial \mu} = 0 \qquad 0 < \mu \leqslant T_S$$

$$\frac{\partial n_2(\mu,t)}{\partial t} + \frac{\partial n_2(\mu,t)}{\partial \mu} = 0 \qquad 0 < \mu \leqslant T_2$$

where $n_1(\mu,t)$, $n_S(\mu,t)$ and $n_2(\mu,t)$ are the cell density functions for the G_1, S and G_2+M phases and T_1, T_S and T_2 are the respective transit time of cells in these phases. Note that

$$T_A = T_1 + T_S + T_2 \tag{25}$$

The first initial condition in (19) is replaced by the set of initial conditions

$$n_1(\mu,0) = f_1(\mu) \qquad n_S(\mu,0) = f_S(\mu) \qquad n_2(\mu,0) = f_2(\mu) \tag{26}$$

where f_1, f_S and f_2 are prescribed functions of μ. In addition, the first two boundary conditions in (20) are replaced by the set of boundary conditions.

$$\begin{aligned} n_1(0,t) &= \alpha\, N_0(t) & n_S(0,t) &= n_1(T_1,t) \\ n_2(0,t) &= n_S(T_S,t) & g(0,t) &= 2n_2(T_2,t) \end{aligned} \tag{27}$$

The steady state behavior of the neutrophil production system can be obtained by disregarding the initial conditions (19) and the time derivatives appearing in (18) In addition, from (15), (16) and (17), it follows that in the steady state

$$\alpha = \beta = \alpha_0 \qquad\qquad \gamma = \gamma_0.$$

Using these facts, the steady solutions of (18), denoted by an overbar, are:

$$\begin{aligned} \bar{n}(\mu) &= n_0 & 0 &\le \mu < T_A \\ \bar{g}(a) &= g_0 e^{-2\alpha_0 a} & 0 &\le a \\ \bar{m}(\mu) &= m_0 & 0 &\le \mu < T_M \\ \bar{r}(a) &= r_0 e^{-\gamma_0 a} & 0 &\le a \\ \bar{b}(a) &= b_0 e^{-\gamma a} & 0 &\le a \end{aligned} \tag{28}$$

Here $n_0, g_0, m_0 r_0$ and b_0 are constants. Substituting (28) into (21) yields the total number of cells in each compartment in the steady state:

$$\begin{aligned} \bar{N}_A &= n_0 T_A & \bar{N}_0 &= g_0/2\alpha_0 & \bar{N}_M &= m_0 T_M \\ \bar{N}_R &= r_0/\gamma_0 & \bar{N}_B &= b_o/\lambda \end{aligned} \tag{29}$$

Hence, the total population

$$\bar{N} = \bar{N}_A + \bar{N}_0 + \bar{N}_M + \bar{N}_R + \bar{N}_B \tag{30}$$

is also a constant. Utilizing (28) and (29) in the boundary conditions (27) results in

$$g_0 = 2n_0 \qquad m_0 = g_0/2 = n_0 \qquad r_0 = g_0/2 = n_0 \qquad b_0 = r_0 = n_o \tag{31}$$

Substituting the constants appearing in (31) into (29) gives

$$(32) \qquad \bar{N}_A = n_o T_A \qquad \bar{N}_o = n_o/\alpha_o \qquad \bar{N}_M = n_o T_M$$

$$\bar{N}_R = n_o/\gamma_o \qquad \bar{N}_B = n_o/\lambda$$

Note that the sizes of the compartmental populations are all expressed in terms of a single constant n_o. This constant represents the steady state production rate of the system which equals the steady state loss from the blood compartment $\lambda\ \bar{N}_B$. From (30) and (32) the total population

$$(33) \qquad \bar{N} = n_o(T_A + \alpha_o^{-1} + T_M + \gamma_o^{-1} + \lambda^{-1}).$$

Finally, utilizing (33) and the last equation in (20 yields the following compatibility condition satisfied by the system parameters:

$$(34) \qquad \bar{N} = \bar{N}_B\lambda(T_A + \alpha_o^{-1} + T_M + \alpha_o^{-1} + \lambda^{-1}).$$

If subcompartments of N_A are introduced as in (24) the steady state solutions are:

$$(35) \qquad \bar{n}_1 = \bar{n}_S = \bar{n}_2 = n_o \qquad \bar{N}_1 = n_o T_1$$

$$\bar{N}_S = n_o T \qquad \bar{N}_2 = n_o T_2$$

Naturally, $\bar{N}_A = \bar{N}_1 + \bar{N}_S + \bar{N}_2$.

The system of equations (18) is nonlinear because of the dependence of α, β and γ on $N(t)$ and $N_B(t)$. In general, an analytic solution of these equations will be difficult to find. By integrating equations (1) over age or maturity, as required, ordinary differential equations can be derived for the total compartmental compartmental populations. These equations can be solved numerically.

The ordinary differential equation for the active compartment is derived as follows. Integrating the first equation in (18) over μ and utilizing (21) yields

$$(36) \qquad \frac{dN_A(t)}{dt} + n(T_A,t) - n(0,t) = 0.$$

The general solution of the first partial differential equation in (18) for the cell density function is

$$(37) \qquad n(\mu,t) = \begin{cases} n(\mu-t,0) & t \leq \mu \\ n(0,t-\mu) & t > \mu \end{cases}$$

according to Fig. 3.5 of Chapter 3.

Using initial condition (19) and boundary condition (20) equation (37) can be rewritten as

$$(38) \qquad n(\mu,t) = \begin{cases} f_A(\mu-t) & t \leq \mu \\ \alpha(t-\mu)\ N_o(t-\mu) & t > \mu \end{cases}$$

Substituting the values of the function n appearing in (36) from (38) leads to the following difference - differential equation

$$(39) \qquad \frac{dN_A}{dt} = \alpha(t)\ N_o(t) - J_A$$

where

$$(40) \qquad J_A = \begin{cases} f_A(T_A-t) & 0 < t \leq T_A \\ \alpha(t-T_A)\ N_o(t-T_A, & t > T_A \end{cases}$$

Similarly, the following equations for the other compartmental populations are readily deduced,

$$(41) \qquad \begin{aligned} \frac{dN_o}{dt} &= -[\alpha(t)+\beta(t)]N_o(t) + 2\ J_A \\ \frac{dN_M}{dt} &= \beta(t)\ N_o(t) - J_M \\ \frac{dN_B}{dt} &= -\gamma(t)\ N_R(t) + J_M \\ \frac{dN_B}{dt} &= -\mu\ N_B(t) + \gamma(t)\ N_R(t) \end{aligned}$$

where $\gamma(t)$ is given by (17) for $t \geq t_R$ and by

$$(42) \qquad \gamma(t) = \begin{cases} \gamma_o+\gamma_1[(\bar{N}_B/F_B(t))^{\nu}-1] & F_B(t) < \bar{N}_B \\ \gamma_o & F_B(t) \geq \bar{N}_B \end{cases}$$

for $0 \leq t < t_R$. Here

$$(43) \qquad J_M = \begin{cases} f_M(T_M-t) & 0 < t \leq T_M \\ \beta(t-T_M)N_o(t-t_M) & t > T_M \end{cases}$$

Suppose that the active compartment is divided into three subcompartments as in in (24). Then, equation (39) and the first equation in (41) are replaced by the following equations obtained by integrating (24) over the maturation or age variables

(44)

$$\frac{dN_1}{dt} = \alpha(t)\ N_o(t) - J_1$$

$$\frac{dN_S}{dt} = J_1 - J_2$$

$$\frac{dN_2}{dt} = J_S - J_2$$

$$\frac{dN_o}{dt} = 2J_2 - [\alpha(t)+\beta(t)]\ N_o(t)$$

where

$$J_1 = \begin{cases} f_1(T_1-t) & 0 < t \leq T_1 \\ \alpha(t-T_1)\ N_o(t-T_1) & t > T_1 \end{cases}$$

$$J_S = \begin{cases} f_S(T_S-t) & 0 < t \leq T_S \\ f_1(T_1+T_S-t) & T_S < t \leq T_1+T_S \\ \alpha(t-T_1-T_S)\ N_o(t-T_1-T_S) & t > T_1+T_S \end{cases}$$

$$J_2 = \begin{cases} f_2(T_2-t) & 0 < t \leq T_2 \\ f_S(T_S-T_2-t) & T_2 < t \leq T_2+T_S \\ f_1(T_A-t) & T_2+T_S < t \leq T_A \\ \alpha(t-T_A)\ N_o(t-T_A) & t > T_A \end{cases}$$

The right hand side in (44) is always the difference between cell influx and cell efflux. Cell division, which occurs when cells leave compartment 2, are accounted for by the factor 2 in the last equation in (44).

The set of equations (39) and (41) can be solved numerically. Their solution requires the initial values of the compartment populations $N_A(0)$, $N_o(0)$, $N_M(0)$, $N_R(0)$, $N_B(0)$, the function F_B appearing in (23) and the initial maturity distribution functions f_A and f_M. If the system (44) is to be utilized, the initial maturity distribution f_A must be partitioned into f_1, f_S and f_2. In this situation $N_1(0)$, $N_S(0)$ and $N_2(0)$ are also needed. Furthermore, the values of the 13 parameters of the system (or 15 if (44) is utilized) must be specified. The values of $\bar{N}_B$, T_A+1/α_o, T_M, $1/\gamma_o$ and λ can be determined from observations in the steady state situation. The total neutrophil population $\bar{N}$ is then determined from the compatibility relation (34). The values T_A and $1/\alpha_o$ were not determined uniquely, but were assigned values compatible with experimental data by Rubinow and Lebowitz [38]. This still leaves the six system parameters α_1, β_1, γ_1, ν, ρ and t_R to be determined. These parameters enter into the homeostatic control elements α, β and γ. Thus, the values of these parameters can only be inferred from experiments which perturb the steady state behavior of the system. As a simplification Rubinow and Lebowitz assumed that

$$\alpha_1 = \alpha_0 \qquad \beta_1 = \frac{1}{2}\alpha_0 \qquad \gamma_1 = \alpha_0 \qquad \rho = \nu.$$

Note that the condition for stability of the system, $\beta_1 < \alpha_1$, is satisfied. The assumption $\rho = \nu$ implies the dynamical control of the normal blood neutrophil population is similar to the dynamical control of the total neutrophil population. There is no biological reason for this assumption. Note that with these assumptions assumptions there are only two dynamical parameters (t_R and ν) to determine. These parameters are determined from experimental data by numerically solving the system of ordinary differential equations (39) and (41) and adjusting the parameter values to produce solutions which compare favorably with the available observations.

As an example of perturbing the steady state behavior of the system leukopheretic studies were simulated. A death term $- D(t)\, b(a,t)$ was added to the last equation appearing in (18). Here

$$D(t) = f_0 \sum_{J=1}^{K} \delta[t-(j-1)t_0]$$

where δ is a Dirac delta function, f_0 and t_0 are constants and K is a positive integer. The term f_0 is the constant fraction of cells removed from the blood per unit time at equal intervals t_0 commencing with $t = 0$ and ceasing after K removals, at time $t = (K-1)t_0$. The equation for $N_B(t)$ now becomes

$$\dot{N}_B(t) = -\lambda N_B(t) + \gamma(t) N_R(t) - D(t) N_B(t).$$

Theoretical predictions are made for the transient behavior of the normalized blood cell population $N_B(t)/\bar{N}_B$ by numerically solving the system of equations for various values of t_R and ν. These are shown graphically in their paper. The predicted response is a damped oscillation of the blood cell population. Such cylic variations of the neutrophil blood populations have been observed in the disease state cylic neutropenia and in dogs in response to acute infections.

As another application of the model the theoretical tritium blood granulocyte specific activity (BGSA) curve was calculated by solving (41) and (44) numerically. It was assumed that at $t = 0$, the cells were in a steady state. Furthermore, all cells in the S phase were labeled with a label value of 1, an arbitrary unit of activity. Each cell is assumed to contribute half of its label value to each of its daughters. All cells whose label value is less than 1/4 are not counted. The tritium BGSA function at time t is determined by summing the total number of counted labeled cells in the blood compartment multiplied by their appropriate label value. There is good quantitative agreement between the computed and corresponding experimental BGSA curves. The theoretical BGSA curves were fitted to the experimental curve by varying T_M.

King-Smith and Morley [22] formulated a computer model to show how a feedback control system simulated cyclic variations of the neutrophil blood population. Although a complete description of their model was not given it appears that they assume that the neutrophil production rate depends on the blood neutrophil population. Recall that in the model of Rubinow and Lebowitz the production rate is governed by the total population of the neutrophils. This last assumption seems more plausible since if the marrow cells alone were destroyed at a given time, the system would immediately start to replenish itself, without waiting for the appearance of a blood deficit. King-Smith and Morley's model predicts that a stable small amplitude oscillation appears in response to marrow failure. The Rubinow-Lebowitz model predicts slowly damping, small amplitude oscillations. This result suggests why cylic oscillations of the neutrophil count is rarely observed in healthy people. It is only detectable after a large perturbation which is sufficiently recent to the time of observation.

9.5 Acute Myeloblastic Leukemia

A model for acute myeloblastic leukemia (AML) was introduced by Rubinow [37]. A more detailed account of the kinetics of neutrophil production in AML appears in Rubinow and Lebowitz [39]. In this paper they assume that the normal neutrophil population and the population of leukemic blood cells coexist. Each population possesses feedback control elements that regulate the total number of cells in the population. However, the leukemic cell possess an aberrant set of kinetic parameters. The leukemic state is assumed to display controlled growth with an abnormal set of control elements and not "uncontrolled growth". This viewpoint was advanced by Clarkson [12] and Metcalf [31].

The main assumptions of the model described above are:

(a) There are two cell populations: the normal neutrophil cell population and the leukemic cell population.

(b) The two populations have their own resting compartment denoted by G_0 and G_0', respectively. A resting cell (normal or leukemic) either passes into an active compartment (A or A' respectively) and subsequently divides or else is not proliferative. In the latter situation a normal cell undergoes a period of maturation, ultimately enters the blood and dies (see Fig. 9.7). If the cell is leukemic, it enters the blood directly and eventually dies.

(c) The decision as to whether a cell proliferates or not depends on the total number of cells in the population.

(d) Each population has qualitatively similar feedback control elements that regulate and control the total number of cells in the population. However, the kinetic parameters of the normal cells differ from those of the leukemic blood cells.

(e) Normal neutrophil precursors in G_0 do not distinguish leukemic cells but regard them as part of the total population.

(f) Cells are only lost from the blood compartment.

(g) Stem cell influx is negligible.

The schematic representation of the model for the leukemic state in AML is shown in Figure 9.8. The resting and actively proliferating leukemic cell pools

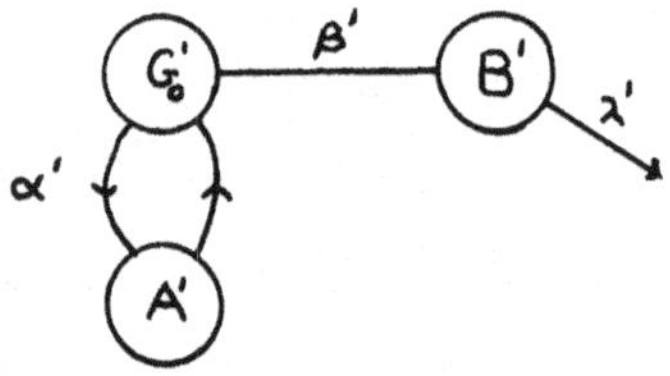

Figure 9.8 Model of the leukemic state in AML.

are represented by the compartments G_0' and A', respectively. Cells leave G_0' at a fractional rates per unit time α' and β' to enter the active and blood compartment G', respectively. Note that there are no maturation and reserve compartments as in the normal neutrophil production scheme of Fig. 9.7. Presumably, the authors have assumed that leukemic cells are immature and are not held in reserve. Cells randomly leave the blood compartment to die, at a fractional rate per unit time λ'.

The mathematical translation of the leukemic state model involves three age-time equations analogous to those appearing in (18). When the cell density functions of the A', G_0' and B' compartments are integrated over age or maturity variables the following equations are obtained for the total number of cells in each compartment:

$$\frac{dN_A'}{dt} = \alpha'(t)\; N_0'(t) - J_A'$$

(46)
$$\frac{dN_0'}{dt} = -[\alpha'(t)+\beta'(t)]\; N_0'(t)+2J_A'$$

$$\frac{dN_B'}{dt} = \beta'(t)\; N_0'(t) - \lambda' N_B'(t) \qquad t > 0$$

where

$$J_A' = \begin{cases} f'(T_A'-t) & 0 < t < T_A' \\ \alpha'(t-T_A')N_0'(t-T_A') & t \geq T_A' \end{cases}$$

Here, N_A', N_0' and N_B' represent the number of leukemic cells in the A', G_0' and B' compartments, respectively. The initial distribution of leukemic cells in the compartment A' as a function of age a is denoted by f'(a). The control function

(47)
$$\alpha' = \alpha_o' + \alpha_1^1\; \ell n \frac{\bar{N}'}{N(t)+N'(t)}$$

$$\beta' = \alpha_0' + \beta_1'\; \ell n \frac{\bar{N}'}{N(t)+N'(t)}$$

where $\bar{N}'$ is the stable population number of leukemic cells and is much larger than $\bar{N}$, the Romeostatic level of the total neutrophil population and its precursors. Here N(t) is the total normal population of neutrophils and their precursors a time t and

$$N'(t) = N_A'(t)+N_0'(t)+N_B'(t)$$

is the total number of leukemic blood cells (LBC). It is assumed that $\alpha_1' > \beta_1'$ to to insure stability about $\bar{N}$!

The equations corresponding to the compartments containing normal neutrophils and their precursors are just (39) and (41) with

(48)
$$\alpha(t) = \alpha_0+\alpha \quad ([\bar{N}/(N(t)+N'(t))]^{\nu}-1)$$
$$\beta(t) = \alpha_0+\beta_1([\bar{N}/(N(t)+N'(t)]^{\nu}-1)$$

where ν is a positive number. Note that now α and β are defined in terms of both N(t) and N'(t) (c.f. (15) and (16) because of (c). Rubinow and Lebowitz also suggested that the unprimed version of (47) would also be suitable control laws. Stability at the homestatic level N requires that $\alpha_1 > \beta_1$.

To study labeling experiments for LBC, the active compartment is partitioned into three subcompartments leading to primed versions of (44).

After specifying the parameters of the model, the initial and boundary conditions, Rubinow and Lebowitz were able to solve the above system of equations numerically. From these solutions the following curves were plotted:

(i) total blood population, LBC population and normal blood neutrophil population versus time,

(ii) total leukemic blood cell population versus time,

(iii) labeling index of leukemic cells versus time;

(iv) marrow cellularity (i.e., the ratio of the total marrow population of normal plus leukemic cells to the normal steady state marrow population $\bar{N}$-$\bar{N}_B$) versus time.

This model had also been utilized previously by Rubinow et al. [36] to analyze data obtained from two adults with AML with the aid of autoradiography.

The qualitative effect on the normal population of the presence of LBC can be intuitively understood without solving the system equations. From (48) it follows that if the total population size is greater than $\bar{N}$, the normal steady-state level, then $\beta > \alpha$. Suppose that the normal population number is at its steady state level $\bar{N}$ and a few leukemic cells are introduced. The total population will then be greater than $\bar{N}$ and so $\beta > \alpha$. Consequently, the normal production of neutrophil precursors is slowed down and operates below a self-sustaining level. On the other hand, the leukemic cells' controller recognizes that the total number of cells is less than the leukemic steady state level N_1' so that $\alpha' > \beta_1'$ Therefore the leukemic population increases. This leads to a further decline in the production of normal cells. A graph illustrating this behavior appears in the paper of Rubinow and Lebowitz [39]. Ultimately the number of leukemic cells exceed a certain critical limit while the number of normal cells fall below another critical value and the patient dies.

After a short transient period, the initial small number of LBC are in steady exponential growth. To verify this assertion analytically let $N'(t)$ be small compared with $N(t)$ and let $N(t) = \bar{N}$, the normal homeostatic level. It follows from (47) that α' and β' are essentially constants, denoted by $\bar{\alpha}'$ and $\bar{\beta}'$, respectively, since $N(t)+N'(t) \approx \bar{N}$. The equations in (46) can be solved by assuming that $N_0'(t)$ is proportional to $e^{\gamma t}$ where γ is determined from

(49) $$\bar{\alpha}' + \bar{\beta}' + \gamma = 2\bar{\alpha}' e^{-\gamma T_A}$$

It follows that the leukemic compartment population is in a state of steady exponential growth with growth rate γ - that is,

(50) $$N'(t) = ce^{\gamma t}$$

where c is a constant. Rubinow and Lebowitz assert that the natural lifetime of the acute myeloblastic state can be estimated as the time τ it takes one cell to grow into 10^{12} cells. Therefore, from (50), with $c = 1$ and $N'(\tau) = 10^{12}$.

$$\tau = \gamma^{-1}\, 12 \,\ln 10 = 27.6\gamma^{-1}$$

The LBC continue to increase until a stationary growth phase appears as N' approaches its steady state level $\bar{N}'$.

The utility of a concrete mathematical model is that specific predictions about the natural history of the disease can be made. These predictions may be susceptible to observational test. For example, the variety of responses seen in humans with AML can be accounted for by assigning different parameters values to different individuals. By appropriately increasing $\bar{N}'$, it is possible to produce marrow cellularities as much as 5 to 10 times normal, as seen in patients with AML in the advanced stage of the disease. Correct predictions made by a model lend support to the assumptions which have not been verified. The Rubinow-Lebowitz model of AML successfully simulates the kinetic behavior of AML in a qualitative manner. Hence, the basic assumption of two distinct cell populations, normal and leukemic, with different control parameters is supported.

9.6 A Chemotherapy Model of AML

Another use of a model of AML is that it can be applied to the simulation of the possible outcomes of various drug protocols. Such simulations can be used to maximize the cytocidal action of the drug on the leukemic population while minimizing toxicity of the drug on the normal population. This application was investigated by Rubinow and Lebowitz [40].

The treatment of acute myeloblastic leukemia and related myeloid forms of this disease has been implemented via the L-6 protocol described in Gee et al. [19] and Clarkson [12,13]. In this protocol drug doses of the cycle specific drug arabinosyl cytosine (ara-C) in combination with 6-thioguanine (6-TG) are administered sufficiently close together in time so that leukemic cells are prevented from completing the S-phase[4] without being exposed to a lethal drug concentration. Treatments are continued until marrow depression or other toxic effects become too great. Then a rest period is instituted, usually of 3 weeks duration, to allow the normal marrow cells time to recover. If necessary, the entire course of treatment is repeated in an attempt to induce remission.

Clinically, a remission means the apparent elimination of leukemic myeloblasts from the blood and the reduction of myeloblasts in the marrow to normal levels. Such an apparent disappearance occurs after an approximately 100-fold decrease in the leukemic cell number. A more precise clinical definition, which depends on the hemopoietic status of other cells can be found in the paper by Gee et al. [19]. For the purposes of their model, Rubinow and Lebowitz conveniently define a state of remission as the reduction of the leukemic cell population to 10^{10} or less while the normal population is at an acceptable level of depletion.

The L-6 protocol produced a remission rate of 56% in 88 previously untreated patients [19]. The median duration of remission was 10 months. The median

---

[4] The average duration of the S-phase for LBC is about 2 h.

survival time of responders was 2 years, whereas prior to 1966, the median survival time of adults with AML was about 4 months. On the other hand, prior to the introduction of the L-6 protocol, of 36 patients treated with daily doses of ara-C plus 6-TG, but without rest periods, 53% had complete or partial remissions. Hence, Hence, the trial and error method of designing protocols did not lead to a significant increase in the incidence of remission. Thus a mathematically designed protocol seems essential.

Rubinow and Lebowitz utilized their model of AML to simulate the effects of drug protocols in the following way. A course of treatment is described by the administration of the cycle-specific drugs (ara-C followed by 6-TG) n times with periodicity P. The regimen consisted of the course of treatment repeated six times. They assume that the introduction of a fixed dosage of the chemotherapeutic agents at time t_i reduces the leukemic population $N_S'(t_i)$ by a fixed fraction f, which was kept at the value 0.9. After perturbation the tumor grows from the new population level $(1-f)\ N_S'(t_i)$. Similar statements hold for the normal population $N_S(t_i)$ since the drug kills both normal and leukemic cells in the S phase. The system of equations describing the growth of normal and leukemic cells is then solved numerically over the interval t_i to t_{i+1} using the initial conditions $(1-f)\ N_S(t_i)$ and $(1-f)\ N_S'(t_i)$ for the normal and leukemic cells in the S phase. The whole process is repeated at time t_{i+1}.

Rubinow and Lebowitz studied the effects of different drug protocols both on slow and fast growing leukemic cell populations by varying the parameters α_1' and β_1' of their model. They also varied the periodicity P, the number of drug administrations n and the resting time τ_R.

Some conclusions of their work are:

(a) Significant changes in the outcome of the treatment can occur with small alterations in scheduling parameters.

(b) The optimum periodicity of drug doses may be equal to that of the S phase interval of leukemic cells.

(c) The more intensive a course of treatment, the longer the rest period following it should be.

(d) The concept of aggressive continuation of chemotherapy should not be abandoned, but only for good responders, that is, individuals who achieve remission status after two courses of treatment.

(e) Treatment could be successful against fast growing leukemia, which had a natural lifetime of about two years, if very large toxic effects were accepta acceptable.

It is interesting to note that (a) and (c) are observed in clinical practise.

Although the above model has made some qualitative predictions which agree with observations, it still has some shortcomings. Disease states such as smoldering leukemia may not be adequately represented by any set of parameter values. The toxic effect on other cells in the body is disregarded. The behavior of the leukemic population following the attainment of a state of remission is not correct since the immune system probably comes into play. The variability of the S-phase and other phases has been neglected. The possibility that ara-C inhibits cells at the G_1-S boundary is not accounted for. Finally, the trial and error method of studying different protocols by varying the drug parameters should be automated and the possibility of using optimal control theory should be investigated.

9.7 Models of Chronic Granulocytic Leukemia (CGL)

Wheldon [43] proposed a simple model for the regulation of blood cell production which incorporates a time delay for cellular development. Such models are prone to develop oscillations and so can be utilized to study oscillatory blood cell production.

Wheldon's model is not as complex as the one of Rubinow and Lebowitz since it only considers the number of cells in the marrow store, x, and the number of cells in the blood, y. The biological background of the model can be found in Wheldon et al. [44]. It is described by equations of the form

$$\frac{dx}{dt} = \frac{\alpha}{1+\beta[x(t-\tau)]^a} - \frac{\lambda\, x(t)}{1+\mu[y(t)]^b} \tag{51}$$

$$\frac{dy}{dt} = \frac{\lambda\, x(t)}{1+\mu[y(t)]^b} - w\, y(t) \tag{52}$$

where the parameters have the following interpretations:

α = maximum rate of cell production when the number of marrow granulocytes is zero

β = coupling constant for cell production loop

a = gain control of cell production loop

λ = maximum rate of release of matured cells from marrow when blood cell number is zero

μ = coupling constant for release loop

b = gain control of release loop

w = rate constant for loss of granulocytes from blood to tissue

τ = control system time delay (it represents the mean time taken for development from the stem cell state to that of the mature cell stage at which the cell is available to be released)

Since the equations are nonlinear and analytically intractable, numerical integration was employed. The parameters of the model were chosen to represent normal granulopoiesis and subjected to perturbations. Steady state values were regained by way of damped oscillations having a period of around 20 days. This is in line with the observed behavior of the real system. A more mathematical discussion of the properties of the model appears in Wheldon [42].

The model was also used to simulate cyclical neutropenia and oscillations in chronic granulocytic leukemia. The simulations qualitatively mimicked the following features of these diseases. In cyclical neutropenia, the mean granulocyte number is low and the number fluctuates with a large aptitude around this mean with a period of about 20 days. On the other hand, the period of the cycle in CGL is considerably longer than that in cyclical neutropenia (40-60 days versus 15-20 days). Moreover, the oscillations in CGL may exhibit divergent tendencies unless controlled by therapy.

The oscillatory pattern in CGL could be produced in two ways, either by an increased cell production rate or by an increased maturation time. Further biological experiments are required to discriminate between these alternative mechanisms.

Another simple explanation for the oscillatory behavior observed in CGL is contained in Mackey and Glass [27]. The oscillations are associated with bifurcations in the dynamics of a first-order differential-delay equation describing the population of mature circulating cells. The delay is required since there is a significant time lag between the initiation of cellular proliferation and the release of mature cells into the blood.

A later paper by Mackey [28] generalizes the differential-delay equation to a system of two such equations which describe the hematopoietic pluripotential stem cells. Based on this more biologically realistic model, the simplest hypothesis for the origin of aplastic anemia and periodic hematopoiesis is that they are both due to the irreversible loss of proliferative stem cells.

It would be interesting to compare the results generated by Wheldon's model, Mackey's model and the Rubinow-Lebowitz model. Note that similar qualitative behavior can be produced by different models.

9.8 A Discrete Mathematical Model of Acute Lymphoblastic Leukemia

The paper by Mauer et al. [30] is based on the biological assumption that the cell population in acute lymphoblastic leukemia (ALL) consists entirely of myeloblasts in which no maturation or differentiation occurs. This population is assumed to be characteristic of the bone marrow at the time of diagnosis or in full relapse. Since the type of the population does not change they were able to model the phase of the ALL cell cycle as shown in Figure 9.9.

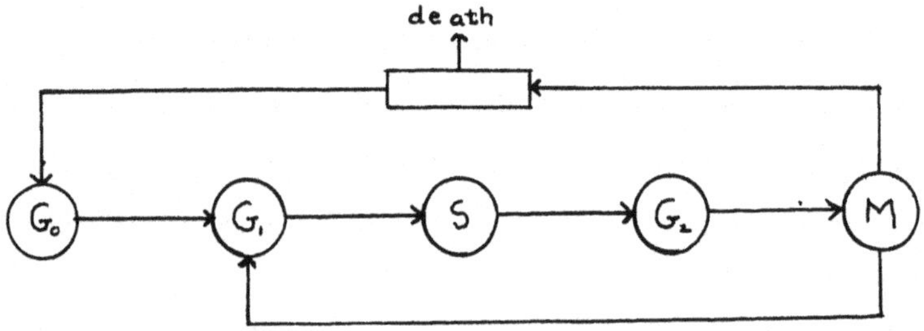

Figure 9.9 Cellular compartments for simulation of ALL.

Cells proceed through the G_1,S,G_2 and M phases. After mitosis a certain percentage of the cells enter the G_0 phase and the remainder reenter the mitotic cycle. The transit time in the S phase was considered to be constant. However, variations in the transit times in the other phases were allowed. Cell death was assigned to occur in some of the cells designated to be G_0 cells. However, not much is known about this aspect of leukemic cell kinetics in human disease. A few animal studies lend support to this hypothesis.

The model was simulated using the General Purpose Simulation System (GPSS) developed for operations research studies by IBM.

An interesting feature of the paper by Mauer et al. [30] is the simulation of the growth of a single leukemic cell. After a time span of three and one-half years a population of 10^{12} cells was reached. This time corresponds to the peak incidence of ALL in children! Thus a mutational event in a single cell could lead to ALL in childhood.

The effects of phase specific drugs could also be simulated and correlated with experimental data via generated labeling and mitotic indices. As an example, they attempted to simulate the action of vincristine by trapping cells in metaphase. The model failed to reproduce the observed effect of vincristine in a patient. To simulate the experimental data, an additional block had to be imposed on the flow of cells from G_0 to G_1. This unsuspected effect of vincristine was later confirmed by experiments. It was found that vincristine blocked the blast transformation of lymphocytes exposed to phytohemagglutinin. This effect was associated with early inhibition of RNA synthesis. Thus computer model may provide insights into drug action which can be checked by the design of new experiments.

Further cell kinetic studies were undertaken by Evert. It was found that early versions of GPSS were inconvenient in establishing initial conditions, it had to be run for some time to establish "steady state" conditions before a drug could be introduced and there was no convenient way to alter the model during the course of an experiment. Some of these difficulties were overcome in some of the more advanced versions of GPSS, but the time required for simulation increased. Another operations research language GASP (General Activity Simulation Language) [33] was also tried and found unsuitable. Evert and Palusinski [18] performed further simulations of ALL using CELLSIM [15], a simulation language for modeling populations of biological cells. This language has graphical capabilities and printouts of the simulation indicate that there are oscillations of the number of cells in each of the S, G_1 and G_0 phases. Evert was not satisfied with CELLISM since it is oriented to batch processing and has substantial memory requirements. Hence, Evert [17] developed his own interactive, simulation program CELLDYN for modeling cell dynamics. The moral of this story is that it is often better to write a custom-designed program rather than using a general purpose program written by another person.

9.9 A Comprehensive Computer Model of Granulopoiesis and Cancer Chemotherapy

A comprehensive deterministic model of granulopoiesis that begins with the division of precursor stem cells in the marrow and ends with the death of spent granulocytes in the tissue space was developed by Blumenson [2]. A full description of the underlying experimental work was presented in an earlier paper by Blumenson [1]. That paper includes estimates of the steady state values of the cellular components for a 70 kg. man.

Such models which closely mirror granulopoiesis can be used to study

(a) quantitative relationships among marrow cellularity, the anti-cancer drug protocol and the temporal pattern of the white blood count,

(b) other diseases such as acute infections via a more detailed submodel of the tissue spaces,

(c) iatrogenic destruction of cells of the granulopoietic system by anti-cancer drugs and the dosage schedule of an antibiotic to prevent death from infection.

(d) marrow transplant experiments;

(e) the breakdown of feedback mechanisms for proliferation and release of cells from the marrow in order to gain insight into the etiology of diseases such as chronic granulocytic leukemia.

The simplified block diagram of the model is shown in Fig. 9.10. Each block is affected through feedback from the other blocks as indicated by the dotted lines. The state of the block(s) at the foot of the arrow at time t-1 is conveyed to the receptor block at the arrowhead in order to control the flow of cells from the receptor block during the time from t-1 to t. A detailed description of the

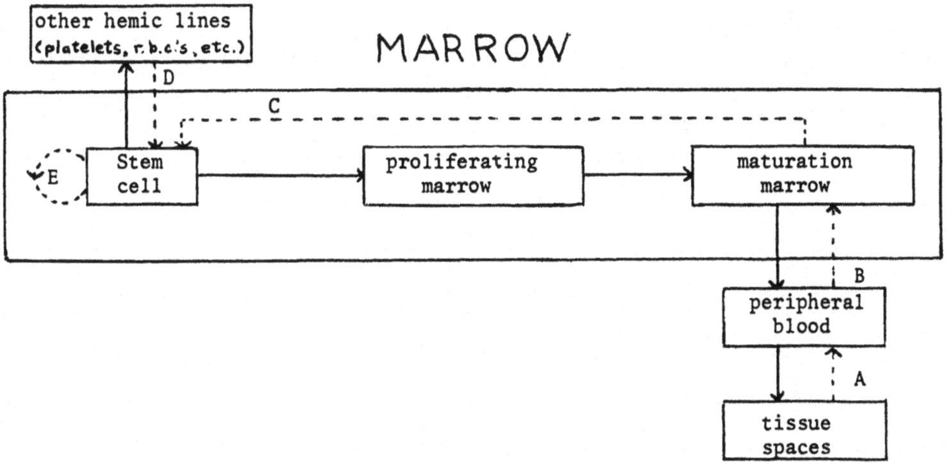

Figure 9.10 Simplified block diagram for granulopoiesis.

internal functioning of these blocks as well as the feedback relationships among them will be described below.

1. Maturating marrow

The maturating marrow consists of 130 one hour age-compartments arranged in a pipeline-like process of maturation. The daughter cells from the division of late myelocytes enter the maturating marrow from the last compartment of the proliferating marrow. These cells are the youngest metamyelocytes and are of age 0. According to the deterministic nature of the model cells of age k at the beginning of an hour become cells of age k+1 at the end of the hour for k = 0,...,129. Cells of age 130 always enter the peripheral blood in accordance with Blumenson's data [1]. If required, younger cells (ages 94-130)can also enter the peripheral blood as described in 2 below. If there are no cells age 94 or greater then no cells enter the blood and it is necessary to wait for younger cells to mature. A decrease in maturing marrow cellularity stimulates stem cell activity as described below and eventually replenishes the lost cells. Cells in the age range 0-40, 41-93 and 94-130 are described as metamyelocytes, bands and segmented, respectively.

2. Peripheral blood

Under normal steady state conditions granulocytes enter the blood at age 131. Cells of age k at the beginning of an hour are of age k+1 at the end of the hour for k = 131,...,159 provided they do not enter the tissues. Cells of age 160 enter the tissues during the next hour. All cells of age less than 160 in the peripheral blood pool have an equal chance 1-f of leaving for the tissue spaces (random loss) during a one hour period. This chance, which is the feedback A, can vary from hour to hour depending on the needs of the tissue spaces. However, in Blumenson's paper f is a constant since the hourly tissue demands were held at the steady state level. Thus there is an associated steady state age frequency distribution of the number of cells of each age in the blood.

The fraction f of cells of age k (k = 131,...,159) remaining in the blood and the frequency distribution of ages of cells in the blood will be calculated in the steady state. Let N(k), k = 131,...,160 be the number of granulocytes of age k in the peripheral blood. If there are N(k) granulocytes of age k at the beginning of an hour, then there will be f N(k) granulocyte of age k+1 for k = 131,...,159. Since during the steady state the number of cells of a given age is the same during any hour

$$f\,N(k) = N(k+1) \qquad k = 131,\ldots,159 \tag{51}$$

The number of granulocytes released from the maturing marrow is

$$N(131) = GTR \tag{52}$$

where GTR is the granulocyte turnover rate. Note that the GTR is the hourly rate at which cells are entering (or equivalently, in the steady state, leaving) the peripheral blood. Methods for calculating the GTR are described in Blumenson [1]. The solution of equations (51) and (52) is the required steady state distribution

$$N(k) = f^{k-131}\ GTR \quad (k = 131,\ldots,160) \tag{53}$$

The constant f can be determined from the fact that the sum of N(k) must be equal to the total number of granulocytes in the peripheral blood or the total blood granulocyte pool (TBGP). Hence

$$(1+f+\ \ldots\ +f^{29}) \times GTR = TBGP$$

or

$$(54) \qquad \frac{1-f}{1-f^{30}} = \frac{GTR}{TBGF}$$

When conditions are not steady state, the age frequency distribution of granulocytes in the peripheral blood will vary from hour to hour. Release of cells from the maturing marrow into the peripheral blood is related to the concentration of granulocytes in the peripheral blood (feedback B). It is assumed that as many segmented cells (ages 94-130) in addition to those of age 130 are released as necessary to elevate the TBGP to its normal steady state value. The older cells are released before the younger ones. The frequency distribution must now account for cells in the blood with ages ranging from 94-160 hours.

To simulate certain pathological conditions such as chronic granulocytic leukemia or severe infection may require that even younger cells (band and metamyelocytes) be released into the blood.

In some situations, such as acute infections, the granulocyte reserves in the marrow may be depleted so that tissue demands for cells may not be met. This depletion initiates feedback C to the stem cells (discussed below) and the marrow attempts to return to normal cellularity by shunting more stem cells into the differentiation track.

The computer program based on the model calculates the following quantities associated with the peripheral blood compartment:

(a) the probability that a granulocyte (age less than 160) leaves the blood during the next hour.

(b) the number of granulocytes of each different age in the blood at the beginning of the hour.

(c) the change dùring the hour in this age distribution due to loss to tissue spaces and gain from the maturing marrow.

Clinically useful information can be obtained from the above computations by relating them to easily measurable quantities such as the hourly granulocyte count (GC). In order to find the GC a more detailed examination of the TBGP is required. At any given time blood granulocytes do not circulate freely, but some adhere to the walls of capillaries and venules. The sticking cells from the marginated granulocyte pool (MGP) and the free cells form the circulating granulocyte pool

(GCP). The number of cells in each of these pools will also be denoted by the same letters as used to denote the pools. Hence,

$$TBGP = MGP + CGP.$$

There is a continual interchange of marginated and free granulocytes. It is estimated that about 56% of the TBGP are marginated. Hence, CGP = .44(TBGP). The granulocyte count for a blood volume V is CG = CGB/V.

3. Tissue spaces

The tissue spaces are "sinks" or "drains" for the granulocytes in the peripheral blood. In the steady state this drain is the GTR. The feedback A (or equivalently the chance (1-f) of a cell of age less than 160 leaving the peripheral blood is calculated as described in 2 above. The tissue demands in Blumenson's paper were held at the steady state level. As mentioned above, the modeling internal to the tissue block can be extended to include variable demands due to various pathological conditions.

4. Proliferating marrow

The proliferating marrow block is divided into 4 compartments representing the various stages of differentiation of committed granulocyte cells. The types of cells in each stage are listed in Table 9.5 in order of increasing maturity from top to bottom (c.f. Fig. 9.1). Each compartment is divided into one hour

TABLE 9.5 SUBDIVISIONS OF THE PROLIFERATING MARROW

Type of cells in a compart.	No. of subcompart. (Cell cycle in hours)	Subcompart. of the S-phase	No. of cells in each subcompart.
myeloblast	18	6-17	1/16 GTR
promyelocyte	24	12-23	1/6 GTR
early myelocyte	52	40-51	1/4 GTR
late myelocyte	52	40-51	1/2 GTR

subcompartments. The number of subcompartments of any compartment is determined by the length of the cell cycle of any cell in the stage of differentiation represented by that compartment.

In addition to maturating, cells in the proliferating marrow can also multiply. A cell in a given stage passes through successively through all the subcompartments of that stage, beginning with compartment 1. Upon completing its cycle of equivalently leaving the last subcompartment of a stage, the cell divides into 2 identical cells of the next type. Both of these cells enter the first subcompartment of the next stage.

The steady-state number of cells in each one hour subcompartment of a given stage is shown in Table 9.5. These numbers are obtained by the following reasoning. The number of cells leaving the proliferating marrow is equal to the steady state GTR. Since each late myelocyte divides to become two metamyeolocytes there are $\frac{1}{2}$ GTR cells in each of the 52 one hour subcompartments of the late myelocyte stage. By working backwards the number of cells in the other subcompartments can be found.

5. Stem cells

The model internal to the stem cell block is depicted in Fig. 9.11 and is an extension of a model of Lajtha et al. [25]. In Fig. 9.11 feedback is denoted by

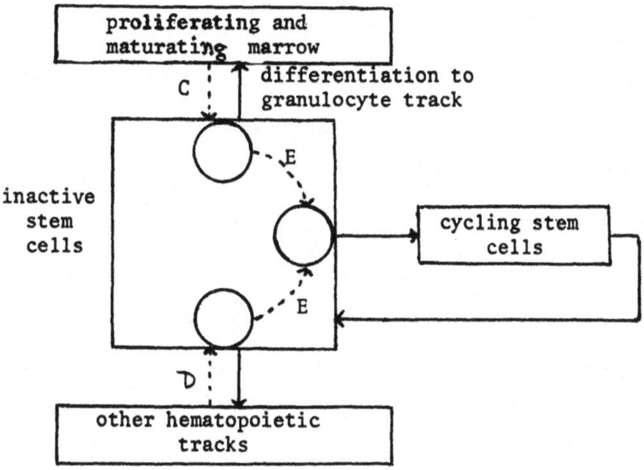

Figure 9.11 Model internal to stem cell block with feedbacks.

dashed arrows, while cellular activity is denoted by solid arrows. Feedback C of the number of granulocytic marrow cells affects the rate of differentiation of inactive stem cells into the granulocyte track. This loss of inactive stem cells is noted by feedback E which affects the rate at which inactive stem cells are triggered into the cell cycle. Feedback D regulates the level of cells in other hemic tracks by affecting the rate of differentiation of inactive stem cells into this track. These feedbacks will be discussed below.

Feedback D

Feedback D is simply the rule that the fraction of noncycling stem cells continually being lost to other tracks each hour is just the steady state level .0075. This loss is independent of the need for granulocytes. The reason for this simple rule is that no internal structure is given for the other hemic lines block. However, the possibility remains of coupling this model of granulopoiesis with multi-block models for the development of other hemic lines.

The steady state loss is calculated from the following facts:

(i) GTR = 4.8×10^9,

(ii) total number of stem cells in the steady state = 5.0×10^{10}

(iii) percentage of hemic marrow cells of the granulocytic series = 56;

(iv) stem cell cycle is 33 hr. with 1 hr. for mitosis.

Recall that the number of stem cells differentiating into the granulocytic track per hour is $\frac{1}{16}$ GTR = 3.0×10^8. From (iii) a rough estimate of the number of stem cells that differentiate into other hemic tracks per hour is $(.44/.56)(3.0 \times 10^8)$ = 2.4×10^8. Hence, 5.4×10^8 stem cells differentiate into some hemic track each hour. This means that the same number of stem cells must be triggered into the stem cell cycle each hour. Moreover, this same number of cells must be in mitosis. Let x be the number of cycling stem cells and assume that these cells are uniformly spread through the cell cycle. It follows from (iv) that $\frac{1}{33}$ x of these cells will be in mitosis and so $\frac{1}{33} x = 5.4 \times 10^8$. Solving for x yields 1.8×10^{10} stem cells in the cell cycle. It follows from (ii) that 3.2×10^{10} stem cells are inactive. Therefore, the fraction of inactive stem cells which differentiate into a non-granulocytic track each hour is, in the steady state, $(2.4 \times 10^8) | (3.2 \times 10^8) = .0075$.

Feedback E

Feedback E, which is internal to the stem cell block, is also quite simple. Let

$$d = \text{[steady state no. of stem cells]} - \text{[no. of stem cells at time t]};$$

If there was a depletion of stem cells ($d > 0$) then during the tth hours an extra fraction $u\,d$ of inactive stem cells are triggered into the cycle in addition to the usual steady state number (5.4×10^8). If d is negative, only the steady state number are triggered into the cycle. Here, u is a constant which satisfies $0 < u \leq 1$. The calculations in Blumenson [2] for the effect of the drug 5-fluorouracil (5-FU) on granulopoiesis were performed for $u = .5$ and $u = 1$. There was little difference in the overall time sequence of marrow cellularity and peripheral granulocyte count for these two values.

<u>Feedback C</u>

Feedback C controls the rate of stem cell differentiation into the granulocyte track. It acts by comparing G(t), the total number of cells in the proliferating and maturating marrow at time t, with the total number of such cells in the steady state, G_s. The rate of stem cell differentiation should also be proportional to S(t), the total number of inactive stem cells at time t. Moreover, this rate should increase (decrease) as G(t) decreases (increases). A proportionality "constant" which varies in such a way that the last condition is satisfied is

$$p(z) = Ce^{-bz}$$

where b anc C are constants which will be determined below and

$$z = \frac{G(t)}{G_s} .$$

Then the number of inactive stem cells at time t which differentiate during the next hour is

$$p(z)\ S(t).$$

In the steady state when $z = 1$,

$$p(1) = \frac{\text{no. of inactive stem cells which differentiate in next hr.}}{\text{total no. of inactive stem cells}}$$

$$= \frac{3 \times 10^8}{3.2 \times 10^{10}} = .0094$$

by using the constants derived in the discussion of feedback D. Since there are two unknown parameters (b,C), the value of p(z) for a second value of $z \neq 1$ is needed. The second value is obtained by comparing the rate of differentiation of stem cells when the number of cells in the granulocyte track is small (z near zero) with the rate when z = 1. Not much is known about this ratio and Blumenson [2] assumed that the value of this ratio would be 130. In other words,

$$\frac{130}{1} = \frac{p(o)}{.0094}$$

Hence, solving for p(o) yields C = p(o) = 1.2. From

$$p(1) = Ce^{-b} = .0094$$

it follows that b = 4.85.

The effect of the drug 5-fluorouracil (5-FU) on the granulopoietic system was simulated by Blumenson. The average steady state values for a healthy 70 kg. man were used to initialize the system. 5-FU is a fluorinated pyrimidine which alters cell function by inhibiting thymidylate synthetase, which causes thymidine deficiency, leading to cellular injury and death. This action was simulated in the model by simply assuming that a certain dose-dependent fraction of marrow cells which are in the S-phase when the drug is present are killed. The drug was assumed only to be effective 3 hours from the time that the injection began. In particular, it was assumed that when the solution prepared with 15 mg/kg was injected 30%, 15% and 7.5% of the marrow cells in the S-phase were killed during the first, second and third hour, respectively. When the 7.5 mg/kg solution was injected the corresponding percentages of marrow cells in the S-phase which were killed during the first three hours were 15%, 7.5% and 3.75%, respectively. These percentages were chosen arbitrarily, but changing them by 20% had little effect on the calculated pattern of granulopoietic recovery.

The particular dosage schedule and protocol simulated was the one actually used in a study of 5-FU on patients with several types of advanced cancer. The dosage schedule was 15 mg/kg of total body weight daily for 5 days and 7.5 mg/kg on alternate days beginning with the 7th day. If toxicity developed (i.e. the white blood count (WBC) was less than 4000), the therapy was discontinued until toxicity disappeared. The trial ran for at least 60 days.

The daily WBC was calculated using the program of the model. The calculated curve was extremely erratic. However, the predicted pattern, including the dip at 10 days and the overshoot at 20 days is qualitatively consistent with the pattern of WBC which has been reported for many of these patients treated with 5-FU.

Blumenson also plotted the hourly WBC, stem cell activity and chance that an inactive stem cell is triggered into the granulocyte track. Comparing the hourly WBC with the daily WBC indicates that only monitoring the daily WBC leads to artifacts which give an incomplete picture of the WBC variation.

9.10 Discussion

Blumenson's model could be generalized in several ways. Instead of considering considering fixed cell cycle times, the cell cycle times could be randonly distributed. The feedbacks could be altered to produce disease states. In addition, a leukemic cell population could be added in the manner suggested by the work of Rubinow and Lebowitz. The other hemic lines block could be broken up into a separate block for each hemic line using the ideas on red cell production from Düchtung [16], Kirk et al. [23,24] and Monot et al. [32] and the ideas on platelet production from Gray and Kirk [20]. Finally, the immune system block might be incorporated by utilizing some of the results and references in Appendix D.

The reader should not be misled into thinking that Blumenson's complex model for granulopoiesis is better than the relatively simpler model of Rubinow and Lebowitz. The genius in modeling lies in knowing which facts are important and which are not. Newton would never have formulated his theory of planetary motion if he had tried to describe each planet realistically with a non-smooth surface and a non-spherical shape. Instead, he reduced each planet to a point. The study of complex computer models may reveal which controls and cell populations are most important.

At present computer models have been used for theoretical studies with average steady-state values of their parameters. The clinical application of such models will require new methods of rapidly estimating the relevant parameters for each patient. The suggested course of therapy will then be computed. The patient will have to be monitored to detect deviations from the predicted course. When these are detected a new course of therapy will have to be planned and so on. Thus, not only will a predictive model be required but methods for detecting and correcting errors are also needed.

An interesting problem would be the application of optimal control theory for finding more efficient drug protocols for treating leukemia using one of the analytic models of leukemia.

A common problem for patients being treated for malignancies is providing appropriate long-term platelet transfusions. Platelet deficiency frequently occurs during a course of cancer chemotherapy. Optimization techniques are identified and applied by Simon [41] for one important aspect of this problem.

A mathematical model for the etiology of leukemia has been developed by Burch [5,6,7,8,9]. It is based on the assumption that a given number of mutations are required for leukemogenesis. Burch inferred, by applying his model to epidemiological data, that various combinations of inherited, somatic mutations and possible nongenetic factors such as viruses will produce the different types of leukemia. For instance, leukemias in childhood generally arise from one inherited mutation, two somatic mutations and a nongenetic factor. On the other hand, chronic lymphatic leukemia in adults develops from one inherited and three somatic mutations. Although the model provides a theory of leukemogenesis, the translation of this theory into biological terms is not evident. It is not known which genes are mutated. Moreover, the biological mechanism by which the combination of different mutations can result in different cell types of leukemia is also unknown.

REFERENCES

1. Blumenson, L.E. A comprehensive modeling procedure for the human granulopoietic system: overall view and summary of data, Blood, 42, 303-313, 1973.

2. Blumenson, L.E. A comprehensive modeling procedure for the human granulopoietic system: detailed description and application to cancer chemotherapy, Math. Biosciences, 26, 217-239, 1975.

3. Bradley, T.R., Metcalf, D. and Robinson, W. Stimulation by leukemic sera of colony formation in vitro by mouse bone marrow cells, Nature (Lond.), 213, 926-927, 1976.

4. Bullough, W.S. Mitotic control in mammalian tissues, Biol. Rev., 50, 99-127, 1975.

5. Burch, P.R.J. A biological principle and its converse: some implications for carcinogenesis, Nature, 195, 241-243, 1962.

6. Burch, P.R.J. Leukemogenesis in man, Ann. N.Y. Acad. Sci., 114, 213-224, 1964.

7. Burch, P.R.J. Natural and radiation carcinogenesis in man I. Theory of initiation phase, Proc. Roy. Soc. (London) B162, 223-239, 1965.

8. Burch, P.R.J. Natural and radiation carcinogenesis in man II. Natural leukaemongenesis: initiation, Proc. Roy. Soc. (London) B162, 240-262, 1965.

9. Burch, P.R.J. Natural and radiation carcinogenesis, Proc. Roy. Soc. (London) B162, 263-287, 1965.

10. Chan, S.H. and Metcalf, D. Inhibition of bone marrow colony formation by normal and leukaemic human serum, Nature (Lond.), 227, 845-846, 1970.

11. Cline, M.J. The White Cell, Harvard University Press, Cambridge, 1975.

12. Clarkson, B.D. Acute myelocytic leukemia in adults, Cancer, 30, 1572-1582, 1972.

13. Clarkson, B.D., Dowling Jr., M.D., Gee, T.S. and Burchenal, J.H. Treatment of acute myeloblastic leukemia, Bibl. Haematol., 39, 1098-1114, 1973.

14. Clarkson, B. and Rubinow, S.I. Growth kinetics in human leukemia, in Growth Kinetics and Biochemical Regulation of Norman and Malignant Cells (Drewinko, B. and Humphrey, R.M.,(editors), Williams and Wilkins, Baltimore, 591-628, 1977.

15. Donaghey, C.E. Cellsim User's Manual, University of Houston, Houston, Texas, 1973.

16. Düchting, W. Computer simulation of abnormal erythropoiesis - an example of cell renewal regulating systems, Biomed. Techn., 21, 34-43, 1976.

17. Evert, Jr., C.F. CELLDYN - a digital program for modeling the dynamics of cells, Simulation, 24, 55-61, 1975.

18. Evert, C.F. and Palusinski, O.A. Application of discrete computer modeling to the dynamics of cell populations, Acta Haemat. Pol., 6, 176-184, 1975.

19. Gee. T.S., Yu, K.P., Clarkson, B.D. Treatment of adult leukemia with arabinosylcytosine and thioguanine, Cancer, 23, 1019-1032, 1969.

20. Gray, W.M. and Kirk, J. Analysis by analogue and digital computer of the bone marrow stem cell and platelet control mechanisms, in Computers for Analysis and Control of Biomedical Systems, I.E.E. Publications, London, 1971.

21. Hirsch, J.G. New light on the mystery of the neutrophil leukocyte, Medical Times, 104, 88-109, 1976.

22. King-Smith, E.A. and Morley, A. Computer simulation of granulopoiesis; normal and impaired granulopoiesis, Blood, 36, 254-262, 1970.

23. Kirk, J., Orr, J.S. and Hope, C.S. A mathematical analysis of red blood cell and marrow stem cell control mechanisms, Brit. J. Haematol., 15, 35-46, 1968.

24. Kirk, J., Orr, J.S., Wheldon, T.E. and Gray, W.M. Stress cycle analysis in the biocybernetic study of blood cell populations, J. Theor. Biol., 26, 265-276, 1970.

25. Lajtha, L.G., Oliver, R. and Gurney, C.W. Kinetic model of bone-marrow stem-cell population, Br. J. Haematol., 8, 442-460, 1962.

26. Lewis, A.E. Principles of Hematology, Appleton Century Croft, New York, 1970.

27. Mackey, M.C. and Glass, L. Oscillation and chaos in physiological control systems, Science, 197 , 287-289, 1977.

28. Mackey, M.C. Unified hypothesis for the origin of aplastic anemia and periodic hematopoiesis, 51, 941-956, 1978.

29. Mathé, G., Belpomme, D., Dantchev, D. and Pouillart, P. Progress in the classification of lymphoid and/or monocytoid leukemias and of lympho and reticulosarcomas (non-Hodgkin's lymphomas), Biomedicine, 22, 177-185, 1975.

30. Mauer, A.M., Evert, Jr., C.F., Lampkin, B.C. and McWilliams, N.B. Cell kinetics in human acute lymphoblastic leukemia: computer simulation with discrete modeling techniques, Blood, 41, 141-154, 1973.

31. Metcalf, D. The nature of leukaemia: neoplasm or disorder of haemopoietic regulation? Med. J. Aust., 2, 739-746, 1971.

32. Monot, C., Najean, Y., Dresch, C. and Martin, J. Models of erythropoiesis and clinical diagnosis, Math. Biosciences, 27, 145-154, 1975.

33. Pritsker, A.A.B. and Kivioat, P.J. Simulation with GASP II, Prentice Hall, Englewood Cliffs, 1969.

34. Robinson, W., Metcalf, D., and Bradley, T.R. Stimulation by normal and leukemic mouse sera of colony formation in vitro by mouse bone marrow cells, J. Cell Physiol., 69, 83-91, 1967.

35. Rubinow, S.I. A simple model of a steady state differentiating cell system, J. Cell Biol., 43, 32-39, 1969.

36. Rubinow, S.I., Lebowitz, J.L. and Sapse, A. Parameterization of in vivo leukemia cell populations, Biophys, J., 11, 175-188, 1971.

37. Rubinow, S.I. Mathematical Problems in the Biological Sciences, SIAM Publication, Philadelphia, 1973.

38. Rubinow, S.I. and Lebowitz, J.L. A mathematical model of neutrophil production and control in normal man, J. of Math. Biology 1, 187-225, 1975.

39. Rubinow, S.I. and Lebowitz, J.L. A mathematical model of the acute myeloblastic leukemic state in man, Biophys. J., 16, 897-910, 1976.

40. Rubinow, S.I. and Lebowitz, J.L. A mathematical model of the chemotherapeutic treatment of acute myeloblastic leukemia Biophys, J., 16, 1257-1272, 1976.

41. Simon, R. Application of opimization methods in hematological support of patients with disseminated malignancies, Math. Biosciences, 25, 125-138, 1975.

42. Wheldon, T.E. Mathematical models of granulopoiesis, Ph.D. Thesis, University of Glasgow, 1973.

43. Wheldon, T.E. Mathematical models of oscillatory blood cell production, Math. Biosciences, 24, 289-305, 1975.

44. Wheldon, T.E., Kirk, J. and Finlay, H.M. Cyclical granulopoiesis in chronic granulocytic leukaemia: a simulation study, Blood 43, 379-387, 1974.

45. Wickramsinghe, S.N. Human Bone Marrow, Blackwell Scientific Publications, Oxford, 1975.

Appendix A

Chemistry of Genes. Protein Synthesis

1. Introduction.

Nearly everyone knows that genes control heredity. However, most people do not realize that genes also control the normal functions of cells. They do this by determining the substances which are synthesized by cells.

A gene is part of a nucleic acid called deoxyribonucleic acid (DNA). DNA automatically controls the formation of another nucleic acid called ribonucleic acid (RNA), which spreads throughout the cell and controls the formation of the different proteins. Some of these proteins are structural proteins from which the various cell organelles are constructed. Others are enzymes that promote the chemical reactions in the cell. These processes will be described in detail later in this appendix after the structures of DNA and RNA have been elucidated.

The DNA of eukaryotic cells is contained in the chromosomes as chromatin. Chromatin is a complex of $\frac{1}{3}$ DNA, $\frac{1}{3}$ histone (small basic proteins) and $\frac{1}{3}$ acidic chromosomal proteins. Five main histones exists, four of which (H_2a, H_2b, H_3 and H_4) fold 200- base-pair sections of double-helical DNA into 125 Å spherical masses or nucleosomes. The remaining histone (H1) may control chromosomal extension and retraction during the mitotic cycle.

2. The Building Blocks of DNA and RNA.

The compounds involved in the formation of DNA and RNA are: phosphoric acid, sugar and nitrogenous bases (purines and pyrimidine). These are described below

(a) phosphoric acid

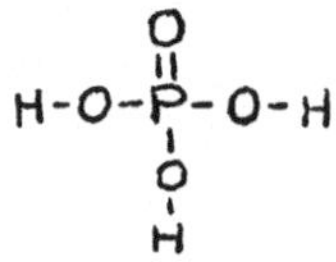

(b) sugars

The sugar in DNA is a pentose (5-carbon) called deoxyribose, differing from the RNA sugar ribose by the lack of one oxygen atom in the 2' position

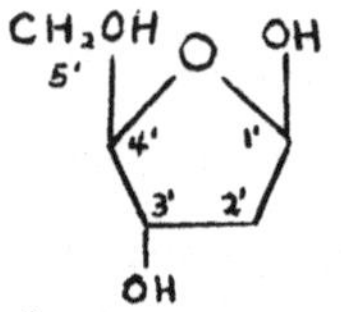

deoxyribose

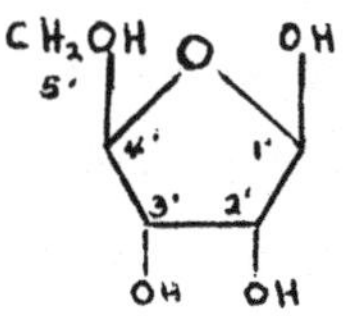

ribose

(c) pyrimidine bases

These are derivatives of pyrimidine.

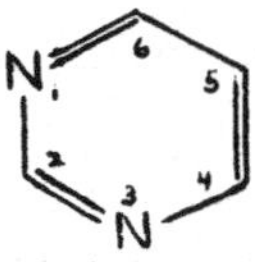

pyrimidine ring

DNA contains the bases cytosine and thymine; RNA contains the bases cytosine and uracil.

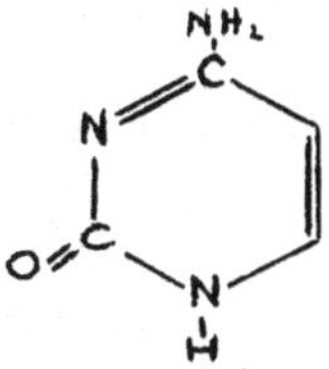

cytosine

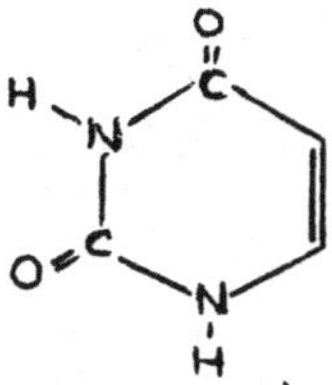

uracil

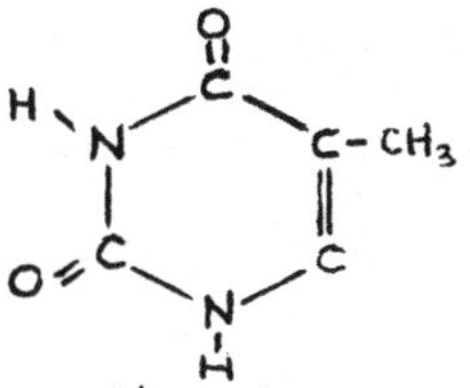

thymine

(d) purine bases

These substances are derivatives of the parent compound purine which contains a pyrimidine and imidazole () ring fused together

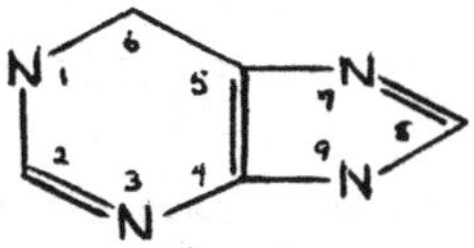

purine

Both DNA and RNA contain the bases adenine and guanine.

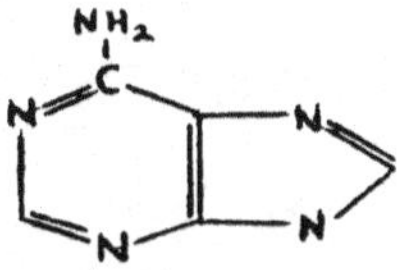

adenine

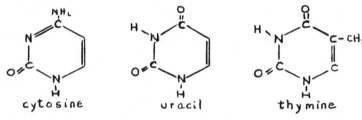

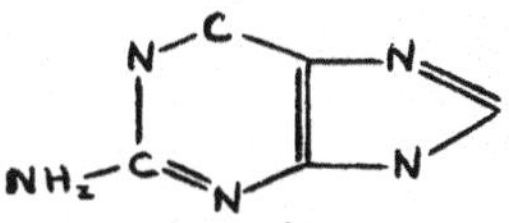

guanine

3. <u>The Chemical Structure of DNA and RNA.</u>

Before describing the chemical structure of DNA the subunits from which it is formed will be defined. These subunits are formed from the basic building blocks.

A nucleoside is the condensation[1] product of a purine (N9) or pyrimidine (N3) with a sugar molecule (joined to 1' of sugar). The basic subunit, called a nucleotide, is formed when a nucleoside is esterified[2] with phosphoric acid. (joined by C5'). For example, the structural formula for adenylic acid is

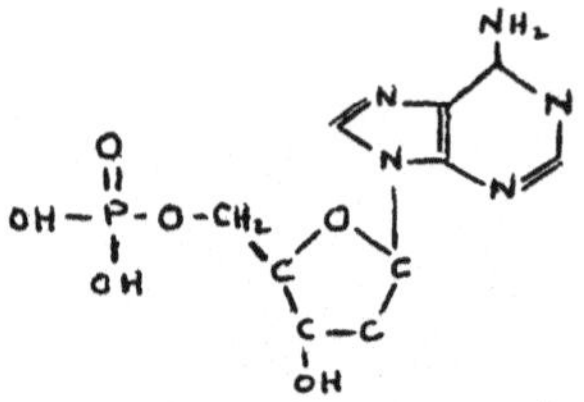

adenylic acid (adenosine 5'-monophosphate)

The three other nucleotides thymidylic, guanylic and cytidylic have a similar structure to adenylic acid with adenine replaced by thymidine, guanine and cytosine, respectively.

The DNA molecule is a double helical structure formed from two strands which are bound together by hydrogen bonds between the bases. Each strand is a long polymer (i.e. a macromolecule composed of a number of similar or identical subunits, monomers, covalently bonded) of thousands of nucleotides as shown in Fig. 1.

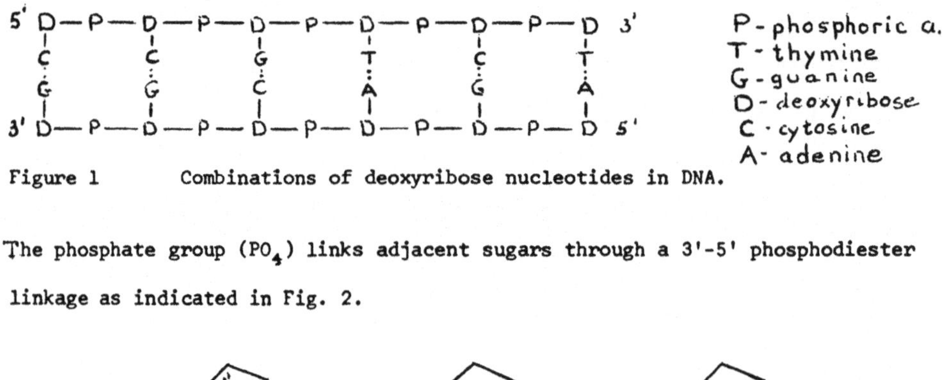

Figure 1 Combinations of deoxyribose nucleotides in DNA.

The phosphate group (PO_4) links adjacent sugars through a 3'-5' phosphodiester linkage as indicated in Fig. 2.

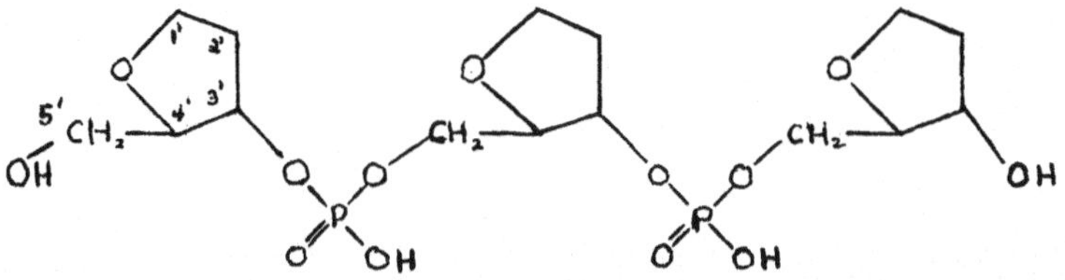

Figure 2 3'-5' phosphodiester linkage.

---

[1] The two molecules are joined by the removal of a molecule of H_2O.

[2] Reaction of an acid with an alcohol or OH-group.

In one chain the linkages are polarized 3'-5'; in the other chain they are in the reverse order 5'-3'. The rungs in the ladder (i.e. the bases connecting one strand of DNA to its polarized complement) can only be formed by purines pairing with pyrimidines and vice versa. Furthermore, the base pairs are linked in such a manner that the number of hydrogen bonds between them is always maximized. For this reason, adenine pairs only with thymine and guanine only with cytosine (See Fig. 3.).

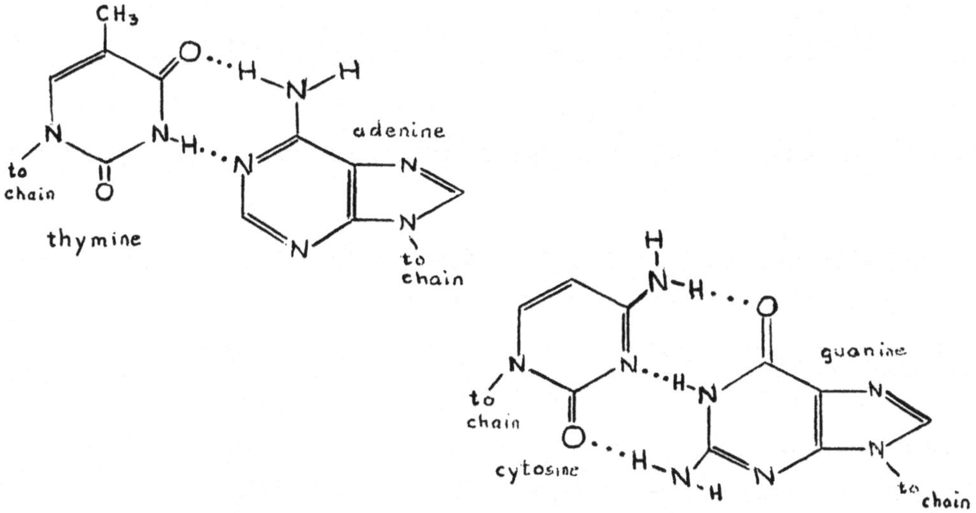

Figure 3 Base pairing in DNA.

Finally, to put the DNA of Fig. 1 into its proper physical perspective, one needs merely pick up the two ends and twist them into a helix. Ten pairs of nucleotides are present in each full turn of the helix.

RNA is different from DNA in the following respects.

(i) RNA is single stranded; DNA is double stranded

(ii) RNA contains ribose sugar; DNA contains deoxyribose sugar

(iii) RNA contains the pyrimidine uracil instead of thymine (Later we shall see that U pairs with A)

(iv) RNA molecules are generally much shorter than DNA molecules.

In some of the plant viruses, DNA is absent and RNA functions as genetic material (see Appendix B).

4. The Replication of DNA

Cellular duplication of DNA occurs by the eventual separation of the two strands of the parent molecule. Each strand acts as a template for the formation of a new strand. The new strands are not identical to the old strand. However, they are complementary in the sense that the sequence of bases along them are uniquely determined by the base pairing (c.f. Fig. 1). The final result is two double stranded DNA molecules which are identical copies of the parent molecule.

Chromosomal replication usually starts at a fixed location (the origin) to produce a replicating bubble containing two replicating forks.[3] The strands do not completely separate before the synthesis of new strands but duplication goes hand-in-hand with separation. Chain growth occurs simultaneously at each replicating fork.

The precursor nucleotides are activated by the addition of two phosphate radicals to form triphosphates of the nucleotides. The energy required is supplied by high energy phosphate bonds derived from adenosine triphosphate (ATP) in the cell. The deoxyribonucleoside-triphosphates now have sufficient energy to be enzymatically added to the 3' ends of short DNA fragments. In turn these fragments become linked to the main daughter strands.

Each cell requires many enzymes for DNA replication. Several different polymerizing enzymes (DNA polymerases) exist. For example, there is a special enzyme for chain elongation and another for filling in of the gaps between the small DNA fragments found at the replicating forks. No known DNA polymerase can initiate DNA chains. This initiation process is thought to start by the synthesis of small RNA primers catalyzed by a form of the enzyme RNA polymerase. Enzymes are also thought to be required for terminating the chain and for cuts required to unravel the parental strands.

Cell DNA polymerases have the capacity to recognize incorrectly incorporated bases and to enzymatically cut them out before addition of the next precursor nucleotide (proofreading). Some of the enzymes involved in DNA duplication also function in the repair of damaged DNA molecules. An unproved conjecture is that

[3] A large part of molecular genetics is obtained from the study of bacteria and viruses.

the bacterial membrane may contain enzymes involved in initiating and terminating DNA synthesis and play an important role in positioning the progeny of DNA synthesis into the daughter cells.

5. The Genetic Code

The importance of the gene lies in its ability to control the formation of other substances in the cell. It does this by producing enzymes which mediate all biochemical reactions. Cell enzymes are proteins. Proteins are polymers of subunits (monomers) called amino acids. Each amino acid has an amino group (NH_2) at one end and a carboxyl group (COOH) at the other end. Each enzyme, therefore, is a precisely ordered sequence of amino acids. This sequence of amino acids is coded by the succession of purine and pyrimidine bases projecting from the side of each DNA strand. It is these projections or bases that form the so-called genetic code.

The number of nucleotides which code for an animo acid is termed a codon. About twenty different amino acids occur naturally in proteins. However, there are only 4 different nucleotides. One nucleotide coding for one amino acid could code for only 4 different amino acids. A doublet code (i.e. two nucleotides coding for one amino acid) yields only 16 different combinations. Therefore, a triplet code (i.e. three nucleotides coding for one amino acid) is the smallest coding unit that will allow 20 amino acids to be represented. Experimental evidence supports this triplet code concept[4]. The code is basically the same for all organisms.

The genetic code is degenerate because more than one codon exists for most amino acids. Only three of the 64 possible are nonsense triplets - that is, codons not coding for any amino acid. These represent signals for terminating protein synthesis. It is not known why three different stop signals are required. Two codons, besides representing amino acids, also serve to initiate protein synthesis. Starting and stopping protein synthesis is especially important when a DNA molecule can cause the formation of two or more different proteins. (Some DNA molecules of viruses can cause the formation of up to 20 different proteins.)

---

4 The first evidence arose from the study of single base pair additions in the single linkage group of bacterial virus (phage T_4).

Almost all DNA is located in the nucleus of the cell. However, most cell functions occur in the cytoplasm. Therefore some means must be available for the genes of the nucleus to control the chemical reaction of the cytoplasm. This is accomplished via ribonucleic acid (RNA) whose formation is controlled by nuclear DNA. The RNA is transported into the cytoplasm where it controls protein synthesis.

At least three types of RNA are important to protein synthesis: messenger RNA (mRNA), transfer RNA (tRNA) and ribosomal RNA (rRNA).

6. Synthesis of RNA

The synthesis of RNA upon DNA templates has many similarities to the DNA duplication process. The two strands of DNA separate. However, only one strand functions in the production of RNA. The activated monomer precursors, the ribonucleoside triphosphates (ATP, GTP, CTP and UTP), are linked to the exposed bases of the DNA strand. A ribose nucleotide base always combines with the complementary deoxyribose bases in the following manner:

DNA base	G	C	A	T
RNA base	C	G	U	A

The enzyme RNA polymerase causes the two extra phosphates on each nucleotide to split away. This liberates enough energy to cause bonds to form between the successive ribose and phosphoric acid radicals. Simultaneously, the RNA strand automatically separates from the DNA strand and becomes a free molecule of RNA.

Synthesis of all chains starts with either adenine or guamine, depending on the specific start signal in the DNA template. Elongation proceeds in the 5' to 3' direction, so the 5' terminal nucleotides always contain a triphsophate group. Synthesis ends upon reaching stop signals. Termination sometimes depends on the presence of a specific protein, the ρ factor. The molecular mode of ρ action is unsolved.

RNA polymerase is constructed from five types of polypeptide chains. One of these chains, σ, easily dissociates from the other four, which form the core enzyme. The core catalyzes the linkage of phophoric acid with the two sugars. The σ polypeptide helps in the recognition of the start signals along DNA molecules. After initiation occurs, σ dissociates from the DNA-RNA core complex, becoming free to attach to another core.

It is important to note that a given sequence on a single DNA strand uniquely determines a complementary sequence on the RNA strand it synthesizes. Thereby, each codon appearing on the DNA template is uniquely transferred as a complementary codon on the synthesized RNA strand. For example,

CODONS ON DNA STRAND	GGC	AGA	CTT
COMPLEMENTARY CODONS ON SYNTHESIZED RNA	CCG	UCU	GAA

This process of transferring the genetic code from DNA to messenger RNA is called transcription.

Table 1 gives the RNA codons for the 20 common amino acids found in protein molecules. Note that several of the amino acids are represented by more than one codon. The nonsense codons represent the signal "stop manufacturing a protein molecule". The signal for "start manufacturing a protein molecule" is usually represented by the codon AUG for the starting amino acids, N-formyl methionine, the same codon as for methionine. GUG which usually codes for valine, also can code for N-formyl-methionine.

TABLE 1 RNA CODONS.

AMINO ACIDS	RNA CODONS					
Alanine	GCU	GCC	GCA	GGG		
Arginine	CGU	CGC	CGA	CGG	AGA	AGG
Asparagine	AAU	AAC				
Aspartic acid	GAU	GAC				
Cysteine	UGU	UGC				
Glutamic acid	GAA	GAG				
Glutamine	CAA	CAG				
Glycine	CGU	GGC	GGA	GGG		
Histidine	CAU	CAC				
Isoleucine	AUU	AUC	AUA			
Leucine	CUU	CUC	CUA	CUG	UUA	UUG
Lysine	AAA	AAG				
Methionine	AUG					
Phenylalanine	UUU	UUC				
Proline	CCU	CCC	CCA	CCG		
Serine	UCU	UCC	UCA	UGG	AGU	AGC
Threonine	ACU	ACC	ACA	ACG		
Tryptophan	UGG					
Tyrosine			UAU	UAC		
Valine	GUU	GUC	GUA	GUG		
Start	AUG	GUG				
Stop	UAA	UAG	UGA			

Once the RNA molecules are formed, they are transported, by unknown means, into all parts of the parts of the cytoplasm where they perform further functions. Three types of RNA are formed as shown in Table 2.

TABLE 2 TYPES OF RNA.

TYPE	LOCATION	PERCENTAGE OF TOTAL IN CELL	STRUCTURE	FUNCTION	MOLECULAR WEIGHT
ribosomal	ribosomes	80	single strand with some coiled branches	unknown	high
transfer		15-20	1. single-stranded, semirigid, L-shaped; about 25 nucleotides 2. 1/2 the bases, part of double helical stems; other 1/2, form loops[5] 3. one loop contains anticodon[6]. Different tRNA for each amino acid. 4. CCA always at 3'-end; G at 5'-end. 5. may contain some unusual bases	1. transfers amino acids to site of protein synthesis 2. positions amino acid in correct position by means of anticodon	25,000
messenger[7]		1	single strand of several hundred to several thousand nucleotides	1. transcription 2. carries genetic information from nucleus to site of protein synthesis 3. serves as template for protein synthesis	very high; several million

5 Stabilized by hydrogen bonds and stacking forces.

6 Groups of 3 bases which base pair to 3 successive template bases (codon)

7 mRNA is very unstable in bacteria and must be continually replaced by newly synthesized molecules.

7. Formation of Proteins

The coded information in DNA is transcribed into the coded sequence of the synthesized mRNA by using one strand of the DNA as a template. The mRNA leaves the nucleus and contacts a ribosome. There as the messenger RNA travels past the ribosome (see Fig. 4) a protein molecule is formed stepwise beginning at the amino terminal end. The codon for each amino acid is read unidirectionally by the ribosome beginning at the 5'-hydroxyl end. The exact process, called translation, by which the ribosome reads the message is unknown.

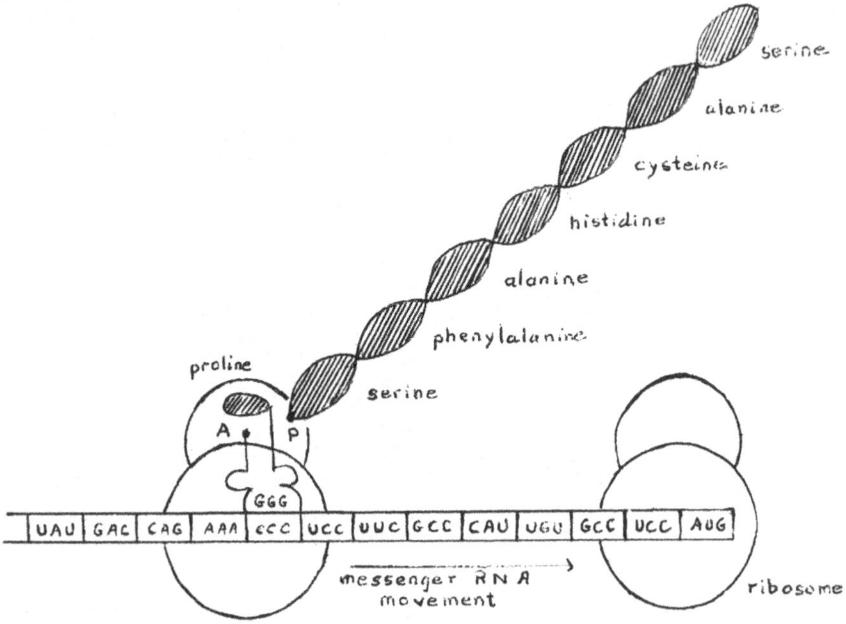

Figure 4 Mechanism for protein synthesis by ribosomes in association with mRNA.

The steps in protein synthesis are:

(i) Each amino acid is activated. ATP combines with the amino acid to form adenosine monophosphate (AMP) complex with the amino acid.

(ii) The activated amino acid, having an excess of energy, then combines with its specific tRNA by its carboxyl group to the 3'-terminal adenosine of the tRNA molecule. This forms an amino acid - tRNA complex (AA ~ tRNA) and releases AMP. There is a specific enzyme (amino-acyl synthetase) for each amino acid. It is the enzyme, not tRNA which recognizes the amino acid!

(iii) The transfer RNA complexed with the amino acid, then comes in contact with
the mRNA molecule in the ribosome (see Remark on ribosomes below). The anti-
codon of tRNA attaches temporarily to its specific codon of the mRNA thus
lining up the amino acids in the proper sequence for forming the protein
molecule.

(iv) The energy from guanine triphosphate[8] (GTP), another high energy phosphate,
is used to form the chemical bonds between successive amino acids. This
process is called peptide linkage and proceeds according to the following
reaction:

$$R_1-\overset{\overset{NH_2}{|}}{C}-\overset{\overset{O}{\|}}{C}-OH \ + \ H-\overset{\overset{H}{|}}{N}-\overset{\overset{R_2}{|}}{C}-COOH \longrightarrow R_1-\overset{\overset{NH_2}{|}}{C}-\overset{\overset{O}{\|}}{C}-\overset{\overset{H}{|}}{N}-\overset{\overset{R_2}{|}}{C}-COOH$$

$$+ \ H_2O$$

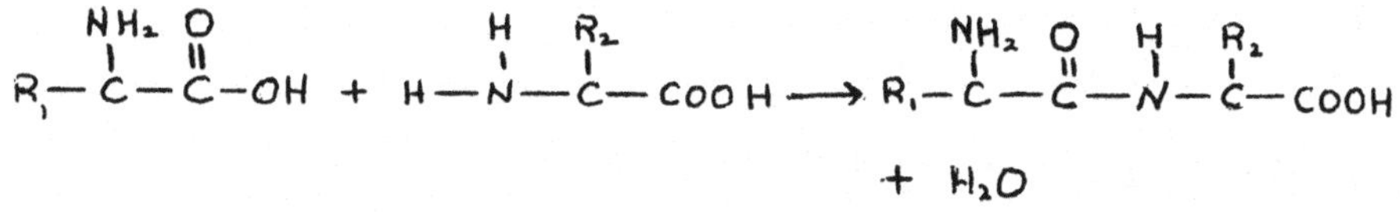

The tRNA is released from its amino acid and from mRNA. The tRNA is then free
to activate another free amino acid of the same species.

(v) When the ribosome reaches the codon for chain termination translation of
the nucleotide code into an amino acid sequence (polypeptide) is complete and
the polypeptide is released. At this stage the polypeptide (protein) possesses its primary structure (the linear sequence of amino acids). At least
some positions of the protein may have secondary structure in the form of an
α-helix.

(vi) The protein chain may then fold back upon itself, forming internal bonds
(including strong disulfide bonds) which stabilize its tertiary structure
into a precisely and often intricately folded protein. This may occur before
release.

(vii) Two or more tertiary structures may unite into a functional quaternary structure. For example, hemoglobin consists of four polypeptide chains, two
identical α-chains and two identical β-chains. A protein does not become an
active enzyme until it has assumed its tertiary or quaternary pattern.

---

[8] This is just guanine 5'-monophosphate with two more phosphate groups attached
to the already present phosphate group.

Remarks on Ribosomes and their Action

Attachment of the growing chain of amino acids to the ribosome occurs through binding of the chains terminal tRNA group to a hole in the ribosomal surface. Each ribosome has two holes: the P (peptidyl) and the A (amino-acyl). Precursor AA ~ tRNA complexes normally first enter the A site, allowing the subsequent formation of a peptide bond to the growing chain held in the P site. This transfers the nascent chain to the A site. Action of the enzyme translocase then moves the nascent chain back to the P site where another cycle can begin. Both the attachment of AA ~ tRNA to the A site and the translocation process require the energy from the removal of a phosphate group from GTP. Movement of the successive codons over the ribosomal surface may occur simultaneously with the translocation step, but nothing is known about how this happens.

On completion of chain synthesis the ribosomes dissociate into large and small ribosomal subunits. Reunion of the subunit occurs only after the initiating AA ~ tRNA has first combined with a mRNA small subunit complex. A given mRNA molecule generally works simultaneously on many ribosomes (polyribosomes or polysomes). Thus, at a given moment many codons of the same template may be read and more than one protein may be constructed.

8. Defining the Gene

The first concept of linked genes was that they were distinct elements analogous to beads on a string. The structure of DNA led to the hypothesis that a gene was one or a sequence of nucleotides. The amount of genetic material was defined by three criteria

(a) the unit leading to phenotypic expression (i.e. any measurable characteristic or distinctive trait possessed by an organism)[9]

(b) the unit of structure which cannot be subdivided by crossing over of non-sister chromatids of homologous chromosomes during meiosis[10].

---

[9] A muton is the smallest unit of genetic material that when changed (mutated) produces a phenotypic effect. May be a single nucleotide or part of a nucleotide.

[10] The recon the smallest unit in which crossovers can occur is believed to involve two adjacent nucleotides.

(c) the unit which can be changes by mutation.

It is now known that these criteria generally do not define the same amount of genetic material.

Another definition of a gene was on the basis of its function. For instance, the structure of an enzyme may be determined by a segment of DNA. It was hypothesized that one gene produces one enzyme. A counterexample to this conjecture is the enzyme tryptophan synthetase. It consists of two structurally different protein chains (A and B). Each chain is produced by an adjacent segment of DNA. Consequently, the "one gene - one enzyme" hypothesis has now been replaced by the "one gene - one polypeptide chain" hypothesis. A gene is a continuous segment of a DNA chain involved in dictating the primary structure of a single polypeptide chain.

REFERENCES

1. Kornberg, A., DNA Synthesis, Freeman, San Francisco, 1974.

2. Lehninger, A.L., Biochemistry, Worth Publishers, Inc., New York, 1970.

3. Watson, J.D., Molecular Biology of the Gene, 3rd ed., W.A. Benjamin, Menlo Park, California, 1976.

Appendix B

Viruses

1. Introduction

Viruses are very small infective agents ranging in size from 20-200nm. Most consists of only 1 molecule of DNA or RNA surrounded by a protein coat (see Table 1). The nucleic acid carries the genetic information necessary for viral multiplication. The protein coat protects the nucleic acid and also confers a host specificity on the virus.

TABLE 1 SOME COMMON VIRUSES.

Virus	Host	% RNA	%DNA	Mol. wt. of nucleic a.	Size
SV40	Vertebrates		13	3×10^6	40 nm diam.
Vaccinia	Vertebrates		3.2	180×10^6	200×300 nm.
T_2	E. coli		50	120×10^6	head 65×95 nm. tail 20×95 nm.
Poliomyelitis	Man	25		2×10^6	30 nm. diam.
Influenza	Man	.9		3×10^6	80 nm. diam.
Tobacco mosaic	Tobacco plants	5		2×10^6	15×300 nm.
Tomato bushy stunt	Tomato plants	15		1.6×10^6	28 nm. diam.
MS_2	E. coli	32		1.05×10^6	17 nm. diam.

Outside the host cell, viruses are inert. They depend on the host cell for energy (ATP) and precursors (e.g., amino acids and nucleotides) necessary for the synthesis of new virus particles.

Most animals, plants and microorganisms are susceptible to viruses. Viruses which infect bacteria are known as <u>bacteriophages</u> or <u>phages</u>.

2. Structure of Viruses

The capsid (coat) of a virus is usually a geometrical (often icosohedral) arrangement of subunits (capsomeres)[1]. Each capsomere is a group of coat protein molecules. In some animal viruses the capsid is surrounded by a membranous

[1] Morphological units when viewed in the electron microscope.

envelope which is formed partly from the membrane of the host cell and partly from the new viral components. The envelope contains viral specific glycoproteins.

Most viruses contain either a molecule of single-stranded RNA or a molecule of double stranded DNA. The nucleic acid is sometimes complexed with basic proteins which may help it adopt a compact conformation necessary for fitting inside the protein coat.

Some viruses contain unusual forms of nucleic acids. Reovirus contains double-stranded RNA which has a helical structure similar to DNA. The mouse virus contains a single-stranded DNA molecule. Human wart virus contains a cyclic DNA molecule.

3. Replication of Viruses

The replication cycle of a virus can be divided into 5 stages:

(a) Absorption

The virus capsid binds to specific receptor sites on the cytoplasmic membrane of a susceptible cell. The membrane envelope of envelope viruses fuses with the cell membrane.

(b) Penetration and Uncoating

The virus particle enters the cell and the protein coat is degraded liberating the nucleic acid.

(c) Synthesis of Virus Components

The RNA of an RNA-virus acts as messenger RNA. The DNA of DNA-viruses is used as a template to form messenger RNA. Virus proteins are synthesized on the cellular ribosomes with the help of virus mRNA. These proteins are of two types: enzymes for the duplication of viral nucleic acid and proteins for the structural components of the virus.

(d) Assembly

The nucleic acid interacts with the structural proteins to form new virus particles.

(e) Release

The cell either dies releasing the virus or the virus particles may "bud off" from the surface of the cell.

Depending on the type of virus, from 10^2 to 10^5 new virus particles are made in an infected cell.

4. Oncogenic Viruses

Many types of viruses can cause cancer. All RNA viruses have essentially the same mode of action. After contacting a susceptible (permissive) cell the usual steps of absorption, penetration and uncoating occur. The 70S[1] molecule of RNA, consisting of two identical 35S chains, is used as a template for the synthesis of a complementary DNA strand. This requires the enzyme reverse transcriptase, several copies of which are contained in every viral particle. In turn, the DNA serves as a template for the synthesis of its complement, becomes circular, and is inserted by crossing over into a host chromosome. The integrated viral specific DNA provirus subsequently becomes transcribed into 35S viral-specific RNA chains. These serve as templates for the various viral proteins. The virus then reproduces by budding.

The oncogenic DNA viruses are a very heterogeneous collection ranging from the small Papova viruses to the very large Herpes and Pox viruses. The most simple DNA viruses that induce cancer belong to the Papova group and include SV40, a monkey virus and Polyoma, a mouse virus. Both SV40 and Polynoma contain a single circular DNA molecule of MW ~ 3×10^6 which may code for as few as 3 genes. Transformation of a cell to a cancer cell is marked by the insertion of the viral DNA into specific host chromosomes. An mRNA molecule is produced which codes for the T (tumor) antigen. Most likely the T antigen induces cancer by turning on the transcription of many gene sets whose functions are normally blocked in the adult animal.

Adenoviruses also induce cancer through insertion of viral genes into host chromosomes. These much larger DNA viruses have linear DNA molecules of MW ~ 25×10^6. Transformation results from the insertion of only a fraction of the adeno genome[2] into the chromosome of a cell. Only the extreme left end

---

[1] This term is found from sedimentation rates in an ultracentrifuge and usually increases with the molecular weight. However, it also depends on the shape of the molecule. S = Svedberg unit = 10^{-13} sec.

[2] the haploid number, which in this situation is all the chromosomes.

(~ 7%) of the adeno chromosome need be inserted. This suggests that only one or two genes function to produce the transformed phenotype.

Transformation by the Herpe viruses is much less well understood. Only a fraction of their DNA genomes are required, but where it is located is unknown.

REFERENCES

1. Fenner, F., McAuslan, B.R., Mims, C.D., Sambrook, J. and White, D.O., The Biology of Animal Viruses, Academic Press, New York, 1974.

2. Tooze, J., The Molecular Biology of Tumor Viruses, Cold Spring Harbor Laboratory, 1973.

Appendix C

Cellular Energy

1. Introduction

The energy in all cells originates from the energy of the sun's light quanta, which is converted by photosynthetic plants into cellular molecules. These plants are then used as food sources by various microorganisms and animals. The animals in turn are eaten by other animals. The principal foodstuffs are carbohydrates, fats and proteins. In the human body these are converted into glucose, fatty acids and amino acids, respectively, before they reach the cell. Inside the cell the foodstuffs react with oxygen. Under the influence of various enzymes that control the rates of reaction, the energy stored within the food molecules (as covalent bonds of reduced carbon compounds) is released by oxidation. All of the released energy is not transformed into heat. More than half is converted into new chemical bonds. Phosphorous atoms play a key role in this transformation. Phosphate esters are formed that have a higher usuable energy content than most covalent bonds. One of the most important compounds for energy storage and also release is ATP.

2. Adenosine Triphosphate (ATP)

The formula for ATP is

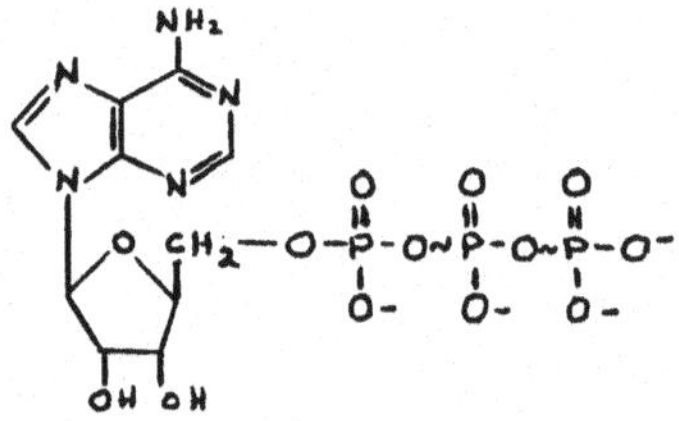

The last two phosphate radicals are connected with the remainder of the molecule by so-called high energy phosphate bonds, which are represented by the symbol ~. Each of these bonds contains about 8000 calories of energy per mole of ATP in the body. This is many times the energy stored in the average chemical bond; hence, the term "high energy" bond.

Whenever energy is required, ATP releases energy by releasing a phosphoric acid radical and forming adenosine diphosphate (ADP). Then, energy derived from the cellular nutrients causes the acceptor molecule ADP and phosphoric acid to recombine to form new ATP. The entire process continues over and over again.

3. Formation of ATP

On entry into the cell glucose is enzymically converted into pyruvic acid (a process called glycolysis)[1]. A small amount of ADP is changed into ATP by the energy released during glycolpis. However, this amount is so slight that it plays only a minor role in the overall energy metabolism of the cell.

The fatty acids and most amino acids are converted into acetoacetic acid. The pyruvic and acetoacetic acid are both converted into the compound acetyl co-A in the cytoplasm. This is transported along with oxygen through the mitochondrial membrane into the matrix of the mitochondria. Here acetyl co-A is acted upon by a series of enzymes and undergoes a sequence of chemical reactions called the Krebs cycle. In the Krebs cycle acetyl co-A is split into hydrogen atoms and carbon dioxide. The carbon dioxide diffuses out of the mitochondria and eventually out of the cell. The hydrogen atones combine with carrier substances and are carried to the surfaces of the shelves which protrude into the mitochondria. Attached to these shelves are the so-called oxidative enzymes. These enzymes participate in a series of oxidation-reduction reactions.

The first step is the reduction of a coenzyme[2] (NAD) by hydrogen to form the reduced coenzyme (NADH). The reduced coenzyme (NADH) is itself oxidized by another molecule (FAD) to yield a new reduced coenzyme ($FADH_2$) and the original coenzyme in the oxidized form (NAD). After several such steps molecular oxygen participates directly to end the oxidation-reduction chain, giving off water. The energy released during this process is used to manufacture tremendous quanti-

---

[1] The majority of carbohydrates undergo glycolyses but some follow the phosphogulconate pathway in which glucose is oxidized and triphosphopyridine nucleotide (TPN) is reduced to TPNH with the release of hydrogen atoms and CO_2, TPNH is used to synthesize fats from carbohydrates

[2] A substance necessary for the activity of the enzyme.

ties of ATP from ADP[3] in some unknown manner. The ATP then diffuses out of the mitochondria into all parts of the cytoplasm and nucleoplasm. There its energy can be used when required by the cell.

REFERENCES

Lehninger, A.L., Biochemistry, Worth Publishers, Inc., New York, 1970.

---

[3] 90% of all the ATP (5% is formed by glycolysis and 5% from the phosphogluconate pathway).

Appendix D

Immunology

1. The Immune System.

The mammalian immune system protects the organism from foreign invaders. A more detailed biological description than appears below can be found in [4,5,6,8, 12]. This system can be divided into two major divisions which can best be understood in terms of their ontogeny.

The two divisions are composed of white blood cells called lymphocytes. These blood cells, like all blood cells, arise in the embryonic yolk sac and ultimately migrate to the bone marrow. Here they differentiate into the various unipotent cells which are in the blood serum. The committed lymphocyte progenitors mature either under the influence of the thymus to become small T cells or under the influence of gut-associated lymphoid tissue (e.g. Peyer's Patches, tonsils and veriform appendix) to become small B cells. [The name "B cells" is a consequence of the fact that, in birds, the differentiation of these cells depends on a hind - gut lymphoid organ termed the Bursa of Fabricius.]

The B and T cells which are produced are relatively small and immature. They are arrested in the G_0 or G_1 phase of the cell cycle. Upon encountering foreign objects called antigens (substances capable of driving lympocytes from their resting state) they enter a metabolically active phase. The small B cells (T cells) mature and increase in size to become large B cells (T cells). This maturation process is irreversible.

The cellular division of the immune system is composed of T cells. T cells react by direct cellular encounter with the stimulating agent. The T cells are responsible for regulation of the B cell response, tissue - graft rejections, tumor-cell destruction and defense against viral and fungal infections.

The humoral division of the immune system is composed of B cells. B cells need not react directly with the agent stimulating the response but can react indirectly through molecular encounter. The B cells are involved in defense against bacterial infections and viral reinfections. Mature antigen - transformed B cells produce antibodies which are protein molecules composed of pairs of bilaterally symmetric heavy and light polypeptide chains. Antibodies circulate in the blood serum and can combine with the inducing antigens leading to its neutralization and elimination from the body. A system of at least fifteen serum proteins called the complement system can be activated by the antigen - antibody reactions. Complement plays an important role in several types of immune reactions (e.g. haemolysis, cytotoxicity, immobilization, opsonisation). Although the end results can be different the identical sequence of biochemical reactions involving the complement system always occur. The differences in observed results depend on differences of target cell structure.

Recent attempts at modelling the immune response by Bell [1,2,3] and Pimbley [16-21] have been based on the clonal selection theory for which Jerne and Burnet received the Nobel Prize. When an antigen encounters a B cell with the proper recognition site for it, the B cell is stimulated to reproduce and a clone is produced. The activated cells can then: (1) differentiate into plasma cells which do not divide but secrete large quantities of antibody specific for the stimulating antigen or (2) differentiate into memory cells which produce no antibody and which remain quiescent or (3) become large lymphocytes which secrete small amounts of antibody. Large lymphocytes divide rapidly and some differentiate into plasma cells. Others may revert back into small lymphocytes, where they probably function as "memory cells". Memory cells can be stimulated into vigorous antibody production when they reencounter the original stimulating antigen. The presence of memory cells may protect a person from getting some diseases a second time. Mathematical models of the lymphocyte - plasma - memory - cell system has been formulated by Perelson et al. [22, 23] using control theory.

A feedback mechanism is thought to prevent excessive proliferation of B cells with a resulting overproduction of specific antibodies. Presumably this overproduction of antibodies inhibits the further reproduction of the clone that produced the antibodies.

Depending on the antigen concentration, the immune system will exhibit different responses. Minute amounts of purified antigens given intravenously will sometimes inhibit the immune response to subsequent (larger) doses of the same antigen. This is called low zone tolerance. Very large doses of certain antigens will result in "paralysis" of the immune system. Maintenance of this type of tolerance requires repeated injections of the antigen. This phenomena is known as high zone paralysis. Mathematical models accounting for these phenomena can be found in Hoffman [19], Marchalonis and Gledhill [14] and Richter [27].

A review of mathematical immunology and further applications of control theory to this area appears in Perelson, [24,25].

2. The Immune System and Cancer.

In this section, evidence will be presented to show that the immune system plays an important role in the early detection, prevention and classification of cancer. This is a complicated and controversial subject. Some hypotheses which try to explain why the immune system does not always mount a decisive attack against tumors will be considered. Finally, some ideas which suggest how the immune system can be harnessed to fight cancer will be discussed. A more detailed account of cancer and its interaction with the immune system can be found in Burnett [5] and a survey article by Carter [7]. A recent account of cancer immunology in mice has been written by Old [15].

Any action of the immune system rests on its ability to distinguish normal from cancerous cells. This ability is a consequence of the fact that every mammalian cell has transplantation antigens on their surface. If a normal cell undergoes oncogenesis, considerable changes occur in the surface antigens. Some of these antigens amy disappear and new antigens (tumor associated antigens) may arise. Hence, it is possible that some type of cancer cells may have antigens unique for that type of cancer and also unique for the mammal in which the cell is growing. The following research supports this last conjecture. Dr. Old and his team at Sloan-Kettering in New York have discovered different antigens among melanoma patients. Some of these antigens are unique to the patient while others are common to all patients.

The following research indicates that cancer classifications may be refined by considering the surface antigens on the cancer cell. Dr. S. Schlossman and his colleagues at Harvard and at the Sidney Farber Cancer Institute have found different antigens among children with acute lymphoblastic leukemia. Blood tests based on these antigens can distinguish between the 70% of the children who respond to drug treatment and the 30% of children who die quickly, often within 15 months.

The fact that new antigens may occur during oncogenesis only means that the tumor may be detectable. The following evidence indicates that immunity places a decisive role in the etiology of a malignancy and in the body's defense against the malignancy.

(a) Statistics show that as the efficiency of the immune system decreases[1] (e.g. in aging and in immunosuppression by drugs or radiation) the frequency of tumors increases.

(b) Nearly all oncogenic viruses are immunosuppressive.

(c) Histological investigation of human tumors show that they will spread less slowly and give rise to less metastases if there are relatively many lymphocytes in the surrounding tissue.

(d) Mitchison showed that an immune response against experimental tumors occurs in the host. He transferred lymphocytes from tumor - bearing or from immunized animals into normal recipients. By this transfer, Mitchison rendered the recipients resistant to tumors which would have grown without this treatment.

(e) Specific antibodies have been detected in the blood serum of animals with experimental tumors and humans with malignancies.

(f) Many abnormal cells occur in a person's body throughout the course of his life. If the immune system did not detect and destroy these cells there would be a far greater incidence of cancer than at present. This supposition of immune surveillance was popular until 1971 when papers offering evidence against this hypothesis began to appear. The conflicting evidence is discussed in De Lisi

---

1. Arguments against this hypothesis are:
   (i) in aging there is more time for mutations, viral infection errors in recognition etc.
   (ii) in immunosuppression only certain tumor frequencies increase.
   (iii) radiation may cause more tumors.

[10] and Swan [28]. All of the above facts indicate that tumors are immunogenic. The answer to the question: Why aren't all tumors rejected? if not known. Some possibilities are:

(1) Tumors escape immune recognition because they have relatively weak tumor associated antigens or the tumor associated antigens are covered by a non-immunogenic mucoid layer.

(2) As a result of the immune response, non-cytotoxic antibodies are produced which react with the tumor but do not kill it. These antibodies may even protect the tumor.

(3) The tumor may grow so fast that the immune system is not able to respond effectively. For example, a small number of tumor cells may escape recognition. By the time there are enough cells to be recognized, the immune system may be overwhelmed.[2]

(4) Suppressor lymphocytes have been discovered recently. These lymphocytes paralyze bone-marrow cells. When the suppressor lymphocytes were removed, the bone marrow cells began growing again. Dr. R. Parkman and his colleagues at the Sidney Farber Institute found more advanced cases of osteogenic sarcoma in patients with suppressor cells than in patients without these cells. It is possible that the immune system may respond to cancer cells by producing suppressor cells as well as killer lymphocytes.

How can the above observations be applied to human malignancies? The answer to this last question is unknown but several approaches are currently being studied.

It has been the dream of immunologists to develop a vaccine against tumors. Considering the great variety of tumor antigens, this possibility seems rather remote. Passive transfer of specific tumor immunity has also been attempted by transferring antibodies or lymphocytes. Unfortunately, the beneficial effects of these treatments were not long lasting.

---

[2] Opponents to this point might ask why this is not also true for disease organisms.

Other immunologists have removed tumor cells surgically. In one experiment, the cells were irradiated while in other experiments the surfaces of the cells were altered biochemically. The altered cells were then reinoculated into the patient. Although the immune response was increased, the results were not encouraging.

The immune response can be enhanced by stimulating the cellular part of the immune system nonspecifically. BCG[3] (microbacterial) treatment of patients combined with chemotherapy is effective in some cases. Hypersensitivity against dinitrobezyl chloride has also been successful in curing certain malignancies. Endotoxins of gram - negative bacteria can enhance nonspecific resistance. Pretreatment with endotoxins renders normal animals resistant to lethal challenges with live tumors. How these results can be applied to humans, without exposing them to the harmful effects of endotoxins, is now being investigated.

Some authorities believe that the shotgun approach of stimulating the entire system may sometimes be harmful. For example, BCG may also increase the number of suppressor lymphocytes which prevent the killer lymphocytes from attacking the tumor.

A recent approach is to try and reduce the number of suppressor cells. Dr. Scholossman's group discovered that if the thymus is removed from a mouse, suppressor cells disappear. A mouse which would normally die of cancer will recover if it undergoes this surgery. The thymus was removed from six children who were afflicted with acute lymphoblastic anemia and who had not responded to drugs. Some of these children have now lived longer than would have normally been expected for children with this condition.

Mathematical models for the interaction between mature lymphocytes and spherical tumors have been developed by De Lisi and Rescigno [10, 26]. A review article on mathematical problems in immunity has been written by De Lisi [11].

---

[3] Bacillus Calmette - Guerin.

REFERENCES

1. Bell, G.I., Mathematical Model of clonal selection and antibody production, J. Theor. Biol., 29, 191-232, 1970.

2. Bell, G.I., Mathematical model of clonal selection and antibody production: II, J. Theor. Biol., 33, 339-378, 1971.

3. Bell, G.I., Mathematical model of clonal selection and antibody production: III, The cellular basis of immunological paralysis, J. Theor. Biol., 33, 379-398, 1971.

4. Bell, G.I., Perelson, A.S., and Pimbley, G.H., (editors). Theoretical Immunology, M. Dekker, New York, 1978.

5. Burnett, F.M., Cancer: biological approach I. Processes of control, Brit. Med. J., 1, 779, 1957.

6. Burnet, F.M., Immunological Surveillance, Pergamon Press, London, 1970.

7. Carter, S.K., Immunotherapy of cancer in man, American Scientist, 64, 418-423, 1976.

8. DeLisi, C., Antigen Antibody Interactions. Lecture Notes in Biomathematics, Vol. 8, Springer-Verlag, New York, 1976.

9. DeLisi, C., Immunological reactions to malignancies: some additional sources of complexity. In environmental Health: Quantitative methods, 149-171, edited by A. Whittemore, SIAM Publications, Philadelphia, 1976.

10. DeLisi, C. and Rescigno, A., Immune surveillance and neoplasia 1. A minimal mathematical model, Bull. Math. Bio., 39, 201-221, 1977.

11. DeLisi, C., Some mathematical problems in the initiation and regulation of the immune response, Math. Biosci., 35, 1-26, 1977.

12. Gordon, B.L. and Ford, D.K., Essentials of Immunology, F.A. Davis Co., Philadelphia, 1973.

13. Hoffman, G.W., A theory of regulation and self-nonself discrimination in an immune network, Eur. J. Immunol., 5, 638-647, 1975.

14. Marchalonis, J. and Gledhill, V.X., Elementary stochastic model for the induction of immunity and tolerance, Nature, 220, 608-610, 1968.

15. Old, L.J., Cancer immunology, Scientific American, 236, May, 62-79, 1977.

16. Pimbley, G.H., On Predator-prey Equations Simulating an Immune Response. In Nonlinear Problems in the Physical Sciences and Biology, 231-243. Lecture Notes in Mathematics, 322, Springer-Verlag, Berlin, 1973.

17. Pimbley, G.H., Periodic solutions of predator-prey equations simulating an immune response I. Math. Biosci., 20, 27-31, 1974.

18. Pimbley, G.H., Periodic solutions of predator-prey equations simulating an immune response II. Math. Biosci., 21, 251-277, 1974.

19. Pimbley, G.H., Periodic solutions of third order predator-prey equations simulating an immune response. Arch. Rat. Mech. Anal., 55, 93-124, 1974.

20. Pimbley, G.H., A direction of bifurcation formula in the theory of the immune response, Bull. Am. Math. Soc., 82, 56-58, 1976.

21. Pimbley, G.H., Bifurcation behavior of periodic solutions for an immune response problem, Arch. Rat. Mech. Anal., 64, 169-192, 1977.

22. Perelson, A.S., Mirmirani, M. and Oster, G.F., Optimal strategies in immunology, J. Math. Biol., 3, 325-367, 1976.

23. Perelson, A.S., Mirmirani, M. and Oster, G.F., Optimal strategies in immunology: II, B memory cell production. J. Math. Biol. 5, 213-256, 1978.

24. Perelson, A.S., Optimal strategies for an immune response, Lectures on Mathematics in the Life Sciences, Vol. 11, Am. Math. Soc., 109-163, 1979.

25. Perelson, A.S., Mathematical Immunology, to be published as: Chapter 5 in Mathematical Models in Molecular and Cellular Biology, L.A. Segel, ed., Cambridge University Press.

26. Reseigno, A. and DeLisi, C., Immune surveillance and neoplasia. II. A two-stage mathematical model, Bull. Math. Biol., #9, 487-497, 1977.

27. Richter, P.H., A network theory of the immune system, Eur. J. Immunol., 5, 350-354, 1975.

28. Swan, G.W., Immunological surveillance and neoplastic development, Rocky Mountain J. Math., to appear.

Appendix E

Mathematical Theories of Carcinogenesis

The development of cancer, carcinogenesis, requires one or more changes in a normal cell. These transformations render the cell capable of unchecked growth leading to a colony (clone) of descendants. When the number of cells in this colony is noticeable, the clone is called a detectable tumor. Mathematical theories of carcinogenesis relate detectable tumor frequency to the concentration and potency of the carcinogen, the susceptibility and age of the host, and other factors. These theories can be tested by comparing their predictions with observed incidences of human cancer or induced animal cancer.

The earliest stochastic model of carcinogenesis was that of Iverson and Arley [1,14]. These researchers assumed that a single interaction between a carcinogen and a normal cell resulted in a hit-that is, mutated the cell into a cancer cell.

The one hit models were unable to explain why the death rates for many forms of human cancer increased proportionately to the fifth or sixth power of age. To interpret this last observation in another way, assume that the time from tumor detection until death is negligible compared to the time required for transformation and growth. Consequently, the death rate from cancer is the same as the rate of appearance of tumors (the incidence rate). In addition, assume that one is exposed to a constant concentration of carcinogen from birth to death. Then the duration of exposure is just the age of the individual. It follows that the incident rate is proportional to the fifth or sixth power of the duration of exposure to a carcinogen of constant concentration.

There are two proposals for explaining this power law relationship. The first due to Fisher and Holloman [9], proposes that six or seven different cells have to be transformed in a single tissue in order to form a tumor. This multicell hypothesis for carciogenesis is not supported by biology for many tumors. The other proposal, suggested by Muller [20] and Nordling [22] is the multi-hit hypothesis. The term "multi-hit" refers to the situation in which more than one hit is required to mutate a normal cell. The number of hits required to produce a cancer cell is called the number of stages. The quantitative deductions from the multi-hit theory were derived by Stocks [24] and by Armitage and Doll [2]. Both Armitage and Doll [3], Fisher [9] and Neyman and Scott [21] modified the multistage theory so that a two or three stage theory could explain the observed data. A recent theoretical and experimental paper concerned with the multi-hit theory is by Emmelot and Scherer [7].

None of the above models consider the division of carcinogenesis into the following two states: initiation, the alteration of the cells in the tissue by an initiator, followed by promotion, the development of initiated cells to detectable tumors by the promoter. A theory developed by Keller and Whittemore [15], which allows alteration of tumor growth rates by carcinogenic exposure, can be used for examining initiation-promotion data.

A two hit theory of retinoblastoma was proposed by Knudson [17]. In the hereditary form of the disease, one mutation occurs in the germinal cells while the other occurs in the somatic cells. Knudson et al [18] developed a probabilistic model for the hereditary form of retinoblastoma. The model predicts the age-dependent incidence of the disease in close agreement with data from patients. Childhood cancer is discussed in Knudson [19].

Recent surveys of the quantitative theories of carcinogenesis appear in Dubin [6], Hethcote [12], Hartley and Sielken [11], Whittemore [25] and Whittemore and Keller [26]. The paper by Schürger and Tautu [23], describing a model for carcinogenesis, contains a detailed bibliography on earlier models.

Biological insight into carciogenesis can be found in the books edited by King [16] and Hiatt et al [13]. The latter book also contains some quantitative material.

Fears et al. [8] present data and simple power relations which support the hypothesis that sunlight leads to skin cancer. In particular, they show that the risk of non-melanoma skin cancer is related to the cumulative lifetime ultra-violet exposure. They also show that the risk of melanoma is related to the annual ultra-violet exposure.

De Lisi [5] demonstrates that a three-stage stochastic model of carcinogenesis leads to an equation which can be fitted to both Oriental and Western age-incidence curves for female breast cancer.

Davison [4] identifies twenty-six chemical reactions occurring within a cell. He describes these reactions by a system of nonlinear ordinary differential equations. Davison also presents a list of perturbations of the initial conditions and the parameters of the cell equations which cause the simulated cell to exhibit tumor-like behavior.

REFERENCES

1. Arley, N. and Iversen, S., On the mechanism of experimental carcinogenesis, III. Further development of the hit theory of carcinogenesis, Acta Path. Microbiol. Scand., 30, 21-53, 1952.

2. Armitage, P. and Doll, R., The age distribution of cancer and a multi-stage theory of carcinogenesis, British J. Cancer, 8, 1-12, 1954.

3. Armitage, P. and Doll, R., A two-stage theory of carcinogenesis in relation to the age distribution of human cancer, British J. Cancer, 11, 161-169, 1957.

4. Davison , E.J., Simulation of cell behavior: normal and abnormal growth, Bull. Math. Biol., 37, 427-458, 1975.

5. DeLisi, C., The age incidence of female breast cancer - simple models and analysis of epidemiological patterns, Math. Biosc., 37, 245-266, 1977.

6. Dubin, N., A Stochastic Model for Immunological Feedback in Carcinogenesis: Analysis and Approximations, Lecture Notes in Biomathematics, Vol. 9, Springer-Verlag, New York, 1976.

7. Emmelot, P. and Scherer, E., Multi-hit kinetics of tumor formation, with special reference to experimental liver and human lung carcinogenesis and some general conclusions. Cancer Res., 37, 1702-1708, 1977.

8. Fears, T.R., Scotto, J. and Schneiderman, M.A., Mathematical models of age and ultraviolet effects on the incidence of skin cancer among whites in the United States, Am. J. Epidemiol., 105, 420-427, 1977.

9. Fisher, J.C., Multiple-mutation theory of carcinogenesis, Nature, 181, 651-652, 1958.

10. Fisher, J.C. and Holloman, J.H., A new hypothesis for the origin of cancer foci, Cancer, 4, 916-918, 1951.

11. Hartley, H.O. and Sielken, R.L., Estimation of "safe doses" and carcinogenic experiments, Biometrics, 33,1-30, 1977.

12. Hethcote, H.W., Mutational models of carcinogenesis. In Environmental Health: Quantitative Methods, edited by Whittemore, A.S., SIAM Publications, Philadelphia, 172-182, 1976.

13. Hiatt, H.H., Watson, J.D. and Winsten, J.A. (editors). Origins of Human Cancer (3 volumes), Cold Spring Harbor Laboratory, Cold Spring Harbor, 1977.

14. Iversen, S. and Arley, N., On the mechanism of experimental carcinogenesis, Acta Path. Microbiol. Scand., 27, 773-803, 1950.

15. Keller, J.B. and Whittemore, A., A theory of transformed cell growth with applications to initiation-promotion data, in Environmental Health: Quantitative Methods, edited by A. Whittemore, SIAM Publications, Philadelphia, 183-196, 1976.

16. King, T.J. (editor). Developmental Aspects of Carcinogenesis and Immunity, Academic Press, New York, 1974.

17. Knudson, A.G., Mutation and cancer: statistical study of retinoblastoma, Proc. Natl. Acad. Sci., 68, 820-823, 1971.

18. Knudson, A.G., Hethcote, H.W. and Brown, B.W., Mutation and childhood cancer: a probabilistic model for the incidence of retinoblastoma, Proc. Natl. Acad. Sci., 72, 5116-5120, 1975.

19. Knudson, A.G., Genetics and the etiology of childhood cancer, Pediat. Res., 10, 513-517, 1976.

20. Muller, H.J., Radiation damage to the genetic material, Science Pregress, 7, 93-492, 1951.

21. Neyman, J. and Scott, E., Statistical aspects of the problem of carcinogenesis, Fifth Berkeley Symposium on Math. Stat. and Proa., University of California Press, 745-776, 1967.

22. Nordling, C.O., A new theory on the cancer inducing mechanism, British J. Cancer, 7, 68-72, 1953.

23. Schürger, K. and Tautu, P., A Markovian configuration model for carcinogenesis, in Lecture Notes in Biomathematics, Vol. 11, Mathematical Models in Medicine, edited by Berger, J., Buhler, W., Repges, R. and Tautu, P., Springer-Verlag, New York, 1976.

24. Stocks, P., A study of the age curve for cancer of the stomach in connection with a theory of the cancer producing mechanism, British J. Cancer, 7, 407-417, 1953.

25. Whittemore, A., Quantitative theories of oncogenesis, Advances in Cancer Research, 27, 55-88, 1978.

26. Whittemore, A. and Keller, J.B., Quantitative theories of carcinogenesis, SIAM Review, 20, 1-30, 1978.

Appendix F

Radiology and Cancer

Radiology plays two roles in the battle against cancer - radiotherapy and detection. Radiotherapy is one of the major therapeutic approaches to cancer problems since it may be effective against many types of solid tumors. Since a patient cannot receive total body irradiation, radiotherapy can only be used when the tumor is localized. The object of radiotherapy is to reduce the tumor burden to the extent that chemotherapy, immunotherapy or the body's natural defences can produce remission.

In order to destroy a large number of malignant cells with minimum damage to surrounding normal tissue radiation dosages must be carefully calculated. Thus, radiology, mathematics, and computers have been intimately related in cancer therapy as evidenced by early models for radiotherapy by Cohen [4], Hall [6] and the book by Fitzgerald et al. [5]. Other more recent applications of mathematics to radiation analysis of cancer therapy problems appear in the book by Alper [2], the works of Kirk et al. [9,10] and Redpath et al. [15].

Hethcote et al. [8] compared five radiation fractionation models. In the first model, the total dose is a power function of the number of fractions. The other four models for the dose-log surviving fraction are described by (1) a multitarget, (2) a product of a single hit and multitarget factors, (3) a quadratic function and (4) a cubic function. Hethcote and coworkers show that the first and last models do not fit specific data and so seem unsuitable for use.

Recently, research has begun on the design of optimal fractionation schedules for the treatment of solid tumors using modern mathematical optimization methods. Hethcote and Waltman [7] studied the reactions of a tumor and normal cell populations to irradiation. Their kinetic cell models consider such factors as repair, reoxygenation and repopulation. The optimal fractionation schedule was derived by means of dynamic programming. This optimization technique for scheduling radiotherapy was also used by Wheldon and coworkers [17,18,19]. The approach of Almquist and Banks [1], which utilizes optimal control theory, has several advantages over the previous work on scheduling radiotherapy. More complex situations can be handled by the proper choice of the performance index and the problem of dimensionality in dynamic programming is alleviated. However, the proper choice of the performance index is not simple. The mathematical difficulty with all of these papers is that a global optimal solution has not been shown to exist.

Besides optimizing the magnitude and timing of the radiation dosage, optimal control theory has also been applied by McDonald and Rubin [13] to optimize the mode of delivery of the external beam. Bjärngard [3] critically reviews this paper.

The C.A.T. (computerized axial tomography) scanner has revolutionized the pinpointing of tumors. Unlike a convential x-ray machine, this device provides cross-sectional photographs of the body. It operates by the use of x-ray transmissions along lines parallel to a large number of different directions. The resulting photographs are pieced together by computer programs to yield a cross-sectional photograph. A non-mathematical review of computerized tomography is contained in Kreel [11]. Mathematical problems in this area can be found in the papers by Thames [16], Orphanoudakis and Strohbehn [14], and the review article by McCullough and Payne [12]. The whole issue of Computers in Biology and Medicine, Vol. 6., No. 4, 1976 is devoted to this topic. There is also a J. of Comp. Tomography and a C.T.J. of Comp. Tomography.

REFERENCES

1. Almquist, K.J. and Banks, H.T. A theoretical and computational method for determining optimal treatment schedules in fractionated radiation therapy, Math. Biosci., 29, 159-179, 1976.

2. Alper, T. (editor) Cell Survival after Low Doses of Radiation: Theoretical and Clinical Implications, J. Wiley, New York, 1975.

3. Bjärngard, B.E. Optimization in radiation therapy, Int. J. Radiat. Oncol. Biol. Phys., 2, 381-382, 1977.

4. Cohen, L. Theoretical "iso-survival" formulae for fractionated radiation therapy, Br. J. Radiol, 41, 522-528, 1968.

5. Fitzgerald, J.J., Brownell, G.L. and Mahoney, J. Mathematical Theory of Radiation Dosimetry, Gordon and Breach, New York, 1967.

6. Hall, E.J. Time dose and fractionation in radiotherapy. A comparison of two evaluation systems in clinical use, Br. J. Radiol., 42, 427-431, 1969.

7. Hethcote, H.W. and Waltman, P. Theoretical determination of optimal treatment schedules for radiation therapy, Radiation Res., 56, 150-161, 1973.

8. Hethcote, H.W., Mclarty, J.W. and Thames, H.D., Jr. Comparison of mathematical models for radiation fractionation, Radiation Res., 67, 387-407, 1976.

9. Kirk, J., Gray, W.M. and Watson, E.R. Cumulative radiation effect. Part VI: simple nomographic and tabular methods for the solution of practical problems, Clin. Radiol., 28, 29-74, 1977.

10. Kirk, J., Cain, O. and Gray, W.M. Cumulative radiation effect. Part VII: computer calculations and applications in clinical practice, Clin. Radiol., 28, 75-92, 1977.

11. Kreel, L. Computerized tomography using the EMI general purpose scanner, Brit. J. Radiol., 50, 2-14, 1977.

12. McCullough, E.C. and Payne, J.T. X-ray-transmission computed tomography, Med. Phys., 4, 85-98, 1977.

13. McDonald S.C. and Rubin, P. Optimization of external beam radiation therapy, Int. J. Radiat. Oncol. Biol. Phys., 2, 307-317, 1977.

14. Orphanoudakis, S.C. and Stroehbehn, J.W. Mathematical model of conventional tomography, Med. Phys., 3, 224-232, 1976.

15. Redpath, A.T., Vickery, B.L. and Duncan, W. A comprehensive radiotherapy planning system implemented in Fortran on a small interactive computer, Brit. J. Radiol., 50, 51-57, 1977.

16. Thames, H.D., Jr. Simulation of many-independent-variable systems. In Computer Techniques in Biomedicine and Medicine, 186-203 edited by E. Haga, R.D. Brennan, M. Ariet and J.G. Llaurado, Auerbach Publishers, Philadelphia, 1973.

17. Wheldon, T.E. and Kirk, J. Mathematical derivation of optimal treatment schedules for the radiotherapy of human tumors. Fractionated irradiation of exponentially growing tumors, Br. J. Radiol., 49, 441-449, 1976.

18. Wheldon, T.E. and Kirk, J. Optimal radiotherapy of tumor cells following exponential-quadratic survival curves and exponential repopulation kinetics, Br. J. Radiol., 50, 681-682, 1977.

19. Wheldon, T.E. Optimal control strategies in the radiotherapy of human cancer, in Cell Kinetics and Biomathematics: Paris 1978, (A.J. Valleron and P.D. McDonald, Eds.) Elsevier Press, Amsterdam, 1978.

Appendix G

Applications of Control Theory to Normal and Malignant Cell Growth

W. Düchting

Biologists have known for some time that cell growth is governed by complex control mechanisms. However, the quantitative application of control theory to the study of cell growth only began in the 1960's.

Iversen and Bjerknes [12] simulated the behavior of the system of epidermal cells by means of an analogue computer.

In [3] Düchting formulated the question: Are cancer diseases nothing but unstable control loops? To answer this question a simple single-loop control circuit of cell growth was developed.

The disadvantages of the rather general statements of [3] have been improved step by step during the following years. The first step was to specify as model for erythropoiesis (the formation of red blood cells). This topic was chosen because experimental data is readily available [2] and a rich experience in this field enables one to understand the actual cell renewal process [17]. In [4,5,6, 7] Düchting developed a multi-loop nonlinear control model of erythropoiesis which could explain the global time behaviour of the red cells in totally different blood diseases (e.g., of an aplastic anemia and of a polycythaemia). This model was too complex for mathematical analysis and could only be thoroughly studied after block-oriented computer languages had been developed. These languages enable one to easily simulate the behavior of complex nonlinear multi-loop control systems by means of a digital computer. It was also possible to introduce external disturbances and to study the effects of parametric and structural changes in the closed-loop control circuit. Other models of blood formation have been studied in [1,13,16,20,21] at the same time.

The macro model described above outputs only the total number of cells in a special compartment. It is impossible to describe the run of a single cell through the different compartments. Furthermore, in this model a single cell can't be destroyed at any arbitrary moment. Therefore the next step [8] was to develop a generalized multi-loop compartment model. In this improved approach each compartment consists of several single cells, each being a multi-input - multi-output system. The system as well as each single cell can be stimulated by parametric and structural alterations. As a simulation result the number of cells in the different compartments are plotted versus the time. The number of cells increase in an oscillatory manner when the conditions for neoplastic malignant cell growth are imposed. The graphs are similar to those observed clinically. Further detailed deterministic and stochastic compartment models can be found in [10, 18].

The next problems which had to be tackled were

- intercellular communication,
- cell systems growing in competition and
- the determination of the exact position of a cell in a tissue.

To study these problems, the comfortable block-oriented language could no longer be used as the number of control-loop elements had reached its maximum limit. Therefore, in [9], a list of specification and formation rules were set up concerning the communication between adjacent cells and competing cell systems with different mean life spans. This idea was stimulated by previous papers about "game life" and automata theory of Gardner, Lindenmayer and Rittgen & Tautu [11, 14, 18]. Steinberg [19] had tried to describe the same phenomena by "adhesion theory".

By means of the above rules, it is possible to describe the intercellular behaviour of the cells by a cell matrix. The computer simulation of the system was performed on a minicomputer. The most interesting results show the time behaviour, the local arrangement and the distribution of disturbed cell systems growing in competition for different initial conditions. The simulation results are similar to those obtained from experimental morphological cuts.

The priority of problems which should be worked on in the future are:

1. Performing designed tests with cell cultures to get suitable experimental data for the validation of the developed models and the formulated hypotheses.

2. Improving the models to minimize the deviation between theoretical and real biological results.
3. Focusing the range of possible causes (parametric and/or structural changes) which are responsible for the instability of malignant cell growth processes.

Such a program will require close cooperation between biologist and engineers or mathematicians.

REFERENCES

1. Blumenson, L.E., A Comprehensive Modeling Procedure for the Human Granulopoietic System: Over-all view and Summary of Data, Blood, 42, 303-313, 1973.

2. Covelli, V., Briganti, G. and Silini, G., An analysis of bone marrow erythropoiesis in the mouse, Cell Tissue and Kinetics, 5, 41-51, 1972.

3. Düchting, W., Krebs, ein instabiler Regelkreis, Versuch einer Systemanalyse, Kybernetik, Band 5, Heft 2, 70-77, 1968.

4. Düchting, W., Entwicklung eines Erythropoese-Regelkreis-modells zur Computer-Simulation, Blut, Band XXVII, 342-350, 1973.

5. Düchting, W., Computersimulationen von Zellerneuerungs - systemen, Blut, Band 31, Heft 6, 371-388, 1975.

6. Düchting, W., Computer simulation of abnormal erythropoiesis - an example of cell renewal regulating systems, Biomedizinische Technik, Band 21, Heft 9, 34-43, 1976.

7. Düchting, W., Control Models of Hemopoiesis, In "Biomathematics and Cell Kinetics", edit. by Valleron, A.-J. and Macdonald, P.D.M., Elsevier/North-Holland, Amsterdam, 297-308, 1978.

8. Düchting, W., A cell kinetic study on the cancer problem based on the automatic control theory using digital simulation, Journal of Cybernetics, Vol. 6, 139-172, 1976.

9. Düchting, W., A Model of Disturbed Self-Reproducing Cell Systems, in Biomathematics and Cell Kinetics, edit. by Valleron, A.-J. and Macdonald, P.D.M., Elsevier/North-Holland, Amsterdam, 133-142, 1978.

10. Fuchs, G. , Compartmentmodelle in der Medizin, Methods of Information in Medicine 11, 137-144, 1972.

11. Gardner, M.: On cellular automata, self-reproduction, the Garden of Eden and the game life, Scientific American, Vol. 224, 112-117, 1971.

12. Iverson, O.H. and Bjerknes, R., Kinetics of Epidermal Reaction to Carcinogens, Norwegian Monographs on Medical Science, Universitetsforlaget, Copenhagen, 1963.

13. King-Smith, E.A. and Morley, A., Computer simulation of granulopoiesis: Normal and impaired granulopoiesis, Blood, 36, 254-262, 1970.

14. Lindenmayer, A., Developmental Algorithms for Multicellular Organisms: A Survery of L-Systems, Journal of Theoretical Biology, 54, 3-22, 1975.

15. May, R.M., Biological Populations, Obeying Difference Equations: Stable Points, Stable Cycles and Chaos, Journal of Theoretical Biology, 51, 511-524, 1975.

16. Mylrea, K.C. and Abbrecht, P.H., Mathematical Analysis and Digital Simulation of the Control of Erythropoiesis, Journal of Theoretical Biology, 33, 279-297, 1971.

17. Rajewsky, M.F., Proliferative parameters of mammalian cell systems and their role in tumor growth and carcinogenesis, Zeitschrift fur Krebsforschung, 78, 12-30, 1972.

18. Rittgen, W. and Tautu, P., Branching models for the cell cycle, Lecture Notes in Biomathematics 11, Springer-Verlag, Berlin , 109-126, 1976.

19. Steinberg, M.S., Adhesion-guided Multicellular Assembly, Journal of Theoretical Biology, 55, 431-443, 1975.

20. Wheldon, T.E., Kirk, J. and Gray, W.M., Mitotic Autoregulation Growth Control and Neoplasia, Journal of Theoretical Biology, 38, 627-639, 1973.

21. Wichmann, H.E. und Spechtmeyer, H. Modelle zur Hämopoese (Computersimulation) in "aktuelle Probleme der Inneren Medizin", Schattauer Verlag Stuttgart, 275-282, 1977.

INDEX

D

E

J

K

L

M

Lecture Notes in Biomathematics

Vol. 26: M. B. Katz, Questions of Uniqueness and Resolution in Reconstruction from Projections. IX, 175 pages. 1978.

Vol. 27: N. MacDonald, Time Lags in Biological Models. VII, 112 pages. 1978.

Vol. 28: P. C. Fife, Mathematical Aspects of Reacting and Diffusing Systems. IV, 185 pages. 1979.

Vol. 29: Kinetic Logic – A Boolean Approach to the Analysis of Complex Regulatory Systems. Proceedings, 1977. Edited by R. Thomas. XIII, 507 pages. 1979.

Vol. 30: M. Eisen, Mathematical Models in Cell Biology and Cancer Chemotherapy. IX, 431 pages. 1979.

Lightning Source UK Ltd.
Milton Keynes UK
UKOW04f1656230317
297375UK00017B/393/P

9 783540 097099